30 Springer Series in Solid-State Sciences
Edited by P. Fulde

Springer
Berlin
Heidelberg
New York
Barcelona
Budapest
Hong Kong
London
Milan
Paris
Singapore
Tokyo

Springer Series in Solid-State Sciences

Editors: M. Cardona P. Fulde K. von Klitzing H.-J. Queisser

Managing Editor: H. K.V. Lotsch

Volumes 1–89 are listed at the end of the book

90 **Earlier and Recent Aspects of Superconductivity**
Editors: J. G. Bednorz and K. A. Müller

91 **Electronic Properties of Conjugated Polymers III** Basic Models and Applications
Editors: H. Kuzmany, M. Mehring, and S. Roth

92 **Physics and Engineering Applications of Magnetism** Editors: Y. Ishikawa and N. Miura

93 **Quasicrystals** Editors: T. Fujiwara and T. Ogawa

94 **Electronic Conduction in Oxides**
By N. Tsuda, K. Nasu, A. Yanase, and K. Siratori

95 **Electronic Materials**
A New Era in Materials Science
Editors: J. R. Chelikowsky and A. Franciosi

96 **Electron Liquids** 2nd Edition By A. Isihara

97 **Localization and Confinement of Electrons in Semiconductors**
Editors: F. Kuchar, H. Heinrich, and G. Bauer

98 **Magnetism and the Electronic Structure of Crystals** By V.A. Gubanov, A.I. Liechtenstein, and A.V. Postnikov

99 **Electronic Properties of High-T_c Superconductors and Related Compounds**
Editors: H. Kuzmany, M. Mehring, and J. Fink

100 **Electron Correlations in Molecules and Solids** 3rd Edition By P. Fulde

101 **High Magnetic Fields in Semiconductor Physics III** Quantum Hall Effect, Transport and Optics By G. Landwehr

102 **Conjugated Conducting Polymers**
Editor: H. Kiess

103 **Molecular Dynamics Simulations**
Editor: F. Yonezawa

104 **Products of Random Matrices** in Statistical Physics By A. Crisanti, G. Paladin, and A. Vulpiani

105 **Self-Trapped Excitons**
2nd Edition By K. S. Song and R. T. Williams

106 **Physics of High-Temperature Superconductors**
Editors: S. Maekawa and M. Sato

107 **Electronic Properties of Polymers**
Orientation and Dimensionality of Conjugated Systems Editors: H. Kuzmany, M. Mehring, and S. Roth

108 **Site Symmetry in Crystals**
Theory and Applications 2nd Edition
By R. A. Evarestov and V. P. Smirnov

109 **Transport Phenomena in Mesoscopic Systems** Editors: H. Fukuyama and T. Ando

110 **Superlattices and Other Heterostructures**
Symmetry and Optical Phenomena 2nd Edition
By E. L. Ivchenko and G. E. Pikus

111 **Low-Dimensional Electronic Systems**
New Concepts
Editors: G. Bauer, F. Kuchar, and H. Heinrich

112 **Phonon Scattering in Condensed Matter VII**
Editors: M. Meissner and R. O. Pohl

113 **Electronic Properties of High-T_c Superconductors**
Editors: H. Kuzmany, M. Mehring, and J. Fink

114 **Interatomic Potential and Structural Stability**
Editors: K. Terakura and H. Akai

115 **Ultrafast Spectroscopy of Semiconductors and Semiconductor Nanostructures** By J. Shah

116 **Electron Spectrum of Gapless Semiconductors**
By J. M. Tsidilkovski

117 **Electronic Properties of Fullerenes**
Editors: H. Kuzmany, J. Fink, M. Mehring, and S. Roth

118 **Correlation Effects in Low-Dimensional Electron Systems**
Editors: A. Okiji and N. Kawakami

119 **Spectroscopy of Mott Insulators and Correlated Metals**
Editors: A. Fujimori and Y. Tokura

120 **Optical Properties of III–V Semiconductors**
The Influence of Multi-Valley Band Structures
By H. Kalt

121 **Elementary Processes in Excitations and Reactions on Solid Surfaces**
Editors: A. Okiji, H. Kasai, and K. Makoshi

122 **Theory of Magnetism**
By K. Yosida

123 **Quantum Kinetics in Transport and Optics of Semiconductors**
By H. Haug and A.-P. Jauho

124 **Relaxations of Excited States and Photo-Induced Structural Phase Transitions**
Editor: K. Nasu

125 **Physics and Chemistry of Transition-Metal Oxides**
Editors: H. Fukuyama and N. Nagaosa

M. Toda R. Kubo N. Saitô

Statistical Physics I

Equilibrium Statistical Mechanics

Second Edition
With 90 Figures

Springer

Professor Dr. Morikazu Toda
5-29-8-108 Yoyogi, Shibuja-ku
Tokyo 151, Japan

Professor Dr. Ryogo Kubo †

Professor Dr. Nobuhiko Saitô
Department of Applied Physics, Waseda University
3-4-1 Okubo, Shinjuku-ku, Tokyo 169, Japan

Series Editors:

Professor Dr., Dres. h. c. Manuel Cardona
Professor Dr., Dres. h. c. Peter Fulde*
Professor Dr., Dres. h. c. Klaus von Klitzing
Professor Dr., Dres. h. c. Hans-Joachim Queisser

Max-Planck-Institut für Festkörperforschung, Heisenbergstrasse 1, D-70569 Stuttgart, Germany
* Max-Planck-Institut für Physik komplexer Systeme, Nöthnitzer Strasse 38
D-01187 Dresden, Germany

Managing Editor:

Dr.-Ing. Helmut K.V. Lotsch

Springer-Verlag, Tiergartenstrasse 17, D-69121 Heidelberg, Germany

Second Edition 1992
Third Corrected Printing 1998

Revised translation of the revised original Japanese edition: *Tokei Butsurigaku*
© Morikazu Toda, Ryogo Kubo, Nobuhiko Saitô, and Natsuki Hashitsume 1978
Originally published by Iwanami Shoten, Publishers, Tokyo (1978).
Englisch translation by Morikazu Toda and Nobuhiko Saitô

ISSN 0171-1873
ISBN 3-540-53662-0 2nd ed. Springer-Verlag Berlin Heidelberg New York
ISBN 3-540-11460-2 1st ed. Springer-Verlag Berlin Heidelberg New York

Library of Congress Cataloging-in-Publication Data. Toda, Morikazu, 1917– [Tökei-butsurigaku. English] Statistical physics / M. Toda, R. Kubo, N. Saitô. –2nd ed. p. cm.–(Springer series in solid-state sciences ; 30–) Rev. translation of: Tökei-butsurigaku. Includes bibiographical references and index. Contents: 1. Equilibrium statistical mechanics.
ISBN 3-540-53662-0 (Berlin : v. 1 : acid-free paper).
1. Statistical mechanics. 1. Kubo, Ryōgo, 1920– . II. Saitô, N. (Nobuhiko), 1919– . III. Title.
IV. Series. QC174.8.T613 1991 530.1'3–dc20 91-165 CIP

Typesetting: Macmillan India Ltd., Bangalore-25
Cover concept: eStudio Calamar Steinen
Cover design: *design & production* GmbH, Heidelberg

SPIN: 10673287 54/3144 – 5 4 3 2 1 0 – Printed on acid-free paper

Foreword to Statistical Physics I and II

The search to discover the ultimate structure of matter, from molecules to atoms, from atoms to electrons and nuclei, to nucleons, to elementary particles, and now to quarks, has formed the mainstream of modern physics. The ultimate structure is still elusive, but the efforts of mankind from the beginning of this century to probe the fundamentals of the physical world have been extremely successful in revealing the grandeur of the order in nature. The glory of this success may even outshine another great endeavor of modern physics, which is, however, equally important. This other endeavor is directed towards the synthesis of analyzed elements into organized systems which are encountered in our more common experience. Analysis and synthesis are the two sides of the evolution of science. They are very often so interconnected that they cannot be separated in any simple way. Briefly stated, statistical physics is the methodology of synthesis. It is the subject of this text.

The construction of macroscopic systems from microscopic elements analyzed at the microscopic level is not only limited to physics. Macrosystems are synthesized from microscopic structure and dynamics in biology, the social sciences, psychology and other sciences as well. This activity of synthesizing is undoubtedly one of the most powerful tools of science. However, we may say that it is best developed in physics. This is, of course, because the objects studied in physics are simpler and more concrete than those in other sciences, and theories can be more easily tested through experiments.

The synthesis of a macroscopic system from microscopic elements is not simply a collecting of fragments. The macroscopic system is an entity characteristically different from that existing at the microscopic level. The most typical example is perhaps given by the second law of thermodynamics. Despite the reversibility of the microscopic dynamics, macroscopic phenomena are indeed irreversible, and entropy always increases.

As is well known, the law of increasing entropy is interpreted in terms of probability. In order to describe a macroscopic system consisting of an enormous number of microscopic elements, the extremely complex motion of the microscopic elements has to be projected, so to speak, onto a much smaller number of macroscopic variables. This projection is necessarily of a statistical character. In this sense, it is statistical physics that synthesizes the microscopic world to the macroscopic world. Statistical physics covers a very large area, from statistical thermodynamics (that is, the statistical mechanics which constructs thermodynamics) to the generalized statistical mechanics of irreversible processes and the kinetic theories which inherit the tradition of the classical kinetic theory of gases. The great breadth of these subjects makes it impossible to treat them all within this text. Fortunately, two volumes of the *Iwanami Series in Fundamental Physics* are also included in this series[1] and are devoted to the physics of condensed matter with various applications of statistical physics. Therefore the emphasis in this book will be placed upon methodological aspects rather than upon specific applications.

The year 1972, during which the first Japanese edition of this book was prepared, was the hundredth anniversary of the proposal of the Boltzmann equation. This equation determines the evolution of the velocity distribution function of molecules in a dilute gas. The stationary solution of this equation gives the Maxwell–Boltzmann distribution law on which the statistical thermodynamics of dilute gases was founded. An even more important aspect of this equation is that it provided a method for calculating the properties of dilute gases in nonequilibrium states. The Boltzmann equation was the prototype of the general kinetic method which treats the temporal evolution of the distribution functions of microscopic elements.

Although the kinetic method is very useful and powerful, generalizing it to apply to denser systems is very difficult. It can hardly be regarded as the general basis of statistical thermodynamics. Here again, Boltzmann made a great contribution and in so doing created statistical mechanics. He recognized that the assumption of equal weights of microscopic states is sufficient to build a general scheme for the statistical mechanics of equilibrium states, namely statistical thermodynamics. The inscription

$$S = k \log W$$

on Boltzmann's gravestone in the central cemetery of Vienna is the essence of this work, by means of which Max Planck summarized Boltzmann's somewhat obscure statements. Statistical mechanics was born through this simple equation. The whole structure was beautifully reconstructed by W. Gibbs some years later. Although there was a difference in viewpoints between Boltzmann and Gibbs, this is no longer so important.

The true mechanics of the microscopic world is quantum mechanics. Thus, statistical mechanics, based on classical mechanics, was doomed to show inconsistencies when applied to real physical problems. The best known case of this was the problem of blackbody radiation, which led Planck to the discovery of energy quanta. Moreover, the quantal structure of nature is reflected in the very existence of thermodynamics. For example, the Gibbs paradox for the extensivity of entropy cannot be resolved by the classical picture. If nature had a different structure, macroscopic thermodynamics would have been totally different from what we know in the physical world. The logical structure of statistical

[1] S. Nakajima, Y. Toyozawa, R. Abe: *The Physics of Elementary Excitations*. T. Matsubara (ed): *The Structure and Properties of Matter*, Springer Ser. Solid-State Sci., Vols. 12 and 28, respectively.

mechanics, particularly that constructed by Gibbs, received the new mechanics as if it had been anticipated. The logic bridging the microscopic and macroscopic worlds does not much depend upon the mechanics governing the former. Here we see the general character of the methods of statistical physics.

At least as far as thermodynamic properties are concerned, quantum-statistical mechanics is the most general scheme for elucidating the properties of a macroscopic system on the basis of its microscopic structure. Developments since 1930 in the modern physics of condensed matter have been theoretically supported by quantum mechanics and quantum-statistical mechanics. Some of the basic problems will be treated in this book, but most of the physical problems had to be left to the two volumes of this series mentioned previously. It should be kept in mind that there is no distinct boundary between statistical physics and condensed matter physics. Indeed, progress in the former was made through efforts in the latter. Theoretical methods were developed in the treatment of real physical problems. It is only in the last decades that statistical physics has grown into quantum-statistical physics, which includes some general aspects of nonequilibrium theories. In these years this progress was truly remarkable and was made hand in hand with developments in solid-state physics and related fields.

This text includes such recent developments in the fundamentals of statistical physics. It is not possible to cover the entirety of these subjects in these few pages. Our intention is to make this an elementary introduction to the subjects on the one hand, and to indicate to the reader the directions of future developments on the other.

The treatment is divided into two volumes. *Statistical Physics I*[2] is on equilibrium theories and *Statistical Physics II*[3] is on nonequilibrium theories. The first three chapters of Volume I form an introduction to statistical thermodynamics, specifically to the statistical mechanics of equilibrium states. The reader is expected to be acquainted only with elementary mechanics, such as the Hamiltonian equation of motion in classical mechanics and the concepts of quantum states in quantum mechanics.

Chapter 1(I) discusses some elements of mechanics and treats some simple problems which use only the concept of the average. These do not depend upon the precise meaning of the average and are therefore very illuminating. Chapter 2(I) reviews the skeleton structure of statistical mechanics. As mentioned earlier, statistical mechanics is based on a probabilistic assumption for microscopic states, namely the principle of equal weight. The logical problem of justifying this principle is discussed later in Chap. 5(I). Here the principle is accepted as a postulate, and we will see how the whole structure of statistical

---

[2] M. Toda, R. Kubo, N. Saitô: *Statistical Physics I*, Equilibrium Statistical Mechanics, Springer Ser. Solid-State Sci., Vol. 30 (henceforth denoted by I).

[3] R. Kubo, M. Toda, N. Hashitsume: *Statistical Physics II*, Nonequilibrium Statistical Mechanics, Springer Ser. Solid-State Sci., Vol. 31 (henceforth denoted by II).

thermodynamics is constructed on this basis. This is a standpoint commonly taken in textbooks, so that the construction in this chapter is not much different from that in other books. A beginner should study this chapter carefully.

Chapter 3(I) is devoted to a few applications. Their number is limited for reasons of space. These problems are basic and will be a good preparation for further study.

Chapter 4(I) treats the problem of phase change within the limitations of statistical thermodynamics. The dynamic aspects are not treated and remain as future problems. Even within this limitation, the problem is the most difficult and fascinating one in statistical physics. The first successful theory of the so-called "order–disorder" problem was the Weiss theory of ferromagnets, from which followed a number of approximate theories. Except for one- and two-dimensional models, no rigorous theory of phase transition exists. Rigorous treatments are discussed in this chapter for examples of lattice gases and low-dimensional Ising models. Approximations for three dimensions are discussed. Recently, more examples of rigorous solutions have been found and their intrinsic relations elucidated. These solutions are highly mathematical and are not treated here. While our treatment is also somewhat mathematical, a beginner need not read this through in detail. The problem of singularities associated with a second-order phase transition has been a central topic of statistical mechanics in recent years and is related to many important aspects of statistical physics. This is briefly touched upon along with the problem of critical indices. The reader is referred to other textbooks on the scaling and renormalization group theories.

Chapter 5(I) is devoted to a fundamental consideration of the mechanical basis of statistical mechanics, namely ergodic problems. We have limited ourselves primarily to classical ergodic problems based on classical mechanics. Quantum-mechanical ergodic theories are only briefly sketched. It is questionable whether classical ergodic theories can really be meaningful as a foundation of statistical mechanics; nevertheless, such theories have their own significance as a branch of physics and have made remarkable progress in recent years. Still, it is indeed a great pity for those engaged in research work in statistical physics that basic principles such as that of equal weight lack rigorous proof. The hope that someone among the readers may someday accomplish this ambitious task is one reason why this chapter was incorporated.

Nonequilibrium processes for which temporal evolution is to be considered explicitly are regarded as stochastic processes. This view is discussed in Chaps. 1 and 2(II). The theory of stochastic processes is an important field of mathematics. We emphasize its physical aspects and introduce the reader to the subject using Brownian motion as an example. Brownian motion is not merely random motion of a very fine particle; in general it is random motion of a physical quantity to be observed in a macrosystem. Such random motion is idealized to ideal Brownian motion in the same way that real gases are idealized to ideal gases in statistical thermodynamics. In this sense, the subject matter of Chap. 1(II) is basic to the whole framework of statistical physics. In particular, the fluctuation–dissipation theorem is the heart of the theory that provides a stepping-stone to the treatments in Chaps. 3–5(II).

To reach the macroscopic level of observation starting from the very fundamental microscopic level, we have to climb up successive levels of coarse graining. In going up each level of this staircase, a certain amount of information gets lost and a corresponding uncertainty is added to the probabilistic description. This is precisely the fundamental theme of statistical physics. However, it is difficult to formulate the program of coarse graining in a general way. Therefore, in Chap. 2(II) we treat a few relatively simple examples to show how this program is carried out, and finally we discuss the derivation of the master equation. Boltzmann's equation is somewhat removed from the main theme of this chapter, but we have included some basic matter on this subject. The Boltzmann equation is also important from a historical point of view, as well as from a conceptual and practical one. Difficult problems still remain in its derivation and generalization, but these are not touched upon here.

Chapters 3–5(II) review the developments in nonequilibrium statistical mechanics which have occurred in the past few decades. Chapter 3(II) is an introduction to these problems from their phenomenological side. It treats relaxation processes from nonequilibrium to equilibrium states and the response of a system near equilibrium to a weak external disturbance. These are linear irreversible processes belonging to a category of physics easily accessible as an extension of the well-founded statistical mechanics of equilibrium states. Theoretical methods for deriving from microscopic physics the relaxation and response functions for such linear processes are discussed in the linear response theory in Chap. 4(II).

Chapter 5(II) treats new developmens in quantum-statistical mechanics on the basis of Chap. 4(II). These developments are applications of the Green's functions and their perturbative calculations. These are among the most remarkable developments of recent years. The Green's function method can join the kinetic approach and equilibrium-statistical mechanics, though to a somewhat limited extent. The formalism takes advantage of the fact that microdynamics is quantum mechanics. The content of this chapter is rather condensed. For further study of the subject the reader is referred to a few well-known textbooks.

The number of pages far exceeded the total originally planned. In spite of this, there are many things which the authors regretfully omitted. We hope that the reader will feel inspired enough to go on and study the subjects at more advanced levels.

R. Kubo and *M. Toda*

Preface to the Second Edition

In this, the second edition of *Statistical Physics*, much new material has been introduced while the general plan and arrangement of the volume remain unchanged.

The subject itself has progressed considerably in recent years, especially in relation to the theory of phase changes and various aspects of the ergodic problems. In order to include recent developments of the theory of phase changes, more than half of Chap. 4 has been rewritten. It is hoped that the inclusion of additional material will elucidate the current point of view and the new methods employed in this fascinating branch of statistical physics. Chapter 5, which is devoted to the ergodic problems, has been fully revised to present contemporary knowledge of the ergodic behavior of mechanical systems, which has been actively investigated in the last few years by means of mathematical analysis, supported by numerical computation.

The authors have also taken advantage of the opportunity to correct typographical errors, and to revise some figures.

On behalf of the authors I thank the staff of Springer-Verlag for valuable assistance during the preparation of this edition.

Tokyo,
May 1991

Morikazu Toda

Contents

1. General Preliminaries

In this chapter, we introduce the subjects to be treated in terms of statistical mechanics. The principles of statistical mechanics, the topics of Chap. 2, are well established for the equilibrium state, while the kinetic theory of gases was developed to interpret the thermal properties of matter in the bulk. Though we will not be involved in kinetic theory in this chapter, we will clarify some important relations which can be derived by the use of averages with respect to configuration and motion of molecules.

1.1 Overview

1.1.1 Subjects of Statistical Mechanics

The macroscopic states of matter are determined by the properties and motion of the microscopic constituents such as molecules, atoms, electrons, and atomic nuclei, and the detailed behavior of electromagnetic fields. The motion of these constituents obey laws of mechanics including electromagnetic theory. However, since the number of the constituents, or the degrees of freedom, is extremely large, the laws of mechanics are not directly reflected in the properties of matter in bulk. Statistical mechanics forms bridges between microscopic and macroscopic worlds.

In the later half of the nineteenth century, there came thermodynamics which systematized general phenomenological laws of heat. Though thermodynamics may exclude considerations of the microscopic structure of matter, the results of thermodynamics can be transferred to statistical mechanics since the subjects to be discussed in both theories overlap in general. The term "statistical thermodynamics" is sometimes used to emphasize this relationship [1.1–12].

Since the number of constituents is extremely large, it is impossible to obtain a complete mechanical description of physical systems though they are microscopically subject to mechanical laws. As far as classical mechanics is concerned, it is impossible to solve the equations of motion exactly because we have as many equations as the number of constituents. Even if we could solve them, we would have to give the initial conditions concerning velocities and positions of all the constituents; but it is practically impossible to determine these initial values experimentally because of the extremely large number of particles involved in a macroscopic substance. In quantum mechanics, even if we could solve the Schrödinger equation for an extremely large number of particles, it is again impossible to experimentally determine all the parameters under given conditions.

There is another situation which makes it impossible to determine a quantum state exactly, even for a system in thermal equilibrium. It is because the eigenstates of a system with a large number of particles are generally very close. As we see for an ideal gas, the number of eigenstates in a given finite energy region increases exponentially with the number of particles in the system so that the energy interval between eigenstates decreases as $10^{-\alpha N}$ (α is of the order unity) with increasing number of particles N. Thus, for a macroscopic system composed of a large number of particles, the energy interval between eigenstates will be much smaller than the interaction energy between the system and its environment.

Further, in order to obtain knowledge of the quantum mechanical state of the system, we have to perform some observations which will give rise to an energy uncertainty ΔE. Since $\Delta E \approx \hbar/\Delta t$ due to the uncertainty principle, for an actual time Δt of observation, the energy uncertainty ΔE will be much larger than the energy interval for a macroscopic system. Therefore, it is impossible to know the eigenstate exactly.

We must note the difference between a system with a small number of degrees of freedom ($N \approx 1$) and a system with extremely large N, and that in statistical mechanics we deal with the limit $N \to \infty$. Thus, we have to give up a detailed mechanical description and investigate the probabilistic behavior of macroscopic systems.

A real substance is always in contact with its surroundings. If it is composed of a small number of constituents, its behavior will be subject to the way it is influenced by the surroundings. However, since a macroscopic system to be dealt with in statistical mechanics consists of an extremely large number of constituents interacting in a complicated fashion with the surroundings, its behavior may be expected to be described by certain probability laws, for the energy intervals between adjacent eigenstates will be much smaller than any disturbance due to the surroundings. In other words, any small disturbance will cause transitions between eigenstates, resulting in a uniform probabilistic distribution.

A macroscopic quantity of a system fluctuates, in general, with time. For example, the pressure of a gas is due to the collision of molecules with the wall of the container, and so it always fluctuates because of irregular molecular motion. In a real measurement, due to the inertia of the piston and other effects, the device will record a certain time average of the pressure. If we take the average over a very long time we will have a result which is independent of the initial conditions of the system. If a measurement was done in a comparatively small interval of time and still yielded practically the same result as the long time average, then the system would be in the state of equilibrium. When a system starts from a state out of equilibrium and is isolated, then it will approach the state of equilibrium in an interval of time called the relaxation time.

1.1.2 Approach to Equilibrium

The kinetic theory of gases demands a detailed analysis of molecular collisions to describe the process of approach to equilibrium. On the other hand, it is interesting to note that the statistical mechanics for the equilibrium state is formulated without any detailed analysis of complicated individual molecular motion. In the statistical mechanics for the equilibrium state, we deal with averages taken over an extremely long time, which is very much longer than the relaxation time, and thus irregular molecular motion is concealed behind the probabilistic nature of the microscopic state of the system.

In Chap. 2, we introduce the principles of the statistical mechanics for the equilibrium state. We will see that the assumption of equal a priori probabilities for the occupation of each quantum mechanical eigenstate (microscopic state) accessible to the system is a basic assumption in statistical mechanics. Since we have no complete proof of this assumption, we shall accept it as a principle, which is called the *principle of equal probability.*

If we refer to the kinetic theory, this assumption should be obtained by tracing the microscopic behavior of the system until it reaches the state of equilibrium. In classical mechanics, a kinetic equation describing the temporal evolution of the distribution of molecules in space and velocity space is called a master equation. The Boltzmann equation in the kinetic theory of gases and its semiquantum-mechanical modification, the Uhlenbeck equation, are master equations. These were derived intuitively and are thought of as approximate equations. Since the equilibrium state is established independently of simplified assumptions and approximations with respect to molecular collisions or so, there might be some model system which would explain the principle of equal probability.

In this respect, a probabilistic simple equation is worth mentioning. We think of a nearly isolated system and let suffixes *i*, *k*, etc., denote its quantum mechanical state. The system is supposed to be interacting in some way or other with the surroundings and therefore, transition between quantum states occurs. If we denote by p_{ki} the transition probability ($k \to i$), then we have a probabilistic equation of the form [1.5]

$$\frac{dw_i}{dt} = \sum_k w_k p_{ki} - \sum_k w_i p_{ik} \,, \tag{1.1.1}$$

where w_i stands for the probability that the system is in the quantum state *i*. We assume that all the quantum states are connected directly or indirectly by the transition probability and that

$$p_{ki} = p_{ik} \tag{1.1.2}$$

is satisfied. If all the w_k are the same, the right-hand side of (1.1.1) vanishes, and so the system does not change at all. We can show that when we start from any arbitrary initial distribution of w_k, after a sufficiently long time, all the w_k become equal, that is, we reach the distribution $w_1 = w_2 = \cdots = w_i = \cdots$. Therefore, if we exclude the nonergodic case, where transition to some particular states is forbidden, we ultimately have equal probability among quantum states. Though this is not an essential proof, it provides a standpoint for understanding the basic principles.

1.2 Averages

1.2.1 Probability Distribution

a) Consider a gas in a container. Even if the gas is not uniform initially, as time passes it will become uniform and the change will stop when the density becomes the same everywhere (Fig. 1.1). However, since the gaseous molecules are moving nearly independently, there must be some fluctuation in density, though we know empirically that the fluctuation is quite small. We shall examine it.

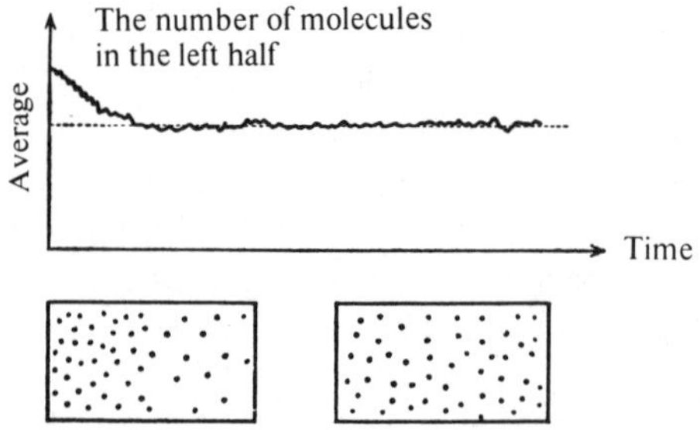

Fig. 1.1. Temporal change in the distribution of molecules in a container

We divide the volume V occupied by a gas into two parts V_1 and V_2, so that $V = V_1 + V_2$. The probabilities for a molecule to be in each part are

$$p = \frac{V_1}{V}, \quad q = \frac{V_2}{V} = 1 - p\,.$$

We have assumed that the molecules are moving independently. In quantum mechanics, this assumption does not hold; we shall discuss this situation later. Now, let the total number of molecules be N. The way of choosing N_1 molecules to be placed in V_1 is $N!/N_1!(N-N_1)!$, and the way of placing N_1 molecules in V_1 and $N_2 = N - N_1$ molecules in V_2 is proportional to $V_1^{N_1}V_2^{N_2}$ or to $p^{N_1}(1-p)^{N_2}$. If we normalize the total probability to unity, the probability that

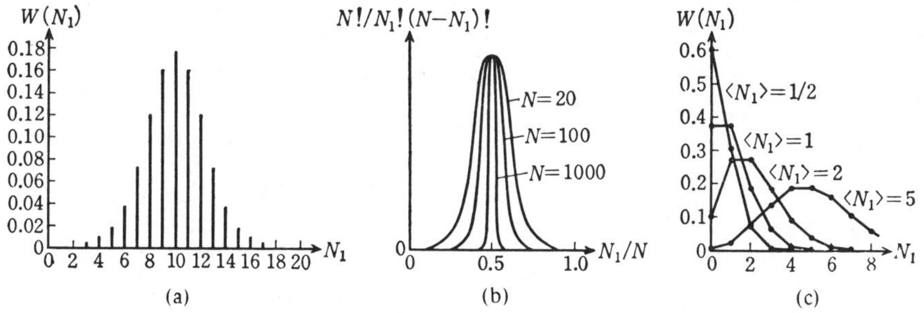

Fig. 1.2 a–c. Binomial distribution. (**a**) $N = 20$, $p = 1/2$. (**b**) Binomial distribution for different N. (**c**) $p \ll 1$, Poisson's distribution Eq. (1.2.1b)

N_1 molecules are in V_1 and N_2 molecules are in V_2 is given by [1.13]

$$W(N_1) = \frac{N!}{N_1!(N-N_1)!}\, p^{N_1}(1-p)^{N-N_1} \,. \tag{1.2.1a}$$

This is called the binomial distribution. When $N \gg 1$, this distribution is represented by a Gaussian error curve (Fig. 1.2) and the location of its maximum gives the average value of N_1. When $p \ll 1$, the binomial distribution can be approximated by Poisson's distribution (Fig. 1.2c)

$$W(N_1) = \frac{\langle N_1 \rangle^{N_1}}{N_1!} \exp(-\langle N_1 \rangle) \,, \tag{1.2.1b}$$

where $\langle N_1 \rangle$ stands for the average of N_1. Average values for the binomial distribution are

$$\langle N_1 \rangle = Np, \quad \langle N_2 \rangle = N(1-p)$$

which are the average number, or the expectation number of the molecules in V_1 and V_2, respectively. The average of the square of deviation can be calculated using (1.2.1a), and we have

$$\langle (N_1 - \langle N_1 \rangle)^2 \rangle = \sum_{N_1} (N_1 - \langle N_1 \rangle)^2 W(N_1) = Np(1-p) \,.$$

Therefore, we have the mean square fluctuation

$$\frac{\langle (N_1 - \langle N_1 \rangle)^2 \rangle}{\langle N_1 \rangle^2} = \frac{1-p}{\langle N_1 \rangle} \,. \tag{1.2.2}$$

The square root of this value is the *relative fluctuation* which is proportional to $1/\sqrt{\langle N_1 \rangle}$. If $\langle N_1 \rangle$ is large, the fluctuation is very small and may be neglected. In general, the fluctuation of a macroscopic quantity is small, since it is inversely proportional to the square root of the number of molecules involved.

The above result also applies to a system of independent spins (Fig. 1.3). Under an external magnetic field, let the probabilities for a spin to be oriented

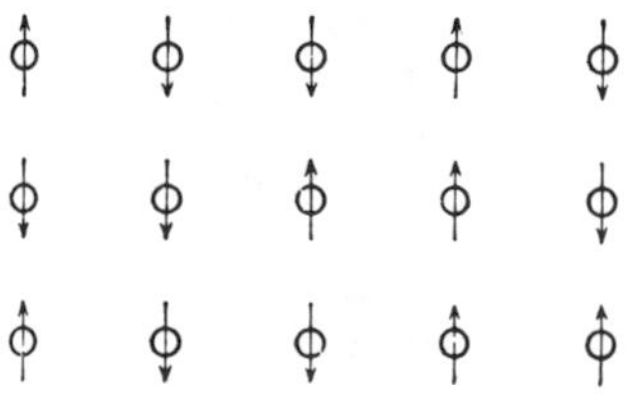

Fig. 1.3. A system of spins

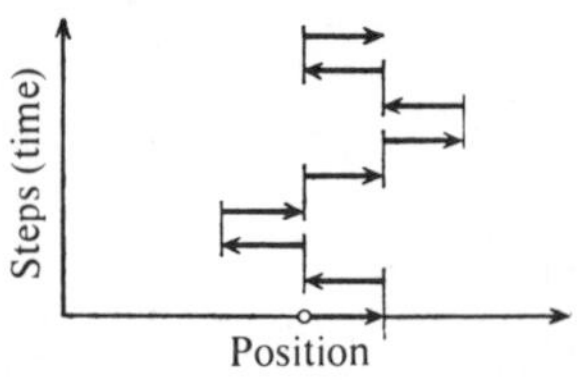

Fig. 1.4. One-dimensional Brownian motion

upwards be p and downwards be $q = 1 - p$. Then, for a system of N independent spins, the probability that there are N_1 up-spins and $N - N_1$ down-spins is given by $W(N_1)$ of (1.2.1).

The same equation also applies to one-dimensional Brownian motion (Fig. 1.4). If p is the probability of a step to be directed to the right and $q = 1 - p$ is that to the left, then $W(N_1)$ gives the probability that the displacement to the right is $N_1 - N_2 = 2N_1 - N$ after N steps. Of course, this is the case where successive steps have no correlation, namely, the case of the Markov process (Chap. 5). The same idea also applies to the calculation of the diffusion of a particle, the extension of a polymer and rubber elasticity.

b) Let us consider the distribution of particles with correlation. Correlation may be due to molecular force. In quantum statistics, the configuration in coordinate space depends on the configuration in momentum space, and thus correlation exists between particles.

Consider a system composed of identical N particles. If $\boldsymbol{r}_j$ $(j = 1, 2, \ldots, N)$ denotes the positions of these particles, then the *number density* at $\boldsymbol{r}$ is given by

$$n^{(1)}(\boldsymbol{r}) = \langle n(\boldsymbol{r}) \rangle, \quad n(\boldsymbol{r}) = \sum_{j=1}^{N} \delta(\boldsymbol{r}_j - \boldsymbol{r}) , \tag{1.2.3}$$

where $\langle \ \rangle$ means the average. It may be a quantum-mechanical expectation value, or a time average, or ensemble average, but at the present moment it is sufficient to suppose that it is an average in a certain sense. $n^{(1)}(\boldsymbol{r})$ is a *one-body density*, while the *two-body density* is

$$n^{(2)}(\boldsymbol{r}, \boldsymbol{r}') = \left\langle \sum_{j \neq k} \delta(\boldsymbol{r}_j - \boldsymbol{r}) \delta(\boldsymbol{r}_k - \boldsymbol{r}') \right\rangle \tag{1.2.4}$$

which expresses the correlation in densities at $\boldsymbol{r}$ and $\boldsymbol{r}'$ (Fig. 1.5a). We write

$$n^{(2)}(\boldsymbol{r}, \boldsymbol{r}') = n^{(1)}(\boldsymbol{r}) n^{(1)}(\boldsymbol{r}') g(\boldsymbol{r}, \boldsymbol{r}' - \boldsymbol{r})$$

and call $g(\boldsymbol{r}, \boldsymbol{r}' - \boldsymbol{r})$ the *pair distribution function*. For an isotropic substance like a gas or a liquid, $\langle n \rangle = n^{(1)}(\boldsymbol{r})$ may be constant except in the vicinity of container walls and we can write

$$n^{(2)}(\boldsymbol{r}, \boldsymbol{r}') = \langle n \rangle^2 g(R), \quad R = |\boldsymbol{r}' - \boldsymbol{r}| . \tag{1.2.5}$$

$g(R)$ is called the *radial distribution function* which can be obtained by X-ray

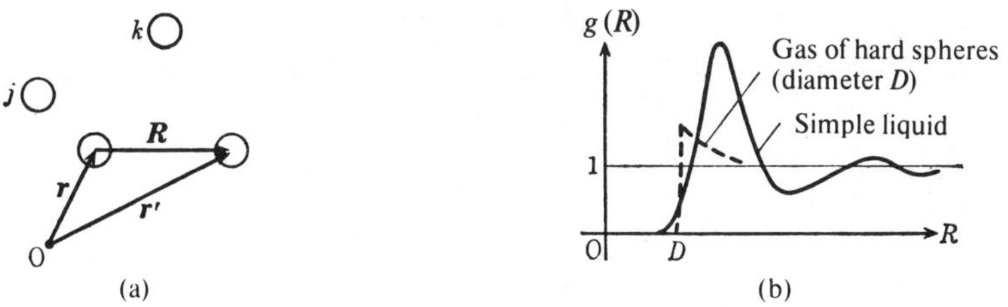

Fig. 1.5 a, b. Radial distribution function. (**a**) Relative position $\boldsymbol{R}$. (**b**) Schematic curve $g(R)$

scattering experiments. When the system is large, the correlation usually vanishes for $R \to \infty$, and we have $g(R) \to 1$ (Fig. 1.5b).

Now, we integrate the two-body density with respect to $\boldsymbol{r}$ and $\boldsymbol{r}'$. The sum with the restriction $j \neq k$ is given as the double sum without the restriction subtracted by a simple sum. Integrating over a volume v, we denote by N_v the number of particles in this volume. Then we have

$$\iint_v n^{(2)}(\boldsymbol{r}, \boldsymbol{r}')d\boldsymbol{r}\,d\boldsymbol{r}' = \langle N_v^2 \rangle - \langle N_v \rangle \,, \tag{1.2.6}$$

where

$$N_v = \int_v \sum_{j=1}^{N} \delta(\boldsymbol{r}_j - \boldsymbol{r})d\boldsymbol{r}$$

and when the volume v is far from the container walls

$$\langle N_v \rangle = \langle n \rangle v \,.$$

On the other hand, when the volume is far from the container walls, we have

$$\iint_v [n^{(2)}(\boldsymbol{r}, \boldsymbol{r}') - \langle n \rangle^2]d\boldsymbol{r}\,d\boldsymbol{r}' = \langle n \rangle^2 v \int_v [g(R) - 1]d\boldsymbol{R}$$

$$= \langle N_v^2 \rangle - \langle N_v \rangle - \langle N_v \rangle^2 \,.$$

Therefore, using the relation $\langle (N_v - \langle N_v \rangle)^2 \rangle = \langle N_v^2 \rangle - \langle N_v \rangle^2$, we have

$$\frac{\langle (N_v - \langle N_v \rangle)^2 \rangle}{\langle N_v \rangle^2} = \frac{1}{\langle N_v \rangle}\left\{1 + \langle n \rangle \int_v [g(R) - 1]d\boldsymbol{R}\right\}. \tag{1.2.7}$$

$g(R)$ usually approaches 1 quite rapidly with increasing R. Thus, when the volume V of the system is sufficiently large and v is a small part of it, then the integral on the right-hand side of (1.2.7) is independent of v (we assume that v is much larger than the size of a molecule).

However, when v is comparable with the total volume V, the above-mentioned integral depends on v. If v is equal to V, then N_v equals N and we have no fluctuation, so that the right-hand side of (1.2.7) must vanish.

In $n^{(2)}(\boldsymbol{r}, \boldsymbol{r}')$, $\langle \delta(\boldsymbol{r}_1 - \boldsymbol{r}) \sum_{k(\neq 1)} \delta(\boldsymbol{r}_k - \boldsymbol{r}') \rangle$ is a term which expresses the probability density at $\boldsymbol{r}'$ when the particle $j = 1$ is at $\boldsymbol{r}$. Therefore, $n^{(1)}(\boldsymbol{r}')g(\boldsymbol{R})$ is the density at $\boldsymbol{r}'$ under the condition that a particle is at $\boldsymbol{r}$. For example, for classical independent particles, when a particle is at $\boldsymbol{r}$, the remaining $N-1$ particles are in the total volume V, and thus

$$\langle n \rangle g(R) = \frac{N-1}{V}, \quad \text{or} \quad g(R) = 1 - \frac{1}{N}.$$

Therefore

$$\langle n \rangle \int_v [g(R) - 1] d\boldsymbol{R} = -\frac{v}{V},$$

where $v/V = p$ is the probability for a particle to be in the volume v. By (1.2.7), the square of the relative fluctuation is thus $(1-p)/\langle N_v \rangle$, which coincides with (1.2.2).

Another extreme case is a system of closely packed hard spheres where fluctuation is nearly impossible, and we must have $\langle n \rangle \int [g(R) - 1] d\boldsymbol{R} \approx -1$.

Taking the Fourier transform of the deviation of the local density $n(\boldsymbol{r})$ from its mean value $\langle n(\boldsymbol{r}) \rangle$, [cf. (1.2.3)], we can calculate the mean square of the Fourier component of the density fluctuation. Thus we obtain

$$\left\langle \left| \int_v [n(\boldsymbol{r}) - \langle n \rangle] \exp(-\mathrm{i}\boldsymbol{f}\cdot\boldsymbol{r}) d\boldsymbol{r} \right|^2 \right\rangle$$

$$= \langle n \rangle v \{1 + \langle n \rangle v \int [g(R) - 1] \exp(-\mathrm{i}\boldsymbol{f}\cdot\boldsymbol{R}) d\boldsymbol{R}\}. \tag{1.2.8}$$

Equations (1.2.7, 8) are relations which connect macroscopic quantities (the left-hand side) to the microscopic radial distribution function $g(R)$ on the right-hand side of these equations.

c) Finally, we shall consider the fluctuation of the intensity of irregular waves. As an example, we consider classical electromagnetic waves and find the intensity of light at an arbitrary point illuminated by many sources. For simplicity, we assume that the frequencies of light from sources and their intensities are the same, and we write the electric field due to the jth source as $a\exp(\mathrm{i}\varphi_j)$. Commutation relations between amplitude and phase are not in question because we are considering a classical electromagnetic field. The intensity of light I is proportional to the square of the electric field $\mathscr{E}$, so that

$$I \propto \mathscr{E}^2 = a^2 \left[\sum_{k=1}^{N} \exp(\mathrm{i}\varphi_k) \right]^* \left[\sum_{j=1}^{N} \exp(\mathrm{i}\varphi_j) \right]$$

$$= a^2 \left[N + 2 \sum_{j>k} \cos(\varphi_j - \varphi_k) \right].$$

Since the phases φ_j are independent, the average gives

$$\langle I \rangle \propto N a^2 \tag{1.2.9}$$

which is proportional to the number of light-sources, as expected. In order to see the fluctuation of light intensity, we must calculate I^2 or $\mathscr{E}^4$. Noting that

$$\left\langle\left[2\sum_{j>k}\cos(\varphi_j-\varphi_k)\right]^2\right\rangle = 4\frac{N(N-1)}{2}\langle\cos^2(\varphi_j-\varphi_k)\rangle$$
$$= N(N-1)\,,$$

we have

$$\langle I^2\rangle \propto (2N^2-N)a^4\,.$$

Or, using the relation $\langle(I-\langle I\rangle)^2\rangle = \langle I^2\rangle - \langle I\rangle^2$, we have for $N \gg 1$,

$$\frac{\langle(I-\langle I\rangle)^2\rangle}{\langle I\rangle^2} = 1\,. \tag{1.2.10}$$

If we compare this with the density fluctuation of particles $\langle(n-\langle n\rangle)^2\rangle/\langle n\rangle^2 = 1/\langle n\rangle$, [cf. (1.2.2)], we see that the fluctuation of light intensity is much more remarkable. The same can be said for waves other than light. If we consider light as a system of photons, we may say that photons exhibit remarkable fluctuation compared to classical particles, so that photons have a tendency to move together. Quantum mechanically, a similar result is obtained for photons (Sect. 3.1.1). In general, in quantum statistics it is shown that Bose particles have such a tendency.

It may be noted that the above treatment is similar to two-dimensional Brownian motion. In a complex plane we may add electric fields $a\exp(i\varphi_j)$ (Fig. 1.6). When the phases φ_j are at random, the result will be the same as for a two-dimensional random walk. If φ_j is restricted to 0 and π, it reduces to one-dimensional Brownian motion with $p = q = 1/2$ in (1.2.1a).

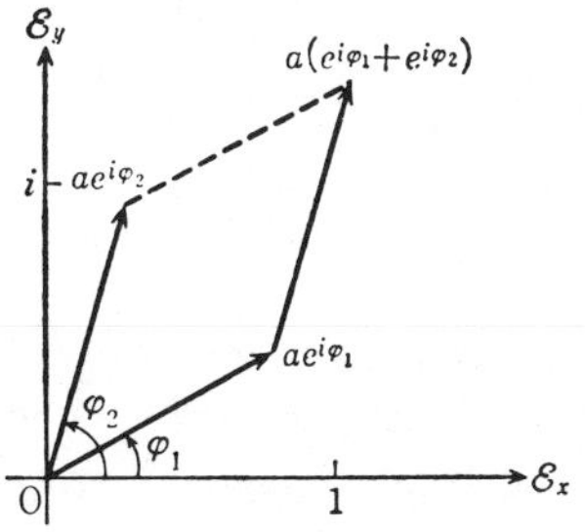

Fig. 1.6. Composition of the electric field of light and its relation to two-dimensional Brownian motion

1.2.2 Averages and Thermodynamic Fluctuation

We have so far taken a microscopic point of view to derive equations related to macroscopic quantities such as density and density fluctuation. If we replace these macroscopic quantities by thermodynamic ones, we will be led to new relations. Thermodynamic equations for fluctuations are derived by making use of the relationship between entropy and probability, which we assume to be known already.

For example, let us consider a small volume v ($\ll V$) of a large system in thermal equilibrium. The fluctuation of the number of molecules in v is related, in thermodynamics, to the isothermal compressibility

$$\kappa_T = -\frac{1}{v}\left(\frac{\partial v}{\partial P}\right)_{T,\langle N_v\rangle} \tag{1.2.11}$$

(T = temperature, P = pressure) by

$$\frac{\langle (N_v - \langle N_v\rangle)^2\rangle}{\langle N_v\rangle^2} = \frac{kT}{v}\kappa_T\,, \tag{1.2.12}$$

where k is the Boltzmann constant. Thus, in view of (1.2.7), we obtain

$$kT\kappa_T = \frac{1}{\langle n\rangle} + \int [g(R) - 1]d\boldsymbol{R} \tag{1.2.13}$$

which is called the *Ornstein-Zernike relation* or the *compressibility equation.* This relation applies to a quantum-mechanical system as well. For usual liquids, compressibility is very small ($kT\kappa_T\langle n\rangle \ll 1$) which means $\langle n\rangle \int [g(R) - 1]d\boldsymbol{R} \approx 1$, as already mentioned. For classical gases subject to Boyle–Charles' law $P = \langle n\rangle kT$. We have $kT\kappa_T\langle n\rangle = 1$ and thus $\langle n\rangle \int [g(R) - 1]d\boldsymbol{R} = 0$. At the critical temperature the compressibility becomes infinitely large showing that correlations between molecules are long-ranged near the critical point.

On the contrary, if we assume that there is no correlation in the positions of molecules, for a small volume v ($v/V \approx 0$), we may neglect $\int [g(R) - 1]d\boldsymbol{R}$ in (1.2.13) to have $kT\kappa_T = 1/\langle n\rangle$, or noting $\langle n\rangle = \langle N_v\rangle/v$,

$$-\frac{kT}{v^2}\left(\frac{\partial v}{\partial P}\right)_{T,\langle N_v\rangle} = \frac{1}{\langle N_v\rangle}\,.$$

Therefore, if we integrate with respect to P and put $P = 0$ for $v = \infty$, we have

$$P = \frac{\langle N_v\rangle kT}{v}\,.$$

Thus Boyle-Charles' law is obtained by simply assuming the absence of correlations for the positions of molecules. However, in quantum mechanics, such correlations exist even when there are no molecular forces, and Boyle-Charles' law does not apply (Chap. 3).

If we assume a classical gas of hard spheres, molecules cannot come closer than their diameter D. For a dilute gas we may approximately put

$$g(R) = \begin{cases} 0 & (R < D) \\ 1 & (R > D)\,. \end{cases}$$

In this approximation we have

$$-\frac{kT}{v^2}\left(\frac{\partial v}{\partial P}\right)_{T,\langle N_v\rangle} = \frac{1}{\langle N_v\rangle} - \frac{4\pi}{3}\frac{D^3}{v}\,.$$

If we integrate with respect to P and put $P = 0$ for $c = \infty$, we obtain

$$P = \frac{kT}{(4\pi/3)D^3} \ln \frac{c}{v - \langle N_v \rangle (4\pi/3)D^3}$$
$$\approx \frac{\langle N_v \rangle kT}{v} \left(1 + \frac{\langle N_v \rangle}{v} \frac{2\pi}{3} D^3 \right) \approx \frac{\langle N_v \rangle kT}{v - b} ,$$

where we have put $b = \langle N_v \rangle 2\pi/3D^3 \ll v$. The last equation expresses the effect of molecular size in the famous van der Waals' equation of state.

It may be noted that such equations for gaseous states are irrespective of the dynamics of molecular motion. For example, Boyle-Charles' law is valid in Newtonian mechanics as well as in relativistic mechanics if there is no correlation of molecular positions, or if the Hamiltonian (Chap. 2) is split into terms, each depending only on the momentum of a molecule.

Similar consideration also applies to magnetization. If χ denotes magnetic susceptibility, the fluctuation of magnetization M is thermodynamically given as

$$\langle (M - \langle M \rangle)^2 \rangle = kT\chi . \tag{1.2.14}$$

If we denote by μ_j the magnetic moment of the jth spin, we have

$$M = \sum_{j=1}^{N} \mu_j$$

$$\langle M^2 \rangle = \sum_j \langle \mu_j^2 \rangle + \sum_{j \neq k} \langle \mu_j \mu_k \rangle .$$

If the spins are mutually independent or without correlation, in the limit of the vanishing external magnetic field, $\langle M \rangle = 0$, we have

$$\langle \mu_j \mu_k \rangle = \langle \mu_j \rangle \langle \mu_k \rangle = 0 \quad (j \neq k)$$

and $\langle M^2 \rangle = N \langle \mu_j^2 \rangle$. Thus (1.2.14) gives the magnetic susceptibility

$$\chi = \frac{C}{T}, \quad C = \frac{N \langle \mu_j^2 \rangle}{k} \tag{1.2.15}$$

which is inversely proportional to the absolute temperature. This is the well-known *Curie's law*. For a ferromagnetic substance, χ diverges which means that $\langle \mu_j \mu_k \rangle$ has a large correlation length.

1.2.3 Averages of a Mechanical System – Virial Theorem

So far we have been concerned with the spacial distribution of constituent particles of a system, and have seen that certain general properties of the system can be derived from the averages and by further applying thermodynamic considerations.

The main purpose of statistical mechanics is to establish a general method of deriving thermal properties from the known Hamiltonian of the system. Thus it is concerned with the distribution of the constituents in coordinate space as well as in momentum space. In Chap. 2, we will see that such a distribution is most generally given in terms of Gibbs' ensemble. But before that, we will extend our consideration of averaging to relations derived from equations of motion for the constituent particles or the Hamiltonian of the sytem.

We shall first consider a perfect gas again. The pressure upon a rigid container wall of a gas arises from forces exerted upon it by those molecules which happen to be undergoing collision with it. If the wall is perpendicular to the x-axis, the pressure is given as the change in momentum in the x-direction of the molecules colliding with the wall per unit area and per unit time (Fig. 1.7). This is twice the momentum flow in the positive x-direction. If we denote by p_x the x-component of the momentum of a molecule with the velocity component $\dot{x}$, and by $n(\dot{x})$ the number of such molecules per unit volume, then the pressure P is given as

$$P = \sum_{\dot{x}>0} 2p_x\dot{x}n(\dot{x}) = \sum_{\dot{x}\gtrless 0} p_x\dot{x}n(\dot{x}) = \langle n\rangle\langle p_x\dot{x}\rangle ,$$

where we have used the averaged number density $\langle n\rangle$ and $\langle p_x\dot{x}\rangle$ is the average of $p_x\dot{x}$.

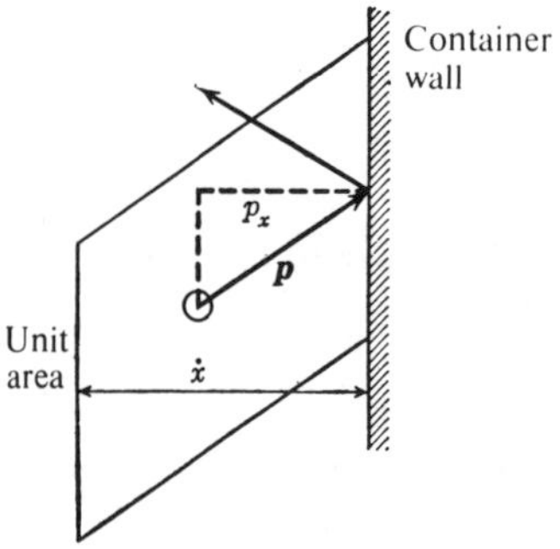

Fig. 1.7. Calculation of gas pressure

Let $q = (x, y, z)$ be the coordinate in general, and let q_j be the coordinate of the jth molecule. If $\mathscr{H}$ denotes the Hamiltonian, we have

$$\dot{q}_j = \frac{\partial\mathscr{H}}{\partial p_j}$$

and we may write

$$PV = \frac{1}{3}\left\langle \sum_{j=1}^{3N} p_j \frac{\partial\mathscr{H}}{\partial p_j}\right\rangle ,$$

where $\sum_j$ denotes the sum over $3N$ degrees of freedom of all the molecules in the container with volume V, and since the sum is extended over y and z as well as x, the coefficient $\frac{1}{3}$ has been introduced. Comparing the above result with

Boyle–Charles' law, we see that

$$\left\langle p_j \frac{\partial \mathscr{H}}{\partial p_j} \right\rangle = kT$$

holds for each degree of freedom. The microscopic meaning of the thermodynamic absolute temperature is implied in this equation, and it is valid both for Newtonian mechanics and relativistic mechanics. The above relation is usually known as the *law of the equipartition of energy*, which holds for any momentum component in classical statistics. We can extend the law to coordinates when they are contained in the Hamiltonian, as will be shown in what follows.

If the Hamiltonian includes a coordinate q conjugate to a momentum p, canonical equations of motion can be written as

$$\dot{q} = (q, \mathscr{H}), \quad \dot{p} = (p, \mathscr{H})$$

both in classical mechanics and quantum mechanics. Here, (A, B) denotes the Poisson bracket in these mechanics. Differentiating the product pq with respect to time, we have

$$\frac{d}{dt}(pq) = p(q, \mathscr{H}) + (p, \mathscr{H})q \, .$$

We assume that the motion is restricted in a region of q by the potential part of the Hamiltonian depending on q. Then the time average of the left-hand side of the above equation vanishes and so does the right-hand side. In classical mechanics, $(q, \mathscr{H}) = +\partial\mathscr{H}/\partial p$ and $(p, \mathscr{H}) = -\partial\mathscr{H}/\partial q$, so that the average over a sufficiently long time yields

$$\left\langle \frac{\partial \mathscr{H}}{\partial q} q \right\rangle = \left\langle p \frac{\partial \mathscr{H}}{\partial p} \right\rangle . \tag{1.2.16}$$

For an ideal gas, the interaction U_w between the wall and the molecules may be included in $\mathscr{H}$. Then $\partial U_w/\partial q_{xj}$ is the force exerted by the jth molecule on the wall perpendicular to the x-axis. Since the sum of such forces gives rise to a pressure, then if we take the sum of the left-hand side of (1.2.16) for a rectangular box of side-length L, we obtain $3PL^2 \cdot L = \sum_j \langle p_j \partial\mathscr{H}/\partial p_j \rangle$, which is an equation we have obtained earlier. The same can be said formally for quantum mechanical systems.

When the particles interact, we rewrite the left-hand side of (1.2.16) separating the part due to U_w from the part due to the Hamiltonian $\mathscr{H}$ which includes the interaction, in such a way that

$$3PV + \sum_j \left\langle \frac{\partial \mathscr{H}}{\partial q_j} q_j \right\rangle = \sum_j \left\langle p_j \frac{\partial \mathscr{H}}{\partial p_j} \right\rangle . \tag{1.2.17}$$

If the interaction can be written as a pair-wise sum of the potential $\phi(r_{jk})$, we have

$$\sum_l \frac{\partial \mathscr{H}}{\partial q_l} q_l = \sum_{j>k=1}^{N} r_{jk} \frac{\partial \phi(r_{jk})}{\partial r_{jk}} .$$

Thus we obtain the pressure equation

$$PV = \frac{1}{3}\left\langle \sum_{j=1}^{N} \boldsymbol{p}_j \cdot \frac{\partial \mathscr{H}}{\partial \boldsymbol{p}_j} \right\rangle - \frac{1}{3}\left\langle \sum_{j>k=1}^{N} r_{jk} \frac{\partial \phi(r_{jk})}{\partial r_{jk}} \right\rangle . \tag{1.2.18}$$

This is called the *virial theorem*, which is valid in classical statistics as well as in quantum statistics. $\frac{1}{2}\langle \sum r_{jk} \partial \phi(r_{jk})/\partial r_{jk} \rangle$ is called the inner virial and $\frac{3}{2}PV$ is called the outer virial. The sum of the inner virial and the outer virial is equal to $\frac{1}{2}\langle \sum \boldsymbol{p}_j \cdot \partial \mathscr{H}/\partial \boldsymbol{p}_j \rangle$ which is, in classical mechanics, equal to the kinetic energy. The virial theorem holds for a stationary motion even if the pressure is zero.

In classical mechanics, the virial theorem can be derived from the Hamiltonian principle which can be written as

$$\delta \int \mathscr{L}\, dt = \int \left(\frac{\partial \mathscr{L}}{\partial q} \delta q + \frac{\partial \mathscr{L}}{\partial \dot{q}} \delta \dot{q} \right) dt = 0 , \tag{1.2.19}$$

where $\mathscr{L}$ is the Lagrange function. For a simple case where the motion is periodic in q, the variation δq may be taken to be proportional to q in such a way that

$$\delta q = \varepsilon q, \quad \delta \dot{q} = \varepsilon \dot{q} .$$

If we denote the potential energy by U, by the property of the Lagrange function we have

$$\frac{\partial \mathscr{L}}{\partial q} = -\frac{\partial U}{\partial q}, \quad \frac{\partial \mathscr{L}}{\partial \dot{q}} = p$$

and thus

$$\varepsilon \int \left(-\frac{\partial U}{\partial q} q + p\dot{q} \right) dt = 0$$

which means that for the time average, $\langle (\partial U/\partial q) q \rangle = \langle p\dot{q} \rangle$. This is the result of the virial theorem (1.2.16) for this degree of freedom. If we take the summation over all the coordinates, we obtain the usual theorem. Further, since

$$\int \left(-\frac{\partial U}{\partial q} q + p\dot{q} \right) dt = \int \left(p \frac{\partial \mathscr{H}}{\partial p} - q \frac{\partial \mathscr{H}}{\partial q} \right) dt = \int \frac{d}{dt}(pq) dt , \tag{1.2.20}$$

if we take the ensemble average in the phase space (pq-space) instead of the time average, we obtain the virial theorem again when the ensemble is stationary with respect to time. Such a steady ensemble is guaranteed by the Liouville theorem to be treated in the next section.

Corresponding to the above variation principle in classical mechanics, we have a quantum-mechanical variation principle

$$\delta(\psi, \mathscr{H}\psi) = 0, \quad (\psi, \psi) = 1 ,$$

where $(\psi, \varphi) = \int \psi^* \varphi \, dx$ is the inner product. The variation is taken with respect to the wave-function ψ. We may confine the variation to that induced by the uniform expansion or contraction of each coordinate. For example, with respect to x_1, we assume the change

$$x_1 \to (1 + \varepsilon) x_1 \quad (|\varepsilon| \ll 1)$$

$$\psi(x_1, x_2, \ldots) \to C\psi[(1 + \varepsilon)x_1, x_2, \ldots] .$$

Then $(\psi, \mathscr{H}, \psi)/(\psi, \psi)$ is transformed into

$$\frac{\int \cdots \int \psi^*[(1+\varepsilon)x_1, x_2, \ldots] \, \mathscr{H}(x_1, x_2, \ldots) \, \psi[(1+\varepsilon)x_1, x_2, \ldots] dx_1 dx_2 \cdots}{\int \cdots \int \psi^*[(1+\varepsilon)x_1, x_2, \ldots] \, \psi[(1+\varepsilon)x_1, x_2, \ldots] dx_1 dx_2 \cdots}$$

$$= \frac{\int \cdots \int \psi^*(y_1, x_2, \ldots) \, \mathscr{H}\left(\frac{y_1}{1+\varepsilon}, x_2, \ldots\right) \psi(y_1, x_2, \ldots) dy_1 dx_2 \cdots}{\int \cdots \int \psi^*(y_1, x_2, \ldots) \, \psi(y_1, x_2, \ldots) dy_1 dx_2 \cdots} .$$

Therefore, the variation principle implies

$$\delta(\psi, \mathscr{H}\psi) = \int \cdots \int \psi^*(y_1, x_2, \ldots) \, \delta\mathscr{H} \psi(y_1, x_2, \ldots) dy_1 dx_2 \cdots = 0 ,$$

where

$$\delta\mathscr{H} = \mathscr{H}\left(\frac{y_1}{1+\varepsilon}, x_2, \ldots\right) - \mathscr{H}(y_1, x_2, \ldots) .$$

For example, if the Hamiltonian is of the form

$$\mathscr{H} = -\frac{1}{2m_1}\frac{\partial^2}{\partial x_1^2} - \cdots + U(x_1, x_2, \ldots) ,$$

then to the first order in ε we have

$$\delta\mathscr{H} = -\varepsilon\left\{\frac{1}{m_1}\frac{\partial^2}{\partial y_1^2} + y_1\frac{\partial U}{\partial y_1}\right\} ,$$

and if y_1 is rewritten as x_1,

$$\int \cdots \int \psi^*(x_1, x_2, \ldots)\left\{\frac{1}{m_1}\frac{\partial^2}{\partial x_1^2} + x_1\frac{\partial U}{\partial x_1}\right\}\psi(x_1, x_2, \ldots) dx_1 dx_2 \cdots$$
$$= 0 .$$

Thus, we obtain the virial theorem for the coordinate x_1:

$$\left\langle -\frac{1}{m_1}\frac{\partial^2}{\partial x_1^2}\right\rangle = \left\langle x_1\frac{\partial U}{\partial x_1}\right\rangle ,$$

where $\langle\ \rangle$ denotes the quantum-mechanical expectation value, that is, $\langle A \rangle = (\psi, A\psi)$.

We may consider the real change of volume of a system in a volume V. For simplicity, take a regular cube of side-length L so that $V = L^3$, and let the Schrödinger equation be

$$\left\{-\frac{1}{2}\sum\frac{\partial^2}{\partial x^2} + U(x)\right\}\psi(x) = E\psi(x) \quad (0 < x < L)\,.$$

When we change the variable x to λx, ψ satisfies

$$\left\{-\frac{1}{2\lambda^2}\sum\frac{\partial^2}{\partial x^2} + U(\lambda x)\right\}\psi(\lambda x) = E\psi(\lambda x) \quad \left(0 < x < \frac{L}{\lambda}\right).$$

If we write $\lambda = 1 + \varepsilon$ and assume that $|\varepsilon| \ll 1$, we have

$$\left[-\frac{1}{2}\sum\frac{\partial^2}{\partial x^2} + U(x) + \varepsilon\left\{\sum\frac{\partial^2}{\partial x^2} + \sum x\frac{\partial U}{\partial x}\right\}\right]\psi(\lambda x) = E\psi(\lambda x)\,.$$

In this equation, the term inside the square brackets can be considered as a new Hamiltonian. E is the eigenvalue, the size of the system being V/λ^3, or

$$V/\lambda^3 = V + dV \quad (dV = -3\varepsilon V)\,.$$

Since the real Hamiltonian of the system is not changed, we let the term in the curly brackets $\{\ \}$ in the above equation be eliminated by adding a term of the same as $\{\ \}$ but with the opposite sign; the change in the eigenvalue is then

$$\begin{aligned} dE &= -\varepsilon\int\psi^*(\lambda x)\left\{\sum\frac{\partial^2}{\partial x^2} + \sum x\frac{\partial U}{\partial x}\right\}\psi(\lambda x)dx \Big| \int\psi^*(\lambda x)\psi(\lambda x)dx \\ &= -\varepsilon\int\psi^*(x)\left\{\sum\frac{\partial^2}{\partial x^2} + \sum x\frac{\partial U}{\partial x}\right\}\psi(x)dx \Big| \int\psi^*(x)\psi(x)dx \end{aligned}$$

to first order in ε. Thus, the change in the eigenvalue due to the volume change is given as [1.14]

$$-\frac{dE}{dV} = \frac{2}{3V}\left\{\left\langle -\frac{1}{2}\sum\frac{\partial^2}{\partial x_j^2}\right\rangle - \left\langle \frac{1}{2}\sum x_j\frac{\partial U}{\partial x_j}\right\rangle\right\}, \tag{1.2.21}$$

where $\langle\ \rangle$ means the quantum-mechanical expectation values. The first term on the right-hand side of the above equation is the kinetic energy and the second is the virial.

When averaged over all the eigenstates with appropriate probability, we will obtain thermodynamic pressure (Sect. 2.3.2). The equation for pressure is thus

$$PV = \frac{2}{3}\left\{\left\langle -\frac{1}{2}\sum\frac{\partial^2}{\partial x_j^2}\right\rangle - \left\langle \frac{1}{2}\sum x_j\frac{\partial U}{\partial x_j}\right\rangle\right\}, \tag{1.2.22}$$

where $\langle\ \rangle$ now represents the statistical average of quantum-mechanical expectation values.

1.3 The Liouville Theorem

1.3.1 Density Matrix

In quantum mechanics, a physical quantity A is given by a Hermitian operator. If a wave-function $\psi(x, t)$ represents the state of a system, then the expectation value of A is given by

$$\langle A \rangle = \int \psi^* A \psi \, dx \, .$$

We expand ψ in terms of the orthogonal functions $\varphi_n(x)$ as

$$\psi(x, t) = \sum_n c_n(t) \varphi_n(x) \tag{1.3.1}$$

and write the matrix representation of A as

$$A_{mn} = \int \varphi_m^* A \varphi_n dx \, ,$$

then the expectation value is given as

$$\langle A \rangle = \sum_{m,n} A_{mn} c_n c_m^* \, .$$

If $\mathscr{H}$ is the Hamiltonian, the time rate of change of ψ is given by

$$i\hbar \frac{\partial \psi}{\partial t} = \mathscr{H} \psi$$

and therefore the time rate of change of c_n is given by

$$i\hbar \frac{\partial c_n}{\partial t} = \sum_l H_{nl} c_l \, ,$$

where $H_{nl} = \int \varphi_n^* \mathscr{H} \varphi_l \, dx$. The time rate of change of the product $c_n c_m^*$ is thus

$$i\hbar \frac{\partial}{\partial t} c_n c_m^* = \sum_l (H_{nl} c_l c_m^* - c_n c_l^* H_{lm}) \, , \tag{1.3.2}$$

where we have used the Hermitian property $H_{ml}^* = H_{lm}$.

Now, we consider many systems, each having the same structure and being under the same macroscopic conditions. We refer to this assembly as a statistical ensemble, or simply as an ensemble. The average of A over the ensemble is given as

$$\langle A \rangle = \sum_{m,n} A_{mn} \langle c_n c_m^* \rangle = \sum_{m,n} A_{mn} \rho_{nm} \, .$$

We write the ensemble average of $c_n c_m^*$ as

$$\rho_{nm} = \langle c_n c_m^* \rangle \, . \tag{1.3.3}$$

$\rho = (\rho_{nm})$ is called the *density matrix*. In general, the sum of the diagonal elements of a matrix M is called the diagonal sum or the trace, $\mathrm{tr}\{M\} = \sum_j M_{jj}$. Using this notation, we can write

$$\langle A \rangle = \mathrm{tr}\{A\rho\} . \tag{1.3.4}$$

If the x-representation of A and ρ are

$$\begin{aligned} A(x, x') &= \sum_{m,n} \varphi_m(x) A_{mn} \varphi_n^*(x') \\ \rho(x, x') &= \sum_{m,n} \varphi_n(x) \rho_{nm} \varphi_m^*(x') , \end{aligned} \tag{1.3.5}$$

then

$$\langle A \rangle = \iint A(x', x) \rho(x, x') dx\, dx' . \tag{1.3.6}$$

In view of (1.3.1, 3), we may write

$$\rho(x, x') = \langle \psi(x, t) \psi^*(x', t) \rangle . \tag{1.3.7}$$

The time rate of change of the density matrix is given as the average of (1.3.2), or

$$\mathrm{i}\hbar \frac{\partial \rho}{\partial t} = \mathscr{H}\rho - \rho\mathscr{H} . \tag{1.3.8}$$

If we use the quantum-mechanical Poisson bracket $(A, B) = (AB - BA)/\mathrm{i}\hbar$, we have

$$\frac{\partial \rho}{\partial t} = (\mathscr{H}, \rho) . \tag{1.3.9}$$

Equation (1.3.8) or (1.3.9) is the quantum-mechanical version of the *Liouville equation.*

When the density matrix is a function of energy, that is, when $\rho = f(\mathscr{H})$, we see that $\partial\rho/\partial t = 0$, which implies that the ensemble is stationary. This is a quantum-mechanical version of the classical Liouville's theorem.

For a stationary ensemble, the density matrix is diagonal in the representation in which $\mathscr{H}$ is diagonal

$$\rho_{nm} = w_n \delta_{nm} . \tag{1.3.10}$$

In this case, the ensemble is determined by $w_n = \langle |c_n|^2 \rangle$. We may define the density operator

$$\rho = \sum_n |n\rangle w_n \langle n|$$

using ket $|n\rangle$ and bra $\langle n|$. Its x-representation is

$$\langle x'|\rho|x\rangle = \sum_n \langle x'|n\rangle w_n \langle n|x\rangle = \sum_n \varphi_n(x) w_n \varphi_n^*(x') .$$

1.3.2 Classical Liouville's Theorem

Though we are generally following quantum mechanics, a classical-mechanical treatment has its particular beauty. There are many things which can be elucidated in terms of classical mechanics, and it will be also interesting to clarify some of the relationships between quantum and classical lines of thoughts.

Hamilton's canonical equations of motion are most suitable for a general description of motion in classical mechanics. Motion with s degrees of freedom is described in a space consisting of s generalized coordinates $q_1, q_2, \ldots, q_s$ and s momenta $p_1, p_2, \ldots, p_s$, which are conjugate to the generalized coordinates. This space, qp space, is called *phase space.*

A microscopic state of a system is represented by a point in its phase space. This is called the *representative point* and its path is called the *orbit* or the *trajectory.* Since motion through a point in the phase space is uniquely determined, trajectories do not cross each other. They go to infinity or describe closed curves.

A set of dynamical systems with the same construction, or an ensemble, is expressed by a set of representative points, each belonging to a system in the ensemble, and the set displays various microscopic states of the same substance. If we consider a very large number of systems, we can speak of the density of representative points in phase space. The assembly of representative points moves with time through phase space like a fluid. It can be shown that the density of representative points does not change if we follow the flow. This is the content of Liouville's theorem in classical mechanics. It may be said that the assembly of representative points moves just like an incompressible fluid in phase space. As the simplest example, such motion for an assembly of free particles on a straight line is shown in Fig. 1.8a. In this case, it is clear that the area of phase space is invariant, and the density of representative points is conserved.

In order to prove the theorem in general, we consider a small volume $\Delta q_1 \Delta q_2 \cdots \Delta q_s \Delta p_1 \Delta p_2 \cdots \Delta p_s$ in phase space. There are

$$f \Delta q_1 \Delta q_2 \cdots \Delta q_s \cdot \Delta p_1 \Delta p_2 \cdots \Delta p_s$$

representative points in this volume. The number of representative points which

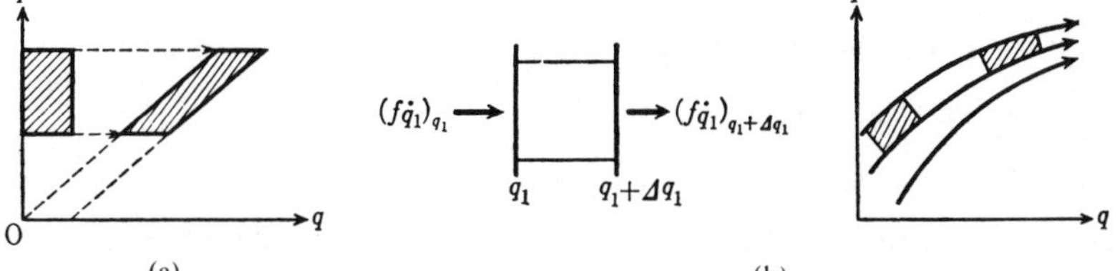

Fig. 1.8 a, b. Motion in phase space. **(a)** Free particles. **(b)** Flow of representative points in phase space

pass through a surface $\Delta q_2 \cdots \Delta q_s$ perpendicular to a coordinate q_1, at q_1 per unit time is

$$f\dot{q}_1 \Delta q_2 \cdots \Delta p_s$$

and the similar flow at $q_1 + \Delta q_1$ is (Fig. 1.8b)

$$(f\dot{q}_1)_{q_1+\Delta q_1} \Delta q_2 \cdots \Delta p_s = \left(f\dot{q}_1 + \frac{\partial f\dot{q}_1}{\partial q_1} \Delta q_1 \right) \Delta q_2 \cdots \Delta p_s \,.$$

Thus, the number of representative points in the volume diminishes per unit time by the amount

$$\frac{\partial (f\dot{q}_1)}{\partial q_1} \Delta q_1 \Delta q_2 \cdots \Delta p_s \,.$$

The same can be said for other surfaces. Therefore, the time rate of change of the density is given by

$$\frac{\partial f}{\partial t} = -\sum_{j=1}^{s} \left(\frac{\partial (f\dot{q}_j)}{\partial q_j} + \frac{\partial (f\dot{p}_j)}{\partial p_j} \right).$$

On the other hand, if $\mathscr{H}(p, q)$ denotes the Hamiltonian, the canonical equations of motion are given by

$$\dot{q}_j = \frac{\partial \mathscr{H}}{\partial p_j}, \quad \dot{p}_j = -\frac{\partial \mathscr{H}}{\partial q_j} \quad (i = 1, 2, \ldots, s) \,.$$

Using $\partial \dot{q}_j / \partial q_j = \partial^2 \mathscr{H} / \partial q_j \partial p_j$ and similar relations we see that

$$\frac{Df}{Dt} \equiv \frac{\partial f}{\partial t} + \sum_{j=1}^{s} \left(\dot{q}_j \frac{\partial f}{\partial q_j} + \dot{p}_j \frac{\partial f}{\partial p_j} \right) = 0 \,. \tag{1.3.11}$$

This is called the Liouville equation. In general, if time and position are shifted a little, the change in f is given by $\Delta f = (\partial f / \partial t) \Delta t + (\partial f / \partial q_1) \Delta q_1 + \cdots$, and $\Delta q_1 = \dot{q}_1 \Delta t$ if we follow the motion. Therefore, the rate of change in f is Df/Dt, which has been shown to vanish. This is the classical Liouville's theorem.

If the x, y, z-axes are used and m_j denotes the mass of a particle j, then $q_j = p_j / m_j$; if there is no magnetic field, we have

$$\dot{p}_j = -\frac{\partial \mathscr{H}}{\partial q_j} = -\frac{\partial U}{\partial q_j} \,,$$

where U denotes the potential energy;

$$\frac{\partial f}{\partial t} + \sum_j \left(\frac{p_j}{m_j} \frac{\partial f}{\partial q_j} - \frac{\partial U}{\partial q_j} \frac{\partial f}{\partial p_j} \right) = 0 \,.$$

The operator appearing in the second term is sometimes called the *Liouville operator*.

When the system is completely isolated or when it is in a constant force field, the energy or the Hamiltonian of the system is kept constant and representative points move on a surface of constant energy

$$\mathscr{H}(p, q) = \text{const.}$$

in phase space.

When the degree of freedom is large, it is nearly impossible to visualize the motion of representative points in phase space. Even for a particle moving in a two-dimensional space, the phase space is four-dimensional. Lissajous' figures, the superposition of perpendicular oscillations in Fig. 1.9, and the motion of a ball in a square (Weyl's billiard) in Fig. 1.10 are examples showing the projection of motion on to coordinate space. These are the cases where the motion in each direction is periodic. The relationship between quantum mechanics and classical mechanics for a periodic motion will be considered in Sect. 1.3.4.

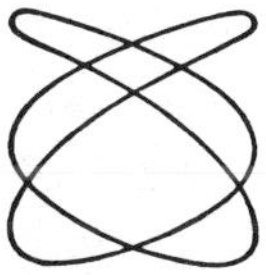

Fig. 1.9. Lissajous' figure

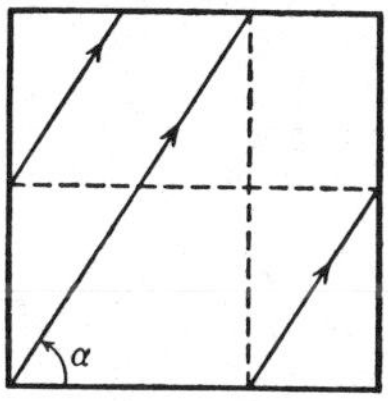

Fig. 1.10. Weyl's billiard. When $\tan\alpha$ is irrational, the ball passes uniformly every point in the square

1.3.3 Wigner's Distribution Function

Classical mechanics is the limit of quantum mechanics as the Planck constant tends to zero. One of the methods to clarify the correspondence is to use Wigner's representation. For simplicity, we shall assume a particle in one-dimensional space. Generalization to a higher degree of freedom is straightforward. We define Wigner's representation A_{w}, for an arbitrary mechanical quantity A ($\hat{A}$ is the corresponding operator) in terms of its coordinate representation $\langle x|A|x'\rangle$ as

$$A_{\mathrm{w}}(p, q) = \int_{-\infty}^{\infty} dr \exp(-ipr/\hbar)\langle x|A|x'\rangle \,, \tag{1.3.12}$$

where

$$q = \frac{x + x'}{2}, \quad r = x - x'$$

or

$$x = q + \frac{r}{2}, \quad x' = q - \frac{r}{2}.$$

That is, the average of x and x' is related to the coordinate q, and the parameter of the Fourier transform with respect to r is related to the momentum p. If A is a function of the coordinate x only, say $U(x)$, then, since

$$\langle x|U|x'\rangle = U(x)\delta(x-x') = U\left(q+\frac{r}{2}\right)\delta(r)\,,$$

(1.3.12) reduces to

$$U_{\mathrm{w}}(p,q) = U(q)\,. \tag{1.3.13}$$

Similarly, if A is a function of the momentum p only, say $K(p)$, then

$$\langle x|K|x'\rangle = K\left(\frac{\hbar}{\mathrm{i}}\frac{\partial}{\partial x}\right)\delta(x-x')$$

and when (x, x') is transformed to (q, r), (1.3.12) gives

$$\begin{aligned} K_{\mathrm{w}} &= \int_{-\infty}^{\infty} dr \exp(-ipr/\hbar) K\left[\frac{\hbar}{\mathrm{i}}\left(\frac{1}{2}\frac{\partial}{\partial \mathrm{q}} + \frac{\partial}{\partial r}\right)\right]\delta(r) \\ &= \int_{-\infty}^{\infty} dr \exp(-ipr/\hbar) K\left(\frac{\hbar}{\mathrm{i}}\frac{d}{dr}\right)\delta(r) = K(p)\,. \end{aligned} \tag{1.3.14}$$

To prove the last equality, it is sufficient to assume a series expansion of $K(p)$ in powers of p, and repeat partial integrations. Equations (1.3.13, 14) are natural consequences which show correspondence between classical and quantum mechanical quantities. When coordinates x and momenta p are mixed, since they are noncommutative, the operator $\hat{A}$, expressed in terms of p and q, will be different from A_{w} in functional forms.

In particular, Wigner's representation for the density matrix

$$f(p,q) = \int_{-\infty}^{\infty} dr \exp(-ipr/\hbar)\langle x|\rho|x'\rangle \tag{1.3.15}$$

is called *Wigner's distribution function*. Its inverse transform is

$$\begin{aligned} \langle x|\rho|x'\rangle &= \frac{1}{h}\int_{-\infty}^{\infty} dp \exp(+ipr/\hbar) f(p,q) \\ &= \frac{1}{h}\int_{-\infty}^{\infty} dp \exp[ip(x-x')/\hbar] f\left(p, \frac{x+x'}{2}\right). \end{aligned}$$

If we use (1.3.6) for a mechanical quantity A averaged in regard to the density matrix ρ, then we obtain the formula

$$\begin{aligned} \langle A\rangle &= \frac{1}{h}\iiint dx\, dx'\, dp \langle x'|A|x\rangle \exp[ip(x-x')/\hbar] f(p,q) \\ &= \frac{1}{h}\iint dp\, dq\, A_{\mathrm{w}}(p,q) f(p,q)\,, \end{aligned} \tag{1.3.16}$$

where we have only changed the integration variables (x, x') to (x', x) and further to (q, r) in order to use the definition (1.3.12) for A_w. This formula expresses the correspondence

$$\hat{A} \leftrightarrow A_w, \quad \rho \leftrightarrow f, \quad \mathrm{tr} \leftrightarrow \frac{1}{h} \int\int dp\, dq \,. \tag{1.3.17}$$

In other words, when we use Wigner's representation, the expectation value in quantum statistical mechanics is given as an average over phase space just as in classical mechanics, but using Wigner's distribution function. Taking a trace (i.e., a sum over discrete states which form the basis of a quantum-mechanical representation) corresponds to an integration over the phase volume h (per degree of freedom) in the Wigner representation.

The equation of motion (1.3.9) for the density matrix can be transformed into an equation of motion for Wigner's distribution function. Let

$$\mathscr{H}(p, x) = \frac{p^2}{2m} + U(x)$$

be the Hamiltonian, then the equation of motion for a density matrix $\langle x|\rho|x'\rangle$ in coordinate representation is given as

$$\mathrm{i}\hbar \frac{\partial}{\partial t} \langle x|\rho(t)|x'\rangle = \left[-\frac{\hbar^2}{2m}\left(\frac{\partial^2}{\partial x^2} - \frac{\partial^2}{\partial x'^2} \right) + U(x) - U(x') \right] \langle x|\rho(t)|x'\rangle \,. \tag{1.3.18}$$

If we change (x, x') to (q, r), the right-hand side is transformed to

$$\left\{ -\frac{\hbar^2}{2m}\left[\left(\frac{1}{2}\frac{\partial}{\partial q} + \frac{\partial}{\partial r} \right)^2 - \left(\frac{1}{2}\frac{\partial}{\partial q} - \frac{\partial}{\partial r} \right)^2 \right] + U\left(q + \frac{r}{2} \right) - U\left(q - \frac{r}{2} \right) \right\} \left\langle q + \frac{r}{2} |\rho| q - \frac{r}{2} \right\rangle$$

$$= \left[-\frac{\hbar^2}{m} \frac{\partial^2}{\partial q \partial r} + U\left(q + \frac{r}{2} \right) - U\left(q - \frac{r}{2} \right) \right] \left\langle q + \frac{r}{2} |\rho| q - \frac{r}{2} \right\rangle .$$

When we make the Fourier transform and note the relations

$$\int_{-\infty}^{\infty} dr \exp(-\mathrm{i}pr/\hbar) \frac{\partial}{\partial r} F(q, r) = \frac{\mathrm{i}}{\hbar} p \int_{-\infty}^{\infty} dr \exp(-\mathrm{i}pr/\hbar) F(q, r) \,,$$

$$\exp(-\mathrm{i}pr/\hbar)\, U\left(q + \frac{r}{2} \right) F(q, r) = U\left(q - \frac{\hbar}{2\mathrm{i}} \frac{\partial}{\partial p} \right) \exp(-\mathrm{i}pr/\hbar) F(q, r)$$

which hold for an arbitrary function $F(p, r)$ (assuming that it vanishes for

$r = \pm\infty$), (1.3.18) is transformed to

$$\frac{\partial}{\partial t} f(p, q, t) = \left\{ -\frac{p}{m}\frac{\partial}{\partial q} + \frac{1}{i\hbar}\left[U\left(q - \frac{\hbar}{2i}\frac{\partial}{\partial p} \right) - U\left(q + \frac{\hbar}{2i}\frac{\partial}{\partial p} \right) \right] \right\} f(p, q, t) . \quad (1.3.19)$$

This is the Liouville equation in Wigner's representation. If we take the classical limit $\hbar \to 0$, since

$$\frac{1}{i\hbar}\left[U\left(q - \frac{\hbar}{2i}\frac{\partial}{\partial p} \right) - U\left(q + \frac{\hbar}{2i}\frac{\partial}{\partial p} \right) \right] \to \frac{\partial U}{\partial q}\frac{\partial}{\partial p} ,$$

the operator on the right-hand side of (1.3.19) reduces to the Liouville operator.

Thus, we have shown that in the limit $\hbar \to 0$, quantum mechanics reduces, in general, to classical mechanics. In this limit, Wigner's distribution function reduces to the classical probability distribution in phase space. However, for finite $\hbar$, Wigner's distribution function takes positive and negative values as well because of the wave character of particles. In this respect, it is unreasonable to consider Wigner's distribution function as a probability distribution function. In spite of this difficulty, it can be used as a distribution function in taking averages as we have done in (1.3.16).

1.3.4 The Correspondence Between Classical and Quantum Mechanics

One of the ways of checking the correspondence between classical and quantum mechanics is given by the correspondence principle in classical quantum theory which was developed prior to quantum mechanics. It is most clearly demonstrated for periodic motion. For simplicity, consider one-dimensional motion. The phase space in this case should really be called a phase plane.

Consider a free particle whose coordinate is q. The region of motion is assumed to be between $q = 0$ and $q = L$, where there are walls and the particle is reflected elastically (Fig. 1.11a). During the collision with the wall, momentum p changes its sign. If the wall is soft, the momentum first diminishes in magnitude from the value p, and after changing to $-p$, it will leave the wall (Fig. 1.11b). In either case, the trajectory will be nearly rectangular. In general, the area in the phase space enclosed by a trajectory of a periodic motion is

$$J = \oint p\, dq , \quad (1.3.20)$$

where J is called the *action variable*. With respect to the present case, if m denotes the mass and E the energy of the particle, the momentum is $\sqrt{2mE}$, and

$$J(E) = 2\sqrt{2mE}L$$

if the wall is perfectly rigid.

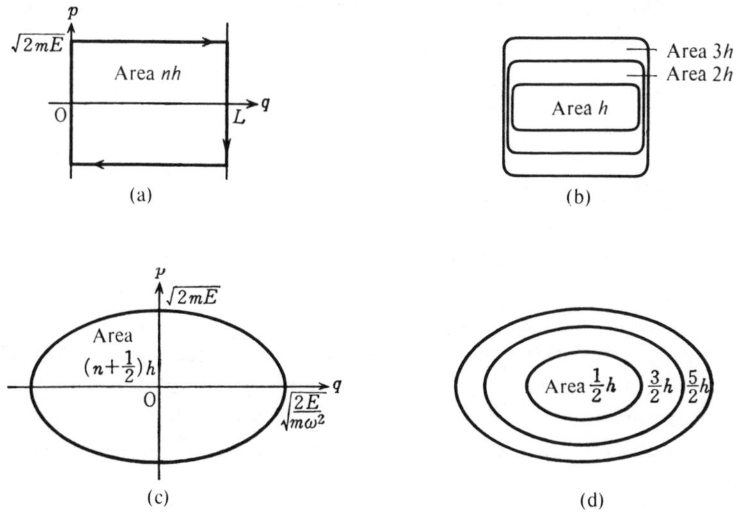

Fig. 1.11 a–d. Examples of trajectories. (**a**) The trajectory of a free particle in a box of size L. (**b**) The trajectory when the walls are soft (the difference between the areas is h). (**c**) The trajectory of a harmonic oscillator. (**d**) A quantum-mechanical harmonic oscillator (the difference between the areas is h)

As is well known in quantum mechanics, the energy eigenvalues for the free particle are given as

$$E_n = \frac{h^2}{8mL^2} n^2 \quad (n = 1, 2, 3, \dots) .$$

Therefore, we have the relation

$$J(E_n) = nh$$

and we see that every time the area in phase space is increased by the amount h, a new state permitted in quantum mechanics appears consecutively. In other words, there is a microscopic state in each area h in qp-space (Fig. 1.11b).

As another example, consider a harmonic oscillator in one dimension. The energy is then

$$\mathscr{H}(q, p) = \frac{p^2}{2m} + \frac{m}{2}\omega^2 q^2$$

and the trajectory is an ellipse in phase space. The radius along the q-axis is $\sqrt{2E/m\omega^2}$ while that along the p-axis is $\sqrt{2mE}$ (Fig. 1.11c). The action variable, or the area enclosed by the ellipse, is

$$J(E) = 2\pi\frac{E}{\omega} = \frac{E}{\nu} ,$$

where $\nu = \omega/2\pi$ is the frequency of oscillation.

On the other hand, in quantum mechanics, the energy of the harmonic oscillator is

$$E_n = (n + \tfrac{1}{2})h\nu$$

and thus

$$J(E_n) = (n + \tfrac{1}{2})h \, .$$

Therefore, there is also a microscopic state in each area h in qp space. In this case, the trajectory is an ellipse, and the representative point moves on it in a clock-wise direction. Each trajectory with area $(n + \frac{1}{2})h$ corresponds to a quantum-mechanical state (Fig. 1.11d).

Since each microscopic state occupies the finite volume h, trajectories may be thought of as having finite thickness, resembling bundles of noodles. A sum over quantum states corresponds in classical mechanics to an integral along the trajectories, rather than a sum over elementary volumes into which the phase space is divided.

2. Outlines of Statistical Mechanics

In this chapter, we start with certain principles and describe the general methods of statistical mechanics [2.1–17]. If we assume that every quantum-mechanical state (microscopic state) has the same weight (the principle of equal probability), then we can establish a standpoint where mechanical laws are combined with probability theory. By considering a system in contact with a larger system, we can describe a system with constant temperature or constant pressure. Thus, we develop the statistical mechanics for an equilibrium state (statistical mechanics in a narrow sense) and we can also find a microscopic interpretation of the laws in thermodynamics.

2.1 The Principles of Statistical Mechanics

2.1.1 The Principle of Equal Probability

If a system is completely isolated, the system will stay forever in a definite state if it is initially in that state. But, as was already pointed out, it is useless to speak of a completely isolated system. We have some uncertainty in the energy of the system because of uncontrollable interaction between the system and the external world.

Nevertheless, we can consider a system which is nearly isolated, and assume the validity of Liouville's theorem during some interval of time. We shall further admit that the time average of a mechanical quantity of a system under a macroscopic equilibrium state is equal to the ensemble average (ergodic hypothesis). This ensemble must be time-independent or stationary. It is a consequence of Liouville's theorem that, if the ensemble is stationary, its density is a function of the energy of the system. Such an ensemble was first clearly mentioned by *W. Gibbs*, and thus it is called *Gibbs' ensemble.* It satisfies the requirement that the statistical ensemble should be compatible with mechanics. The requirement is fundamental to statistical mechanics.

Thus, energy plays an important special role in statistical mechanics, and it is usually assumed that there is no invariant other than energy conservation. In mechanics, there are total momentum and total angular momentum as conserved quantities. However, for a system confined in a box, we have no momentum conservation, and if some asymmetry of the shape of the box is introduced, the total angular momentum will no longer be conserved.

From the classical Liouville's theorem, we conclude that the weight is proportional to the volume of the portion of phase space for a stationary statistical ensemble. The correspondence with quantum mechanics leads to the assertion that every quantum state of the same energy E has the same weight $w(E)$. This is the fundamental principle, which is called the *principle of equal probability* or the assumption of equal *a priori* probability. In short, every quantum state is considered on equal footing. In other words, the a priori probability for a system to be in a particular energy level is the same for all levels.

The principle that the time-average is the same as the ensemble-average and the principle of equal a priori probability, are two basic principles of statistical mechanics. After adopting these principles, we only have to construct the general probabilistic theory.

The ergodic problem aims at deriving the above principles from mechanics. This problem has been usually approached from a mathematical point of view. In Chap. 5, we shall discuss the ergodic problem in some detail from a physical viewpoint.

2.1.2 Microcanonical Ensemble

Consider a system in the energy interval between E and $E + \delta E$, where δE is the order of uncertainty. Each quantum state in this range is assumed to have the same probability. Thus we have an ensemble characterized by the equal probability w_n of each state n, or

$$w_n = w = \text{const.}, \quad E < E_n < E + \delta E \tag{2.1.1}$$

which expresses the principle of equal probability as it is. This is called an ensemble with constant energy or a *microcanonical ensemble*, which can be illustrated by a constant distribution in the small range δE (Fig. 2.1). Of course we may assume a distribution of the error function type, but since it only complicates the matter without any advantage, we take the simple distribution of the step function type.

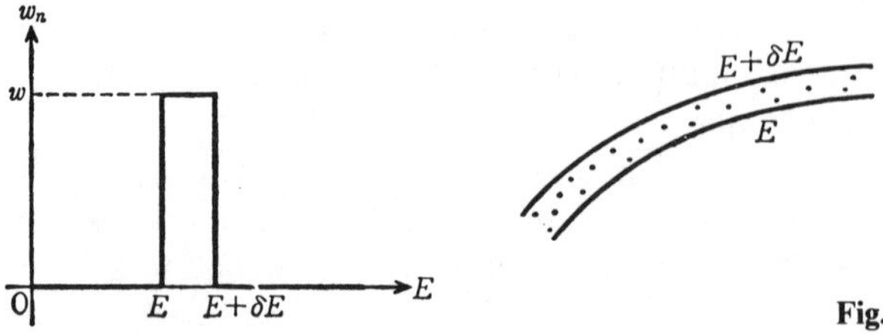

Fig. 2.1. Microcanonical ensemble

If we denote by φ_n a quantum state in the energy range δE, and by s the total number of states in this range, then the density matrix can be written as

$$\rho(x, x') = w \sum_{n=1}^{s} \varphi_n(x)\varphi_n^*(x') . \tag{2.1.2}$$

If the functional set $\{\varphi_n\}$ is transformed by a unitary transformation as

$$\chi_k = \sum_{n=1}^{s} u_{kn}\varphi_n ,$$

we have

$$\sum_{n=1}^{s} \varphi_n(x)\varphi_n^*(x') = \sum_{k=1}^{s} \chi_k(x)\chi_k^*(x') .$$

Therefore, $\rho(x, x')$ is determined by s-dimensional space and is independent of the bases of this space.

For a macroscopic system, energy eigenvalues can be usually seen to be continuous or quasicontinuous. Then, we may formally let $\delta E \to 0$, and the probability of a state with energy $\mathscr{E}$ may be written

$$dw = (\text{const.}) \cdot \delta(\mathscr{E} - E)\, d\mathscr{E} \tag{2.1.3}$$

for a microcanonical ensemble, where δ stands for Dirac's δ-function.

Because of the principle of equal probability, the number of states is of special importance. Let $j(E)$ be the number of states below the energy E. Considering the energy eigenvalue as a continuous variable, we let

$$dj = \frac{dj(E)}{dE}\, dE = \Omega(E)\, dE \tag{2.1.4}$$

express the number of states between E and $E + \delta E$. In general, $j(E)$ is a step function. But on the right-hand side this is written as a differentiable function. $\Omega(E)$ is called the *density of states* (Fig. 2.2).

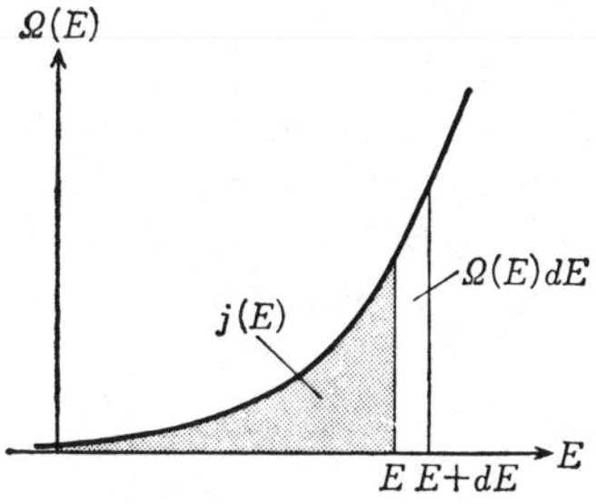

Fig. 2.2. Density of states

2.1.3 Boltzmann's Principle

The number of microscopic states accessible to a macroscopic system is called the *statistical weight* (*thermodynamic weight*). When macroscopic conditions such as the volume of the system are assigned, the statistical weight W of a system in the energy interval between E and $E + \delta E$ may be written as

$$W = \Omega(E)\delta E\ . \tag{2.1.5}$$

If we define *entropy* S by

$$S = k \ln W, \tag{2.1.6}$$

this is very convenient for deriving the thermodynamic relations from the statistical mechanical point of view. It will be shown afterwards that the above entropy is the same as the thermodynamic entropy. The fact that the thermodynamic entropy is related to the number W of microscopic states accessible to the system is called *Boltzmann's principle.* The constant k in the above equation is determined if S is identified with the thermodynamic entropy. k is called the *Boltzmann constant,* and frequently written as k_B.

We shall note some of the considerations which lead to Boltzmann's principle, though it is out of the scope of formal discussion. First we note the theorem of increasing entropy, a theorem of thermodynamics, which states that the entropy of an adiabatic system never decreases and increases in irreversible processes. Here the notion of entropy is somewhat extended to include the nonequilibrium state. On the other hand, as a gas expands into vacuum, irreversible changes occur, in general, towards an increasing microscopic state. Thus, entropy and the number of microscopic states are in a functional relation in that when one increases, the other also increases (if the above equation holds, this means that $k > 0$).

Again, if we consider macroscopic states of a system under hypothetical restrictions, the state of maximum entropy or statistical weight is realized as the equilibrium state. For example, a gas expanding into vacuum cannot stop halfway.

Secondly, if we denote by W_1 and W_2 the statistical weights of two independent systems, the weight of the combined system is the product $W = W_1 W_2$, while the thermodynamic entropy is given by the sum $S = S_1 + S_2$. The relation $S = f(W)$ which satisfies these requirements is obviously $S \propto \ln W$. This is nothing but (2.1.6).

Now, by (2.1.5) the definition of entropy (2.1.6) has an uncertainty, because of the energy width δE, which is difficult to deal with. However, it can be shown that this uncertainty is negligibly small for a system of macroscopic size including a large number of particles. To show this we rewrite (2.1.6) as

$$S = k\left[\ln \Omega(E)\varepsilon + \ln\left(\frac{\delta E}{\varepsilon}\right)\right]. \tag{2.1.7}$$

For ε we take, for example, the energy per particle, or E/N. Even if we assume the largest value E for δE, the second term on the right-hand side is of the order $\ln N$, which is extremely small compared with N (for example, for an ordinary gas, N is of the order 10^{22} per cm^3). On the contrary, the smallest possible value of δE would be given by the uncertainty principle which yields $\delta E = h/t$, where t

is the observation time. If we take, thus, $|\ln(\delta E/\varepsilon)| \sim |\ln(h/\varepsilon t)| \sim N \sim 10^{22}$, then t would be much larger than the life of our universe. In either limit, and in general, the second term is always negligible. We may also note that ε in the first term can be of any value. In the next section it will be shown that, for a system of particles with kinetic energy, we have

$$j(E) \approx \left(\frac{E}{N}\right)^{\alpha N},$$

where α is of the order of unity. Then we see that no numerical change in S occurs if we take

$$S = k \ln j(E) \tag{2.1.8}$$

in place of (2.1.6). In such cases where $j(E)$ and $\Omega(E)$ increase extremely rapidly with E, (2.1.8) is frequently used in place of (2.1.6). But for a spin system, to be discussed in the following section, the total energy is limited from above and in such cases we cannot use (2.1.8) as a substitute.

2.1.4 The Number of Microscopic States, Thermodynamic Limit

We shall show some simple examples of the number of microscopic states $j(E)$ and the state density $\Omega(E) = dj/dE$.

a) *A Free Particle*

First, consider a free particle. The energy is

$$\mathscr{H} = \frac{1}{2m}(p_x^2 + p_y^2 + p_z^2) . \tag{2.1.9}$$

If we take cyclic boundary conditions with the period L (volume is $V = L^3$), then the eigenvalues of momentum components are (Sect. 3.2.1)

$$p_x = \frac{h}{L} n_1, \quad p_y = \frac{h}{L} n_2, \quad p_z = \frac{h}{L} n_3 , \tag{2.1.10}$$

where h denotes the Planck constant, and $n_1, n_2, n_3 = 0, \pm 1, \pm 2, +3, \ldots$ are

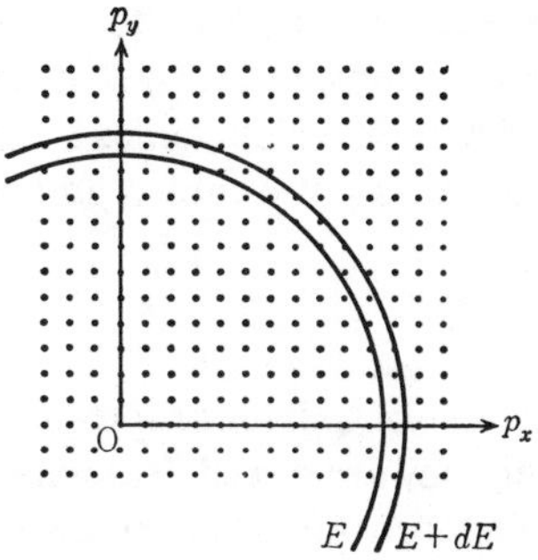

Fig. 2.3. Quantum states of a free particle

quantum numbers. When L is sufficiently large, the number of states $j(E)$ below the energy E is given as the volume of a sphere of radius $p = \sqrt{2mE}$ divided by the unit $(h/L)^3$ (Fig. 2.3). Thus we have

$$j(E) = \frac{4\pi}{3}(2mE)^{3/2}\frac{V}{h^3} . \tag{2.1.11}$$

b) *An Ideal Gas*

Next, let us consider a system of N free particles, assuming no interaction between the particles. This is a model for an ideal gas. The energy is

$$\mathscr{H} = \frac{1}{2m}\sum_{j=1}^{N}(p_{xj}^2 + p_{yj}^2 + p_{zj}^2) \tag{2.1.12}$$

which is assigned by $3N$ quantum numbers. Let us first assume that we can distinguish each particle (classical particle). Then each integer point in the multidimensional space composed of $3N$ quantum numbers represents a microscopic state of the system. If the boundary conditions are periodic, we have

$$\mathscr{H} = \frac{h^2}{2mL^2}\sum_{j=1}^{3N} n_j^2 \leqq E , \tag{2.1.13}$$

where positive and negative integers $n_1, n_2, n_3, \ldots, n_{3N}$ are quantum numbers. The total number $j(E)$ of quantum states with energy less than E is given by the volume of the $3N$-dimensional sphere of radius $p = \sqrt{2mE}$ divided by the unit $(h/L)^3$. This volume is

$$\frac{\pi^{3N/2}}{\Gamma(3N/2+1)}p^{3N} \approx \left(\frac{2e}{3N}\right)^{3N/2}\pi^{3N/2}p^{3N} , \tag{2.1.14}$$

where $\Gamma(\alpha)$ stands for the Γ-function and the right-hand side of (2.1.14) is the asymptotic value for $N \gg 1$. Thus, the number of states below E might be expected to be

$$j'(E, V) = \left(\frac{2e}{3}2\pi m\frac{E}{N}\right)^{3N/2}\frac{V^N}{h^{3N}} , \tag{2.1.15}$$

but this is not correct as can be seen from the following consideration.

Suppose that by using $j'(E, V)$ we deduce entropy by the formula $S' = k \ln j'(E, V)$. If we change N, keeping E/N and V/N constant, then S' does not change proportionally to N because of the term $kN \ln N$ which comes from $kN \ln V$. Therefore, if two such systems are combined, S' is not doubled and is increased by the amount

$$2kN \ln 2N - 2kN \ln N = 2kN \ln 2 .$$

This fact is called *Gibbs' paradox*. If two systems consist of different molecular species it is natural that the entropy is increased when they are mixed. But when two gases consisting of the same kind of molecules are mixed at uniform

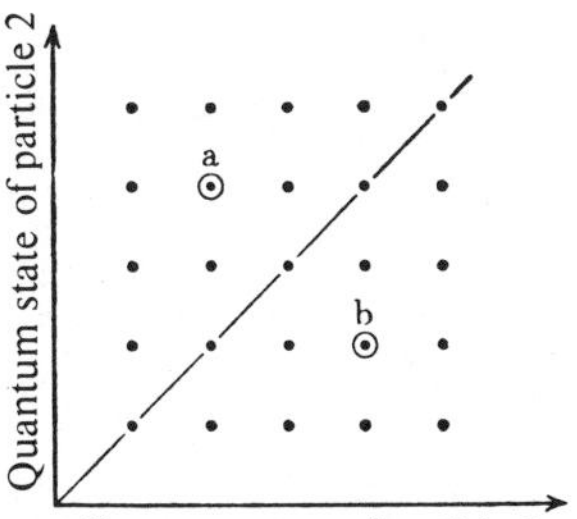

Fig. 2.4. States of a two-particle system

temperature and pressure, the total entropy must stay the same; this assertion is contrary to the above increment of energy by mixing.

Figure 2.4 represents the phase space of the quantum states of a system of two identical particles. It is to be noted that, for example, the states b and a are identical because they differ only in that the particle labels are interchanged. Thus, when the quantum numbers of particle 1 and 2 are different, the corresponding state is duplicated in Fig. 2.4 as well as in phase space. Similarly for an N-particle system, when the roles of the particles are all different, the number of ways of interchanging the roles is the number of permutations $N!$. When the total volume of a gas is very large, the energy levels of the eigenstates for each particle are so dense and the difference between them is so small compared with the total energy, that the chance that two or more particles have the same set of quantum numbers will be negligibly small. Thus, we may say that there are $N!$ equivalent points in the quantum-number space (detailed calculation of the number of states leads to quantum statistics; Chap. 3). Therefore, when the system consists of N identical particles, the number of states is given as

$$j(E, V) = \frac{j'(E, V)}{N!} .$$

In the next chapter we deal with quantum statistics; it will be shown there that in the limit of low density, division by $N!$ is valid for a Bose gas and for a Fermi gas. The above approximation is called classical statistics or Boltzmann statistics. Using Stirling's formula $N! \approx (N/\mathrm{e})^N$, we have

$$j(E, V) = \left(\frac{4\pi}{3} m \frac{E}{N}\right)^{3N/2} \left(\frac{V}{N}\right)^N \frac{\exp(5N/2)}{h^{3N}} \tag{2.1.16}$$

for an ideal gas. Now, $S = k \ln j(E, V)$ is proportional to N when E/N and V/N are kept constant and Gibbs' paradox is eliminated.

Equation (2.1.16) leads to the density of states

$$\Omega(E, V) = \frac{\partial j(E, V)}{\partial E} = \frac{3N}{2E} j(E, V) . \tag{2.1.17}$$

Thus, since we may neglect the small term $\ln(N/E)$ which is not proportional to N, we see that $\ln j$ is practically equal to $\ln \Omega$. Further, when E is decreased by a

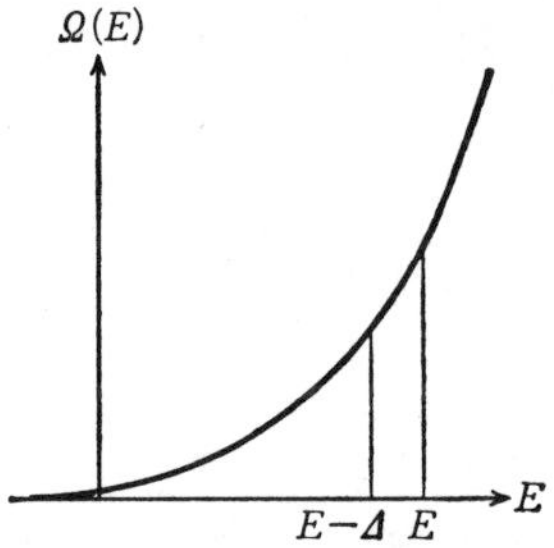

Fig. 2.5. Density of states $\Omega(E)$ for an ideal gas

small amount Δ, the change in Ω is given as (Fig. 2.5)

$$\frac{\Omega(E-\Delta)}{\Omega(E)} = \left(1 - \frac{\Delta}{E}\right)^{3N/2-1} \approx \exp(-\beta\Delta)\,, \tag{2.1.18}$$

where $\beta = 3N/2E$, and $1/\beta$ is of the order of the average energy E/N. Therefore, if the total energy is decreased by the order E/N, then $\Omega(E)$ diminishes by the factor 1/e. In other words, $\Omega(E)$ increases very rapidly with increasing macroscopic energy E.

c) *Spin System*

Consider a system of N independent spins with magnetic moment μ. We assume that the spin takes up and down orientations and the system is in an external magnetic field H, so that the energy of an up-spin is $-\mu H$ and that of a down-spin is $+\mu H$. When there are n up-spins and $n' = N - n$ down-spins, the total energy is

$$E = -(n - n')\mu H = -(2n - N)\mu H\,. \tag{2.1.19}$$

We assume that these spins are localized. Then a microscopic state of the system is determined by specifying the configuration of n up-spins and $N - n$ down-spins, and macroscopic state is determined by the magnetic moment $(n - n')\mu$.

Therefore, a macroscopic state involves

$$W(n) = \frac{N!}{n!(N-n)!} \sim \left(\frac{N}{n}\right)^n \left(\frac{N}{N-n}\right)^{N-n} \tag{2.1.20}$$

microscopic states. Since

$$n = \frac{N}{2} + \frac{E}{2\mu H}\,, \quad n' = \frac{N}{2} - \frac{E}{2\mu H}\,, \tag{2.1.21}$$

we have

$$\ln W = N \ln 2 - \frac{N}{2}\left(1 + \frac{E}{\mu H N}\right) \ln\left(1 + \frac{E}{\mu H N}\right) - \frac{N}{2}\left(1 - \frac{E}{\mu H N}\right) \ln\left(1 - \frac{E}{\mu H N}\right). \tag{2.1.22}$$

Here $\delta E = 2\mu H \delta n$, and $\ln(\Omega \delta E) = \ln W$. It should be noted that $S = k \ln W$ is not necessarily an increasing function of E/N.

In actual fact, $\ln W = 0$ at the minimum energy $E = -\mu HN$ and at the maximum energy $E = +\mu HN$, and $S = k \ln W$ takes its maximum value at $E = 0$ (Fig. 2.6). Thermodynamically, the *absolute temperature* T is defined by $T = (\partial S/\partial E)^{-1}$. In statistical mechanics, absolute temperature is defined by this equation (Sect. 2.2). For the spin system we are considering, we have $\partial S/\partial E > 0$ or $T > 0$, when $E < 0$, but when $E > 0$ we have $\partial S/\partial E < 0$ or $T < 0$. Thus, the energy of a state at a negative temperature is higher than a state at $T = \infty$.

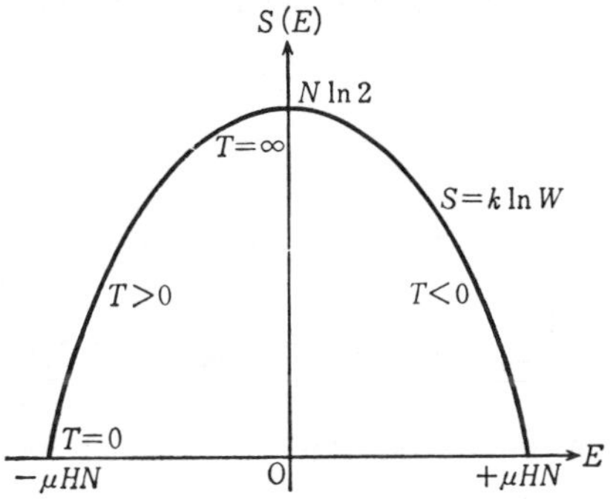

Fig. 2.6. The entropy of a system of free spins

d) *The Thermodynamic Limit*

We have seen that entropy or $\ln W$ increases in proportion to the particle number N for both gas and spin systems when E/N and V/N are kept constant. Variables which are proportional to the quantity of the substance are called extensive variables. Energy and volume are extensive variables, while variables such as temperature and pressure are called intensive variables.

We know empirically that the thermodynamic state of a usual substance is specified by a small number of parameters such as $n = N/V$ and $\varepsilon = E/N$ and is independent of the shape of the container, system size and so forth. When we can consider such a limit as

$$N \to \infty \text{ keeping } n = N/V = \text{const.}, \quad \varepsilon = E/N = \text{const.}\,, \tag{2.1.23}$$

this is called the thermodynamic limit. When the limit exists, we may speak of entropy as an extensive variable. In statistical mechanics we assume the existence of thermodynamic limit.

2.2 Temperature

2.2.1 Temperature Equilibrium

It is observed that when two systems can exchange energy, they are in equilibrium if their temperatures are equal. Let us examine how this equilibrium is established and the meaning of temperature from a statistical-mechanical point of view.

Consider two systems, I and II, and let the energy of system I be E_{I} and that of system II be E_{II}. We assume they are exchanging energy through a small interaction, whose energy can be neglected. Then the total energy is

$$E_{\mathrm{I}} + E_{\mathrm{II}} = E = \text{const.} \tag{2.2.1}$$

If $\Omega_{\mathrm{I}}(E_{\mathrm{I}})$ and $\Omega_{\mathrm{II}}(E_{\mathrm{II}})$ denote the state density of each system, then the number of quantum states (microscopic states) of the combined system, in which system I has the energy between E_{I} and $E_{\mathrm{I}} + dE_{\mathrm{I}}$, and system II has the energy between E_{II} and $E_{\mathrm{II}} + dE_{\mathrm{II}}$, is given as

$$\Omega_{\mathrm{I}}(E_{\mathrm{I}})\,dE_{\mathrm{I}}\,\Omega_{\mathrm{II}}(E_{\mathrm{II}})\,dE_{\mathrm{II}}\,. \tag{2.2.2}$$

Therefore, the number of states of the combined system with energy between E and $E + dE$ is (Fig. 2.7)

$$\iint_{E<E_{\mathrm{I}}+E_{\mathrm{II}}<E+\delta E} \Omega_{\mathrm{I}}(E_{\mathrm{I}})\,\Omega_{\mathrm{II}}(E_{\mathrm{II}})\,dE_{\mathrm{I}}\,dE_{\mathrm{II}} = \delta E \int \Omega_{\mathrm{I}}(E_{\mathrm{I}})\,\Omega_{\mathrm{II}}(E - E_{\mathrm{I}})\,dE_{\mathrm{I}}\,, \tag{2.2.3}$$

where δE denotes the energy width of the microcanonical ensemble of the combined systems.

The integrand on the right-hand side of (2.2.3) represents the number of states of a combined system specified by the macroscopic variable E_{I}. At equilibrium, the value of E_{I} with the maximum number of microscopic states is obtained. When Ω_{I} and Ω_{II} are both rapidly increasing functions of energy, then $\Omega_{\mathrm{I}}(E_{\mathrm{I}})\,\Omega_{\mathrm{II}}(E - E_{\mathrm{I}})$ will have a very sharp maximum which gives the most probable value of E_{I} (Fig. 2.8). Under the condition $E_{\mathrm{I}} + E_{\mathrm{II}} = E = \text{const.}$, we seek a value of E which satisfies

$$\ln \Omega_{\mathrm{I}}(E_{\mathrm{I}}) + \ln \Omega_{\mathrm{II}}(E_{\mathrm{II}}) = \max. \tag{2.2.4}$$

This condition may be written as

$$\frac{\partial \ln \Omega_{\mathrm{I}}(E_{\mathrm{I}}^*)}{\partial E_{\mathrm{I}}^*} = \frac{\partial \ln \Omega_{\mathrm{II}}(E_{\mathrm{II}}^*)}{\partial E_{\mathrm{II}}^*}\,, \tag{2.2.5}$$

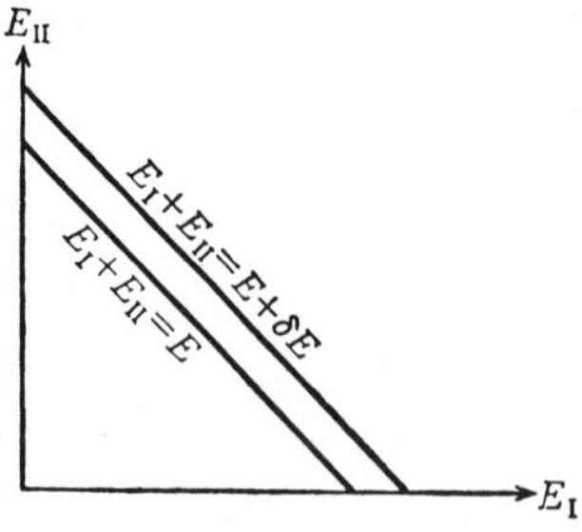

Fig. 2.7. Combined system

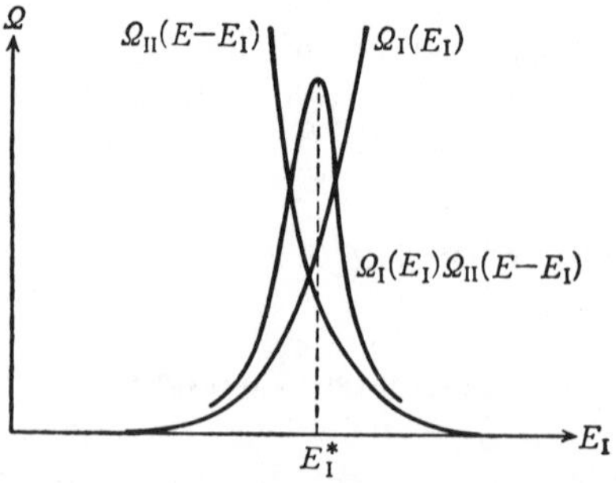

Fig. 2.8. State density of a combined system

where $E_{\rm I}^*$ and $E_{\rm II}^*$ denote the most probable values. For each system we write

$$\frac{\partial \ln \Omega(E)}{\partial E} = \frac{1}{kT}, \tag{2.2.6}$$

and call T the *absolute temperature.* k is the Boltzmann constant. The equilibrium condition may be written as

$$T_{\rm I}(E_{\rm I}^*) = T_{\rm II}(E_{\rm II}^*) \tag{2.2.7}$$

which means that temperatures are equal for systems in equilibrium.

In order that the extremum of $\Omega_{\rm I}\,\Omega_{\rm II}$ is really a maximum (in the following we omit *)

$$\frac{\partial^2 \ln \Omega_{\rm I}(E_{\rm I})}{\partial E_{\rm I}^2} + \frac{\partial^2 \ln \Omega_{\rm II}(E_{\rm II})}{\partial E_{\rm II}^2} < 0\,. \tag{2.2.8}$$

Namely,

$$\frac{\partial T_{\rm I}(E_{\rm I})}{\partial E_{\rm I}} + \frac{\partial T_{\rm II}(E_{\rm II})}{\partial E_{\rm II}} > 0\,.$$

For an ideal gas $T = \frac{2}{3}E/Nk$, we have $\partial T/\partial E \propto 1/N$. Thus, if system II is an ideal gas, assuming the number of molecules to be extremely large compared with system I, we have $\partial T_{\rm II}/\partial E_{\rm II} \to 0$, so that $\partial T_{\rm I}/\partial E_{\rm I} > 0$ for system I in equilibrium with system II. Thus, in order that equilibrium is established irrespective of the relative system size of systems I and II, it is necessary that

$$\partial T/\partial E > 0 \tag{2.2.9}$$

holds for each system. In other words, temperature should be an increasing function of energy. When this condition is satisfied, if two systems at different temperatures are brought into contact, one of the systems will lose energy with decreasing temperature while the other will gain energy with increasing temperature, and equilibrium is established at some intermediate temperature. It is also concluded that the temperature of the system which loses energy must be higher than the other which gains energy. Empirically, this is well known, which means that the above condition is satisfied in general.

The same can be said with respect to the condition for three systems I, II, and III to be in equilibrium. If Ω is an increasing function of E and $\partial T/\partial E > 0$, then we can determine $E_{\rm I}$, $E_{\rm II}$, and $E_{\rm III}$ such that I and II are in equilibrium when II and III are also in equilibrium. In this case I and III with energies $E_{\rm I}$ and $E_{\rm III}$ are also in equilibrium. Conversely, in order that such equilibrium is possible among three systems, the relation $\partial T/\partial E > 0$ must hold for each of them.

Though the condition $\partial T/\partial E > 0$ or $\partial^2 \ln \Omega/\partial E^2 < 0$ might look rather restrictive, usual systems should satisfy this condition because thermal equilibrium is generally established among them. The condition is easily verified with regard to an ideal gas and a spin system. In the above discussion we have assumed that there is only one maximum of $\Omega_{\rm I}\,\Omega_{\rm II}$. There can be two or more sets of E_{I}^{*} and E_{II}^{*} which give extrema. However, if one of them is exceedingly large compared with the others, we only have to account for the largest one. But, there can be a situation in which the extremum extends into a plateau. This means that $\partial T/\partial E = 0$, or that the temperature does not rise even if energy is supplied to the system. The so-called first-order phase transitions, such as condensation and melting, correspond to such cases.

2.2.2 Temperature

If we replace one of the systems, I or II, by an ideal gas, we can easily show that the absolute temperature defined in the preceding section is really identical to the absolute temperature defined according to a gas-thermometer. The absolute temperature of an ideal gas is clearly positive. So, for any substance which can be in equilibrium with an ideal gas, we have

$$T > 0 . \qquad (2.2.10)$$

However, for certain systems with an upper energy bound, such as a perfect spin system, the absolute temperature can be negative. In a two-level spin system, at low temperatures nearly all the spins are on the lower level and with increasing temperature, spins are excited to the upper level (Fig. 2.9). At $T \to \infty$ we have equal numbers of spins on the lower and upper levels. To increase the energy of the system further, we must lift more spins to the upper level. The maximum energy is attained when all the spins are lifted. The highest state, therefore, consists of only one microscopic state. Thus with increasing energy the system passes over to the region where $d\Omega/dE < 0$ or $T < 0$, which means that negative temperature implies a higher temperature than $T = \infty$. The same can be said for any system with an upper energy bound.

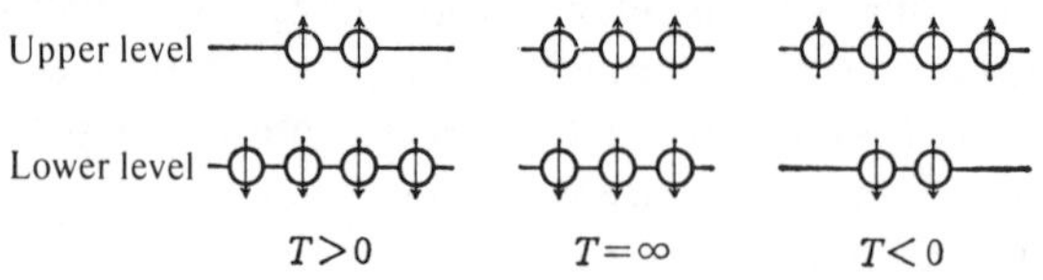

Fig. 2.9. The temperature of a spin system

If there were only such systems, some sort of equilibrium at a negative temperature might be possible. However, even in a spin system, spins are themselves electrons which can move through the crystal and they are also in thermal contact with lattice vibration or external gas, etc. whose energy has no upper bound, and correspondingly, the temperature is positive finite. Then the circumstance is that a system at $T < 0$ is in contact with another system at $T > 0$, and energy flows from the system at $T < 0$ to the system at $T > 0$ to establish finally a state of equilibrium at some positive temperature. In this sense also, a state at $T < 0$ is at a higher temperature than the state at $T = \infty$. Thus in real situations, a negative temperature cannot be a state of equilibrium. However, under certain steady conditions which are not in equilibrium, we can attain negative temperature; the best example will be found in a laser. The concept of negative temperature is very useful in such cases.

2.3 External Forces

2.3.1 Pressure Equilibrium

When two systems I and II are on two sides of a movable barrier, the condition of equilibrium is that the pressures of these systems are the same. We may discuss the pressure equilibrium in a similar way as we did the temperature equilibrium in the preceding section. Let the volumes of the two systems partitioned by a piston be V_{I} and V_{II}, assuming that the total volume is kept constant (Fig. 2.10a):

$$V_{\mathrm{I}} + V_{\mathrm{II}} = V = \text{const.} \tag{2.3.1}$$

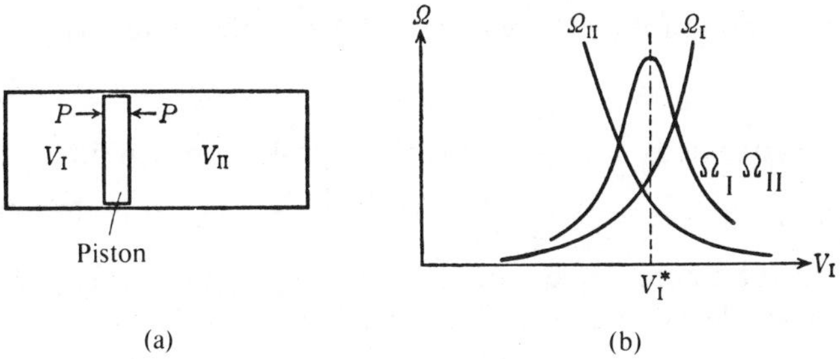

Fig. 2.10. **(a)** Pressure equilibrium. **(b)** The condition of maximum total state density as a function of volume

The state densities of the systems are functions of energy and volume and they can be written as $\Omega_{\mathrm{I}}(E_{\mathrm{I}}, V_{\mathrm{I}})$ and $\Omega_{\mathrm{II}}(E_{\mathrm{II}}, V_{\mathrm{II}})$, respectively. The condition of maximum $\Omega_{\mathrm{I}}\Omega_{\mathrm{II}}$ for a variation of V_{I} is (Fig. 2.10b)

$$\frac{\partial \ln \Omega_{\mathrm{I}}}{\partial V_{\mathrm{I}}^*} = \frac{\partial \ln \Omega_{\mathrm{II}}}{\partial V_{\mathrm{II}}^*}, \tag{2.3.2}$$

where V_{I}^* and V_{II}^* are the most probable values of the volumes of the systems (in the following we shall omit *). If we introduce Boltzmann's constant k and put

$$\frac{\partial \ln \Omega}{\partial V} = \frac{P}{kT}, \tag{2.3.3}$$

then P is the pressure as we shall see immediately below. Since the systems exchange energy through the piston, temperature equilibrium is set up and we have the condition of equilibrium such that $P_{\mathrm{I}} = P_{\mathrm{II}}$, namely, the pressures of both systems are the same.

In order to show that the pressure P is actually the mechanical pressure, let us put system I in a cylinder closed by a piston with a weight on the top (Fig. 2.11). We assume that heat does not penetrate the cylinder and the piston. Then the total energy is kept constant:

$$E_{\mathrm{I}} + wx = E = \text{const.}\,, \tag{2.3.4}$$

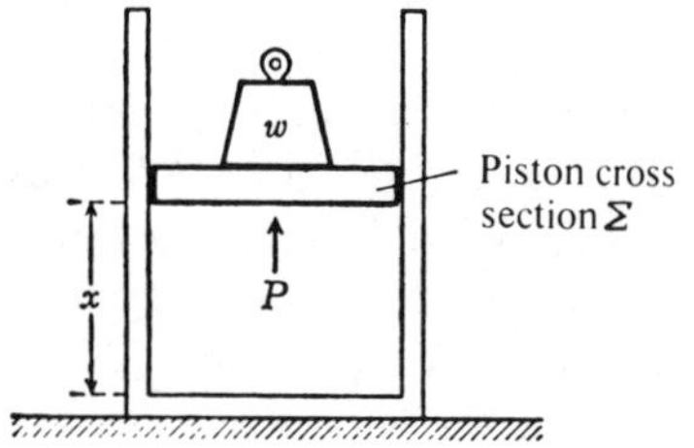

Fig. 2.11. A system with a weight

where w denotes the weight and x its height, so that wx is the potential energy of the weight. We may take as x the length of the system in the cylinder. If we denote by Σ the cross section of the cylinder, the volume of the system is

$$V_{\mathrm{I}} = x\Sigma\,. \tag{2.3.5}$$

Since the energy of the system is $E_{\mathrm{I}} = E - wx$, the state density can be written as

$$\Omega_{\mathrm{I}} = \Omega_{\mathrm{I}}(E - wx, x\Sigma)\,. \tag{2.3.6}$$

In the equilibrium state, the macroscopic state x has the largest state density Ω_{I}. Therefore, the most probable value of x is given by $\partial \ln \Omega_{\mathrm{I}}/\partial x = 0$, or

$$-w\frac{\partial \ln \Omega_{\mathrm{I}}}{\partial E_{\mathrm{I}}} + \Sigma\frac{\partial \ln \Omega_{\mathrm{I}}}{\partial V_{\mathrm{I}}} = 0\,. \tag{2.3.7}$$

The pressure is the force per unit area so that $P_{\mathrm{I}} = w/\Sigma$ and we also have the equation $\partial \ln \Omega_{\mathrm{I}}/\partial E_{\mathrm{I}} = 1/kT_{\mathrm{I}}$. Therefore, we have

$$\frac{\partial \ln \Omega_{\mathrm{I}}}{\partial V_{\mathrm{I}}} = \frac{w}{\Sigma}\frac{\partial \ln \Omega_{\mathrm{I}}}{\partial E_{\mathrm{I}}} = \frac{P_{\mathrm{I}}}{kT_{\mathrm{I}}}\,. \tag{2.3.8}$$

Thus P in (2.3.3) is the same as the mechanical pressure.

If we use the entropy $S = k \ln \Omega$, we have the relation

$$\frac{\partial S}{\partial V} = \frac{P}{T}\,. \tag{2.3.9}$$

The condition that (2.3.7) really gives the maximum can be written as

$$w^2\frac{\partial^2 S}{\partial E^2} - 2w\Sigma\frac{\partial^2 S}{\partial E\,\partial V} + \Sigma^2\frac{\partial^2 S}{\partial V^2} < 0\,. \tag{2.3.10}$$

If we further rewrite (2.3.10) as we usually do in thermodynamics, we obtain

$$\left(\frac{\partial P}{\partial V}\right)_S < 0 \tag{2.3.11}$$

as the condition of the maximum Ω, or of the stability. In (2.3.11), the subscript S implies that the entropy is kept constant when the derivative is taken. This is an adiabatic change (Sect. 2.3.2). Therefore, when the system is in equilibrium, the pressure decreases with increasing volume. It can be shown, by further transformation, that $(\partial P/\partial V)_T < 0$ for isothermal change.

Instead of using a weight to show that P is actually the pressure, we may replace system II by an ideal gas. Then the gas pressure P_{II} is due to the force exerted by molecular collision against the piston, and therefore, P_{I} of system I in equilibrium with the gas is seen to have the meaning of pressure. This is the same as the method we apply to show that the absolute temperature introduced in statistical mechanics coincides with the temperature defined by a gas thermometer. However, there is no reason to reject negative pressure; this is a point which differs from the case of temperature. We may exert negative pressure on the system by using an appropriate spring, etc., and see that the above discussion also applies.

2.3.2 Adiabatic Theorem

a) *Adiabatic Change*

There are external variables like volume V, external magnetic field H, and so forth, which specify the state of the system. However, for brevity, we shall assume that only one of the external variables changes. Extension to a change of many variables is straightforward. When we shift an external variable, the quantum-mechanical levels shift also, and therefore the number of states j below the energy E depends on the external variable a. Thus we write

$$j = j(E, a)\,. \tag{2.3.12}$$

We consider an ensemble of systems with energy below E. When the external variable a is changed, each system will experience some change accordingly. If the change in a happens quite abruptly, a system may be transferred from one level to another, and, as a result, the ensemble will be dispersed in the space of quantum states. Such a change of ensemble corresponds to the law of thermodynamics which says that entropy is increased in general by an abrupt change.

If such a thermodynamic consideration is allowed, we may proceed a little further. That is, we expand the rate of change in the entropy S in powers of $\dot{a}$, the rate of change in a. Then the series will begin with the second power, since S must increase irrespective of the sign of $\dot{a}$. Thus $dS/dt \propto \dot{a}^2$ or $dS/da \propto a$ for a small value of $\dot{a}$. Therefore, when the rate of change $\dot{a}$ approaches 0, the change in S due to the change in a, or dS/da, also approaches 0. We see that in the limit of small $\dot{a}$, the process (quasistatic process in thermodynamics) gives no change in entropy and is shown to be reversible. In this consideration, the system was assumed to have received no influence other than the change in the external variable. That is to say, the system had no transition between levels which might be induced by thermal disturbance from the outside world. Such a process, in which the system is thermally insulated and external variables change sufficiently slowly, is called adiabatic change. During an adiabatic change, the entropy of the system is kept constant.

In quantum mechanics, it is known that no transition occurs and that a system stays in the same quantum state if external variables change sufficiently slowly. This is the adiabatic theorem in quantum mechanics.

In general, the rate of change of the expectation value $\langle \mathscr{H} \rangle$ of the Hamiltonian can be written as

$$\begin{aligned} \frac{d}{dt}\langle \mathscr{H} \rangle &= \frac{d}{dt}(\psi, \mathscr{H}\psi) \\ &= \frac{\mathrm{i}}{\hbar}[(\mathscr{H}\psi, \mathscr{H}\psi) - (\psi, \mathscr{H}^2\psi)] + \left(\psi, \frac{\partial \mathscr{H}}{\partial t}\psi\right), \end{aligned}$$

where the first two terms on the right-hand side cancel due to the Hermitian character of $\mathscr{H}$. If $\mathscr{H}$ is a function of time through an external variable a, we have thus

$$\frac{d}{dt}\langle \mathscr{H} \rangle = \left\langle \frac{\partial \mathscr{H}}{\partial a}\dot{a} \right\rangle . \tag{2.3.13}$$

Let the system be in a state k initially at $t = 0$. Then, if a changes sufficiently slowly, the system stays in the same state and we should take the expectation value for this state. In order to calculate $\partial \mathscr{H}/\partial a$, we differentiate $\mathscr{H}\psi_k = E_k\psi_k$ with respect to a, to have

$$\left(\frac{\partial \mathscr{H}}{\partial a} - \frac{\partial E_k}{\partial a}\right)\psi_k = (E_k - \mathscr{H})\frac{\partial \psi_k}{\partial a} .$$

However, if we multiply ψ^* from the left to calculate the expectation value, the terms on the right-hand side cancel because of the fact that $\psi_k^*\mathscr{H} = E_k\psi_k^*$, yielding

$$\left\langle \frac{\partial \mathscr{H}}{\partial a} \right\rangle_k = \frac{\partial E_k}{\partial a} . \tag{2.3.14}$$

When a changes sufficiently slowly, in (2.3.13) we may consider $\dot{a}$ as a constant and take it out of the bracket. Thus we see that the change in the energy of the system $\Delta\langle \mathscr{H} \rangle$ is given by

$$\Delta E = \frac{\partial E_k}{\partial a}\Delta a . \tag{2.3.15}$$

Since ΔE is the energy change due to the change Δa of the external variable,

$$A_k = -\frac{\partial E_k}{\partial a} \tag{2.3.16}$$

means the force exerted by the system when it is in the state k. If a is the volume, A corresponds to the pressure P.

If we refer to a microcanonical ensemble, then disturbances from the external world or some complicated internal interaction neglected in the Hamiltonian give rise to transitions between different states and each system will experience all the states of the ensemble (the ergodic hypothesis). Therefore, the force actually observed is the average

$$A = \langle A_k \rangle = -\left\langle \frac{\partial E_k}{\partial a} \right\rangle, \tag{2.3.17}$$

where the average is to be taken over the microcanonical ensemble. Since the number of states increases enormously with energy, the average may be taken over all the states below the energy E. In the transition during the slow change, the system is always maintained in the state of equilibrium (a slow change during which equilibrium is maintained is called a *quasistatic process*).

b) *Adiabatic Theorem in Statistical Mechanics*

The change in the energy of a system is given by the average

$$\Delta E = \left\langle \frac{\partial E_k}{\partial a} \right\rangle \Delta a = -A\,\Delta a \tag{2.3.18}$$

which is the work done by the external force A. For this value of ΔE,

$$j(E + \Delta E,\, a + \Delta a) = j(E, a) \tag{2.3.19}$$

or

$$\int_0^{E+\Delta E} \Omega(E, a + \Delta a)\,dE = \int_0^{E} \Omega(E, a)\,dE \tag{2.3.20}$$

holds, as we shall see below. In statistical mechanics, this relation is called the adiabatic theorem. The number of states below E is not altered if E is shifted by the amount of work ΔE due to the slow change of external variables. In other words, the entropy $S = k \ln j$ of the system does not change during an adiabatic change.

To show the above relation, we rewrite it in differential form as

$$\Delta j = \frac{\partial j}{\partial E}\Delta E + \frac{\partial j}{\partial a}\Delta a = 0\,. \tag{2.3.21}$$

Since $\Delta j/\Delta E = \Omega$, the first term can be written as

$$\frac{\partial j}{\partial E}\Delta E = \Omega(E, a)\left\langle \frac{\partial E_k}{\partial a} \right\rangle \Delta a\,. \tag{2.3.22}$$

The second term is the increase in the number of states below the energy surface $\mathscr{H} = E$ when the external variable is changed by the amount Δa. In other words, the number of states below E is decreased every time an energy eigenvalue exceeds E. To calculate this, we note that Ω is the number of states in a unit energy range just below $\mathscr{H} = E$, and let r be the numbering of these states arranged in order according to the magnitude of $\partial E_r/\partial a$. Let $\Omega^{(r)}$ be the number of states with the same value of the derivative $\partial E_r/\partial a$, so that

$$\sum_r \Omega^{(r)} = \Omega \, . \tag{2.3.23}$$

Now, when a is increased by the amount Δa, for each r the number of states whose energy eigenvalue exceeds $\mathscr{H} = E$ is $\Omega^{(r)}(\partial E_r/\partial a)\Delta a$ (Fig. 2.12). Therefore, the second term of (2.3.21) is

$$\frac{\partial j}{\partial a}\Delta a = -\sum_r \Omega^{(r)}\frac{\partial E_r}{\partial a}\Delta a \, . \tag{2.3.24}$$

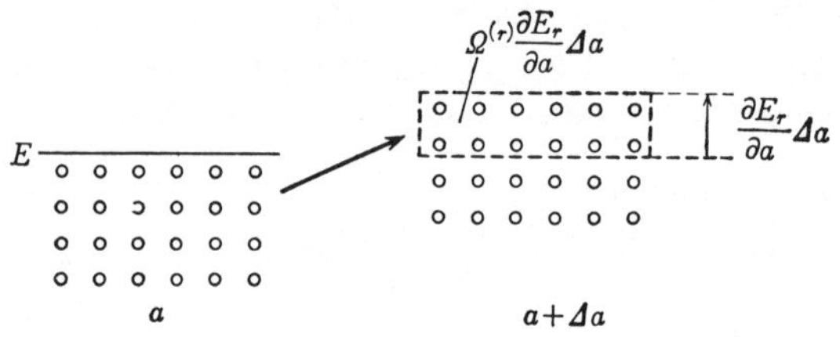

Fig. 2.12. *Circles* represent states. Diagram shows the increase in the number of states above E as a result of an increase in an external variable a

However, since

$$\left\langle \frac{\partial E_k}{\partial a} \right\rangle = \frac{1}{\Omega}\sum_r \Omega^{(r)}\frac{\partial E_r}{\partial a} \, , \tag{2.3.25}$$

we see that the second term just cancels the first term of (2.3.21) and thus $\Delta j = 0$, which proves the theorem.

c) *Adiabatic Theorem in Classical Mechanics*

In classical mechanics a closed trajectory corresponds to a quantum-mechanical stationary state. When an external variable a is changing, the trajectory may not close (Fig. 2.13). But we need to think of a closed trajectory for the fixed value of

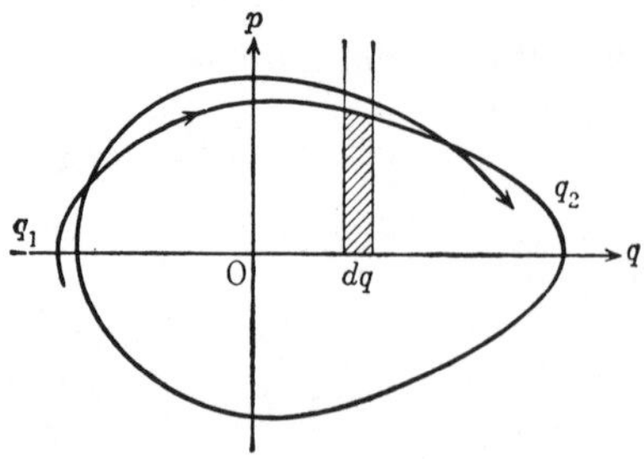

Fig. 2.13. Adiabatic change and trajectory

a that $a(t)$ happens to take at some instant of time. The energy can be written as $E = \mathscr{H}(q, p, a)$, and solving for p, we see that the area enclosed by the trajectory is the action variable given as

$$J(E, a) = \oint p(q, E, a)\,dq \,. \tag{2.3.26}$$

Since our Hamiltonian is of the form $\mathscr{H} = p^2/2m + U(q, a)$, which is not altered by a change of the sign of p, the trajectory is symmetric with respect to the line $p = 0$. Let q_1 and q_2 be the turning points of q ($p = 0$). Then we have

$$J(E, a) = 2 \int_{q_1}^{q_2} p(q, E, a)\,dq \,. \tag{2.3.27}$$

and q_1 and q_2 are functions of E and a. But, since they are roots of $p = 0$, the change in J does not include terms due to Δq_1 and Δq_2, so that

$$\frac{1}{2}\Delta J = \int_{q_1}^{q_2} \left(\frac{\partial p}{\partial E}\Delta E + \frac{\partial p}{\partial a}\Delta a \right) dq \,.$$

For the first term of the integrand of the above equation we use the canonical equation of motion

$$\frac{dq}{dt} = \frac{\partial \mathscr{H}}{\partial p} = \frac{1}{\partial p/\partial E} \,.$$

For the second term, we calculate the derivative $\partial p/\partial a$, where q and E are kept constant, using $\mathscr{H}(q, p, a) = E$:

$$\frac{\partial \mathscr{H}}{\partial p}\frac{\partial p}{\partial a} + \frac{\partial \mathscr{H}}{\partial a} = 0 \quad \text{or} \quad \frac{dq}{dt}\frac{\partial p}{\partial a} = -\frac{\partial \mathscr{H}}{\partial a} \,.$$

Thus, we have

$$\frac{1}{2}\Delta J = \int \left(\Delta E - \frac{\partial \mathscr{H}}{\partial a}\Delta a \right) dt \,. \tag{2.3.28}$$

We have defined J for a virtual motion with a fixed value of a. However, for the actual motion $E = \mathscr{H}(q, p, a(t))$, so we have

$$dE = \left(\frac{\partial \mathscr{H}}{\partial q}\dot{q} + \frac{\partial \mathscr{H}}{\partial q}\dot{p} \right) dt + \frac{\partial \mathscr{H}}{\partial a} da \,.$$

Since the canonical equations of motion hold even when $\mathscr{H}$ depends on time through $a(t)$, the terms in the parentheses on the right-hand side cancel to give

$$dE = \frac{\partial \mathscr{H}}{\partial a} da \,.$$

If the change in a is very slow, the virtual motion will be very close to the actual motion, and we may replace ΔE and Δa by their actual values dE and da.

Therefore, for a very slow change of external variables, we have

$$\frac{dJ}{dt} = 0 , \tag{2.3.29}$$

that is, the area in phase space enclosed by the trajectory of a periodic motion is invariant for a slow change of external variables. This is the adiabatic theorem in classical mechanics.

The above proof is restricted to one-dimensional systems. We shall present a more general proof, since the actual phase space is multidimensional. In view of (2.3.20), it will suffice in classical statistical mechanics to show that the volume $\tau(E, a)$ of phase space below the energy E is equal to the volume $\tau(E + \Delta E, a + \Delta a)$ below $E + \Delta E$. Except that we now deal with a continuous phase space, the proof is essentially the same as that of Sect. 2.3.2b for the quantum-mechanical case. We shall show that the change in the volume $\tau(E, a)$ of phase space vanishes for a sufficiently slow shift of an external variable. We divide the change into two steps.

The first change is a geometrical one, which comes from the change in shape of the energy surface $\mathscr{H} = E$ through the variation of an external variable a. For this change, E is kept constant so that

$$\Delta\mathscr{H} = \sum_j \left(\frac{\partial \mathscr{H}}{\partial q_j} \Delta q_j + \frac{\partial \mathscr{H}}{\partial p_j} \Delta p_j \right) + \frac{\partial \mathscr{H}}{\partial a} \Delta a = 0 . \tag{2.3.30}$$

Let $\Delta\nu$ be the distance between the energy surfaces $\mathscr{H}(a) = E$ and $\mathscr{H}(a + \Delta a) = E$. The gradient of $\mathscr{H}$ in qp space has the value $|\mathrm{grad}\,\mathscr{H}| = \sqrt{(\partial\mathscr{H}/\partial q)^2 + (\partial\mathscr{H}/\partial p)^2}$, and the first term in the above equation can be written as $|\mathrm{grad}\,\mathscr{H}|\Delta\nu$. Thus we have

$$|\mathrm{grad}\,\mathscr{H}|\Delta\nu = -\frac{\partial \mathscr{H}}{\partial a}\Delta a . \tag{2.3.31}$$

The change in volume of $\tau(E)$ for this step is therefore given as (Fig. 2.14a)

$$(\Delta\tau)_1 = \int \Delta\nu \, d\sigma = -\Delta a \int \frac{\partial\mathscr{H}/\partial a}{|\mathrm{grad}\,\mathscr{H}|} d\sigma , \tag{2.3.32}$$

where $d\sigma$ denotes the surface element of the energy surface.

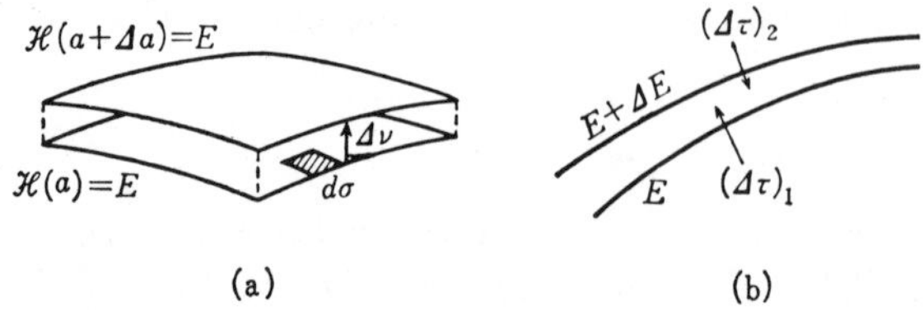

Fig. 2.14 a, b. Adiabatic change

The second change is due to the shift of E, and is

$$(\Delta\tau)_2 = \frac{d\tau}{dE}\Delta E\,, \tag{2.3.33}$$

where we may write

$$\frac{d\tau}{dE} = \int_{\mathscr{H}=E} \frac{d\sigma}{|\mathrm{grad}\,\mathscr{H}|}\,.$$

Therefore, if ΔE is equal to the average of $(\partial\mathscr{H}/\partial a)\Delta a$, or

$$\Delta a \int \frac{\partial\mathscr{H}/\partial a}{|\mathrm{grad}\,\mathscr{H}|}d\sigma \bigg/ \int \frac{d\sigma}{|\mathrm{grad}\,\mathscr{H}|} = -\langle A\rangle_E \Delta a\,, \tag{2.3.34}$$

then the changes in these two steps cancel each other (Fig. 2.14b):

$$\Delta\tau = (\Delta\tau)_1 + (\Delta\tau)_2 = 0\,. \tag{2.3.35}$$

In (2.3.34), $\langle A\rangle_E$ is the average of the force over the microcanonical ensemble. By the ergodic hypothesis that the ensemble average is equal to the time average, $\langle A\rangle_E$ is the force exerted by the system during the quasiprocess. Thus, for an adiabatic, quasistatic process, the energy change of the system is $\Delta E = -A\Delta a$, and we have

$$\tau(E + \Delta E, a + \Delta a) = \tau(E, a)\,, \tag{2.3.36}$$

that is, the volume of phase space is kept constant.

2.3.3 Thermodynamic Relations

If the system has f degrees of freedom, the number of states is given by $j = \tau/h^f$. We have seen that for an adiabatic, quasistatic process, $\tau(E, a)$ is invariant which means that the entropy $S = k \ln j$ is kept constant during such a process. Therefore, for an adiabatic, quasistatic process, the entropy is invariant.

If we consider two energy surfaces $\mathscr{H} = E$ and $\mathscr{H} = E - \delta E$, the numbers of quantum states below these surfaces are invariant for an adiabatic, quasistatic process, and the number of states between these surfaces is also invariant. Thus, if we define entropy by $S = k\ln(\Omega\delta E)$, it is also invariant for an adiabatic, quasistatic process. This result is clear since entropy may be defined either way.

Though we may use either j or $\Omega = \partial j/\partial E$, let us differentiate (2.3.20) with respect to E to have

$$\Omega(E + \Delta E, a + \Delta a) = \Omega(E, a) \tag{2.3.37}$$

for an adiabatic, quasistatic process. Consider the volume V as an external variable. For small $\Delta a = \Delta V$ and ΔE, we may rewrite (2.3.37) as

$$-\frac{\partial \ln\Omega}{\partial E}\Delta E = \frac{\partial \ln\Omega}{\partial V}\Delta V\,. \tag{2.3.38}$$

Using $\partial \ln \Omega/\partial E = 1/kT$, and (2.3.3) for the right-hand side to write it as $(P/kT)\Delta V$, we have

$$-\Delta E = P\Delta V \, .$$

If ΔE is inserted from (2.3.18), we obtain

$$P = -\left\langle \frac{\partial E_k}{\partial V} \right\rangle . \tag{2.3.39}$$

Thus, we see that the force defined by (2.3.17) is equivalent to the pressure defined by (2.3.3).

The energy E of the system is given as the average of E_k, or

$$E = \langle E_k \rangle \, . \tag{2.3.40}$$

During an adiabatic process, volume change is performed without transition between quantum states, and the ensemble over which the average is to be taken does not change. This can also be characterized as a process in which entropy is kept constant. Therefore, (2.3.39) can be written as

$$P = -\left(\frac{\partial E}{\partial V} \right)_S . \tag{2.3.41}$$

This is one of the formulas for pressure.

E is what is called *internal energy* in thermodynamics and is a quantity specified by the macroscopic state of the system. This means that the internal energy is a *quantity of state* and as such a function of macroscopic quantities such as entropy, volume and so on. In (2.3.41), E is regarded as a function of S and V, namely as $E = E(S, V)$. Differentiating, we have

$$dE = \left(\frac{\partial E}{\partial S} \right)_V dS + \left(\frac{\partial E}{\partial V} \right)_S dV.$$

If we put $dV = 0$, we see that

$$\left(\frac{\partial E}{\partial S} \right)_V = 1 \Big/ \left(\frac{\partial S}{\partial E} \right)_V = T \, .$$

Thus, in view of (2.3.41) we have

$$dE = T\,dS - P\,dV \, , \tag{2.3.42}$$

where dE is the energy change of the system, and $-P\,dV$ is the work done upon the system. Therefore, by the law of conservation of energy, we see that $T\,dS$ is the energy given to the system as a quantity of heat. This is the content of the first law of thermodynamics, though absolute temperature and entropy is introduced in thermodynamics by using the second law. Since we have thus derived the fundamental laws of thermodynamics from statistical mechanics, we may also derive all the thermodynamic relations from them.

2.4 Subsystems with a Given Temperature

2.4.1 Canonical Ensemble

A practical method of keeping the temperature of a system constant is to immerse it in a very large material with a large heat capacity. If the material is very large, its temperature is not changed even if some energy is given or taken by the system in contact. Such a heat reservoir serves as a thermostat. In order to apply the method developed in the preceding sections to the present problem, we assume that the system is in thermal contact with another very large system. In other words, we consider the system as being a comparatively small part of a large closed system. We call the small part a subsystem.

Let I be the subsystem, and II the remaining large system (Fig. 2.15). We can use the results of Sect. 2.2. If the total energy is E, the probability that system I has the energy E_{I} is proportional to

$$\Omega_{\mathrm{I}}(E_{\mathrm{I}})\Omega_{\mathrm{II}}(E - E_{\mathrm{I}})\,dE_{\mathrm{I}}\,. \tag{2.4.1}$$

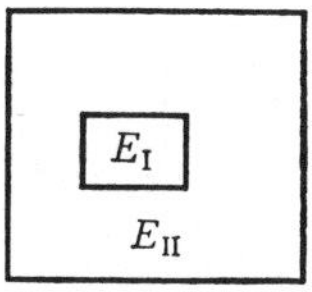

Fig. 2.15. Subsystem

If we take an ideal gas as system II, then as a function of E_{I} we have $\Omega_{\mathrm{II}}(E - E_{\mathrm{I}}) \propto \exp(-\beta E_{\mathrm{I}})$, cf. (2.1.18). In general, we assume that $\Omega_{\mathrm{II}}(E_{\mathrm{II}})$ is a rapidly increasing function of E_{II}, and expand the entropy of system II as

$$k \ln \Omega_{\mathrm{II}}(E - E_{\mathrm{I}}) = S_{\mathrm{II}}(E - E_{\mathrm{I}})$$

$$= S_{\mathrm{II}}(E) - \frac{\partial S_{\mathrm{II}}(E)}{\partial E} E_{\mathrm{I}} + \frac{1}{2}\frac{\partial^2 S_{\mathrm{II}}(E)}{\partial E^2} E_{\mathrm{I}}^2 + \cdots\,. \tag{2.4.2}$$

If we may express the largeness of system II by its number of molecules N, then we see that $S_{\mathrm{II}} \propto N$, $E \propto N$, and these are extensive variables. It follows that $\partial S_{\mathrm{II}}/\partial E \sim 1$, $\partial^2 S_{\mathrm{II}}/\partial E^2 \sim 1/N, \ldots$. Now, we make system II sufficiently large, keeping the energy per molecule, E_{II}/N, finite. Then, in the above expansion the third and higher terms can be neglected. If we write

$$\frac{\partial S_{\mathrm{II}}(E)}{\partial E} = \frac{1}{T} = k\beta\,, \tag{2.4.3}$$

T is the temperature of the heat reservoir, and β is the reciprocal temperature. Thus for $E_{\mathrm{I}} \ll E$, we have the asymptotic form of the state density of system II as

$$\Omega_{\mathrm{II}}(E - E_{\mathrm{I}}) \propto \exp(-\beta E_{\mathrm{I}})\,. \tag{2.4.4}$$

Since every quantum state has equal a priori probability (the principle of equal probability) when the total system is in equilibrium, the probability that system I is in a quantum state with energy E_{I} is proportional to $\exp(-\beta E_{\mathrm{I}})$, and the probability that it is in the range between E_{I} and $E_{\mathrm{I}} + dE_{\mathrm{I}}$ is proportional to

$$\Omega_{\mathrm{I}}(E_{\mathrm{I}})\exp(-\beta E_{\mathrm{I}})\,dE_{\mathrm{I}}\,. \tag{2.4.5}$$

Therefore, the representative points of system I are distributed with the density proportional to $\exp(-\beta E_{\mathrm{I}})$. This is called the *canonical ensemble*, and this distribution of representative points is called the *canonical distribution* (Gibbs' distribution), or *isothermal distribution*. The factor $\exp(-\beta E)$ is often referred to as the *Boltzmann factor*.

Since every quantum state j has a probability proportional to $\exp(-\beta E_j)$, the average of any physical quantity A_j is given by

$$\langle A \rangle = \frac{\sum\limits_j A_j \exp(-\beta E_j)}{\sum\limits_j \exp(-\beta E_j)}\,. \tag{2.4.6}$$

For example, the average of energy is

$$E = \frac{\sum E_j \exp(-\beta E_j)}{\sum \exp(-\beta E_j)}\,, \tag{2.4.7}$$

where we have omitted the bracket $\langle\ \rangle$ because E is a thermodynamic quantity. Similarly, pressure is the average of $-\partial E_j/\partial V$, so that

$$P = -\frac{\sum(\partial E_j/\partial V)\exp(-\beta E_j)}{\sum \exp(-\beta E_j)}\,. \tag{2.4.8}$$

In general, $-PdV$ is the work done upon the system, and

$$dQ = dE + PdV$$

is the quantity of heat given to the system. If we differentiate E and P given by (2.4.7, 8) we see that dQ, multiplied by $k\beta = 1/T$, is the total differential of

$$S = k\beta E + k\ln\sum_j \exp(-\beta E_j)\,. \tag{2.4.9}$$

That is to say,

$$k\beta\, dQ = dS\,.$$

In thermodynamics, the reciprocal of the factor $k\beta$ which, when multiplied, makes dQ the total differential is called the absolute temperature. Thus, $1/k\beta$ is the absolute temperature and S is the entropy in thermodynamics. That this is equivalent to the entropy defined for a microcanonical ensemble is shown as follows.

First we note that the probability of a state is

$$w_n = \frac{\exp(-\beta E_n)}{\sum_j \exp(-\beta E_j)} \tag{2.4.10}$$

and that entropy (2.4.9) for a canonical ensemble can be written as

$$S = -k\langle \ln w_n \rangle = -k\sum_n w_n \ln w_n \, . \tag{2.4.11}$$

On the other hand, the entropy for a microcanonical ensemble is given by (2.1.6) or

$$S_{mc} = k \ln W \, .$$

W is the total number of accessible microscopic states and therefore the probability of each state is the same, namely,

$$p_1 = p_2 = \cdots = p_W = \frac{1}{W} = p \, .$$

Thus we may write

$$S_{mc} = -k \ln p = -k \sum_{k=1}^{W} p_k \ln p_k \, . \tag{2.4.12}$$

In either case, the entropy is the (negative of the) average of the probability of each state.

Since there are $\Omega(E)\,dE$ states in the energy range dE, the entropy (2.4.11) for a canonical ensemble may be written as

$$S = -k\int \{w(E)\ln w(E)\}\,\Omega(E)\,dE \, ,$$

where

$$w(E) = \exp(-\beta E)/\int \exp(-\beta E)\,\Omega(E)\,dE \, .$$

2.4.2 Boltzmann-Planck's Method

One of the distinctive features of a canonical distribution is that it is the only solution which satisfies the requirements of probability theory for independent events. Suppose that two systems I and II are in thermal contact with a heat bath at temperature T. If the probability for system I to be in the quantum state with energy $E_{\mathrm{I}j}$ is designated by $\Pr^{(\mathrm{I})}(E_{\mathrm{I}j})$, and the probability for system II to be in the state with energy $E_{\mathrm{II}\alpha}$ by $\Pr^{(\mathrm{II})}(E_{\mathrm{II}\alpha})$, then by the assumption that these systems are independent, the joint probability for the systems to be simultaneously in the states j and α is given as

$$\Pr^{(\mathrm{I})}(E_{\mathrm{I}j})\Pr^{(\mathrm{II})}(E_{\mathrm{II}\alpha}) = \Pr(E_{j\alpha}) \, ,$$

where

$$E_{j\alpha} = E_{\mathrm{I}j} + E_{\mathrm{II}\alpha}$$

and Pr is the probability for the combined system. The distribution law is assumed to depend only on the energy, being independent of the nature of each system. Then, $\mathrm{Pr}^{(\mathrm{I})}$, $\mathrm{Pr}^{(\mathrm{II})}$ and Pr, as functions of energy, must be of the same functional form. As a law of statistical mechanics, we may demand such a property which is satisfied by

$$\mathrm{Pr}(E) \propto \exp(-\beta E) ,$$

where β is a coefficient which is determined by the environment or the temperature. This probability distribution is the canonical distribution.

If such a consideration is applied to a large system consisting of systems of the same structure, for each system, the others play the role of a heat reservoir. In Boltzmann-Planck's method, we consider the ways of distributing N total systems among states with energies E_j. Let N_j be the number of systems in E_j (Fig. 2.16). We assume that the interaction energy between the systems can be neglected. Furthermore, let E be the total energy:

$$\begin{aligned} N_1 + N_2 + \cdots &= N \\ N_1 E_1 + N_2 E_2 + \cdots &= E . \end{aligned} \tag{2.4.13}$$

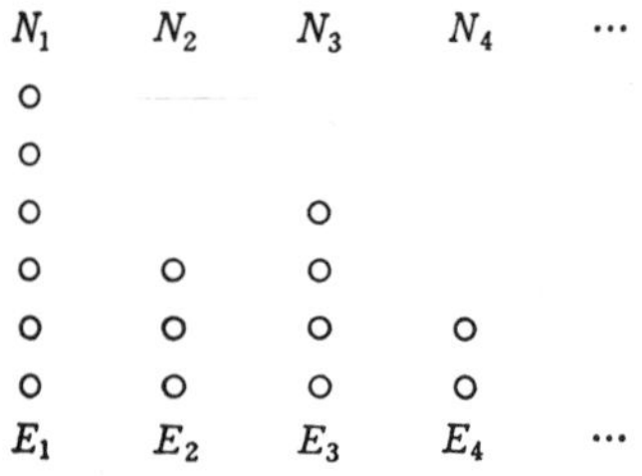

Fig. 2.16. A distribution of N systems

The number of ways of distributing N systems, so that N_1 systems are in E_1 and N_2 systems are in E_2, and so forth, is given as

$$W(N_1, N_2, \ldots) = \frac{N!}{N_1! N_2! \ldots} , \tag{2.4.14}$$

where $W = W(N_1, N_2, \ldots)$ is the number of events specified by $N_1, N_2, \ldots$. For fixed values of N and E, the total number of different distributions, or of microscopic states, is

$$W(E) = \sum_{N_1, N_2, \ldots} W(N_1, N_2, \ldots)$$

and its logarithm gives the entropy of the total system. However, we may replace

$W(E)$ by the maximum term $W(N_1, N_2, \ldots)$ with respect to the change of $N_1, N_2, \ldots$. We assume that N is sufficiently large, so that all the N_j's are also very large. Then we can use Stirling's formula

$$\ln N! = N \ln N - N + \frac{1}{2}\ln(2\pi N) + O\left(\frac{1}{N}\right) \tag{2.4.15}$$

to have

$$\ln W = -N \sum_n w_n \ln w_n \tag{2.4.16}$$

with

$$w_n = \frac{N_n}{N}, \tag{2.4.17}$$

which is subject to the supplementary conditions

$$\begin{aligned} &\sum_n w_n = 1 \\ &\sum_n w_n E_n = \frac{E}{N} = \text{const.} \end{aligned} \tag{2.4.17'}$$

Treating w_n as continuous variables, we have the variational equation

$$\delta\left(\frac{1}{N}\ln W + \lambda \sum w_n + \mu \sum w_n E_n\right) = 0$$

which gives w_n for the maximum W. Here λ and μ are Lagrange's indeterminate multipliers. Thus, we obtain

$$\ln w_n = \lambda + \mu E_n$$

or

$$w_n = C \exp(-\beta E_n), \tag{2.4.18}$$

where β and C are coefficients which replace λ and μ. When E_n goes to infinity, β should be positve. The coefficient C is determined by the condition (2.4.17′), and it is positive.

With the above w_n,

$$S_N = k \ln W = -Nk \sum_n w_n \ln w_n \tag{2.4.19}$$

is the entropy of the total system composed of N systems. Therefore, the entropy of each system is $-k\sum w_n \ln w_n$ as we have already seen, cf. (2.4.11). In this treatment, we have first calculated the most probable value of the entropy of the combined system of N systems in a microcanonical ensemble, and then by dividing by N, we have obtained the entropy of a system in a canonical ensemble.

Strictly speaking, the energy of the combined system, i.e., the microcanonical system, should not be assumed to have a definite value; instead it should be in some interval, say between E and $E + \delta E$. If it were not so, the set of N_j which satisfies the supplementary conditions (2.4.13) would be extremely rare. For example, if E is a rational number and E_j's take irrational values, then there is no set of integral values of N_j. This kind of difficulty is overcome by some small allowance of energy, as we have already indicated in Sect. 2.1.

2.4.3 Sum Over States

The sum over all the accessible states j with energy E_j

$$Z(\beta, V) = \sum_j \exp(-\beta E_j) \tag{2.4.20}$$

is called the *sum over states*, or the *partition function*. Since E_j is a function of volume V, Z is a function of V. If there are other external variables, Z is a function of these variables.

Using the sum over states, energy and pressure are written as

$$E = -\frac{\partial}{\partial \beta} \ln Z(\beta, V) \tag{2.4.21}$$

$$P = k\frac{\partial}{\partial V} \ln Z(\beta, V)\,, \tag{2.4.22}$$

respectively, the entropy is given by

$$S = k\beta E + k \ln Z(\beta, V)\,. \tag{2.4.23}$$

Thus all the thermodynamic quantities can be expressed by using the sum over states, and the knowledge of the sum over states is sufficient to understand all the macroscopic properties of the system. Therefore, as far as the equilibrium state is concerned, the application of statistical mechanics means the calculation of the sum over states.

In statistical thermodynamics, *Planck's characteristic function*

$$\Psi = k \ln Z(\beta, V)$$

is sometimes used. However, we usually use *Helmholtz's free energy*

$$F = E - TS = -kT \ln Z(\beta, V)\,. \tag{2.4.24}$$

We may write

$$\exp(-\beta F) = Z(\beta, V) = \sum_j \exp(-\beta E_j) \tag{2.4.25}$$

or, using the state density Ω,

$$\exp(-\beta F) = Z(\beta, V) = \int_0^\infty \Omega(E, V) \exp(-\beta E)\, dE\,, \tag{2.4.26}$$

where the lower limit of energy is put equal to zero. The last equation means that the sum over states is a Laplace transform of

$$\Omega(E, V) = \exp(S/k) .$$

Since $\Omega(E, V)$ is a rapidly increasing function of E, $\Omega(E, V)\exp(-\beta E)$ has a sharp maximum at some value of E (Sect. 2.2.1). If E^* is this value of E and S^* is the corresponding value of S, we may replace the integral of (2.4.26) by the maximum value of the integrand in such a way that

$$\exp(-\beta F) \approx \exp[-\beta(E^* - TS^*)] ,$$

for, in taking the logarithm, the important term is proportional to the size of the system. Thus, we recover the relation $F = E - TS$.

2.4.4 Density Matrix and the Bloch Equation

We define the density operator (density matrix) by

$$\rho = Z^{-1}\exp(-\beta\mathscr{H}) , \tag{2.4.27}$$

where Z is the sum over states . In x-representation we have

$$\rho(x, x') = \sum_n \varphi_n^*(x')\exp[\beta(F - \mathscr{H})]\varphi_n(x) . \tag{2.4.28}$$

In terms of this operator, the average of any quantity A can be written as, cf. (1.3.4, 6),

$$\langle A \rangle = \mathrm{tr}\{A\rho\} = \iint A(x', x)\rho(x, x')\,dx\,dx' . \tag{2.4.29}$$

Noting

$$\sum_n \varphi_n^*(x')\varphi_n(x) = \delta(x - x') \tag{2.4.30}$$

and

$$\mathscr{H}(p, x) = \mathscr{H}\left(\frac{\hbar}{\mathrm{i}}\frac{\partial}{\partial x}, x\right) , \tag{2.4.31}$$

we may write

$$\rho(x, x') = \exp[\beta(F - \mathscr{H})]\,\delta(x - x') . \tag{2.4.32}$$

$\{\varphi_n\}$ may not be the system of eigenfunctions of $\mathscr{H}$, but it suffices if it is a complete set. For example, the basis functions φ_n can be plane waves. The above density matrix is normalized in such a way that

$$\mathrm{tr}\{\rho\} = 1 .$$

We also use the density matrix

$$\rho = \exp(-\beta\mathscr{H}) \tag{2.4.33}$$

which is not normalized. In terms of this density matrix, the sum over states is written as

$$Z = \mathrm{tr}\{\rho\} . \tag{2.4.34}$$

In x-representation

$$\rho(x, x') = \langle x'|\exp(-\beta\mathcal{H})|x\rangle = \sum_n \varphi_n^*(x')\exp(-\beta\mathcal{H})\varphi_n(x) .$$

In particular, if $\{\varphi_n\}$ is the set of eigenfunctions of $\mathcal{H}$, we have

$$\rho(x, x') = \sum \exp(-\beta E_n)\varphi_n(x)\varphi_n^*(x') . \tag{2.4.35}$$

In general, we may write

$$\rho(x, x') = \exp(-\beta\mathcal{H})\delta(x - x') . \tag{2.4.36}$$

For a many-particle system, δ-functions in the above equations should be those for many particles. We must also take the statistics obeyed by the particles into account. As a result, the wave function must be symmetrized or antisymmetrized according to the statistics, considering also the spin coordinates. For a system of N identical particles, we express the coordinates and spin coordinates of all the particles by $\boldsymbol{q}$ and represent the permutation of particles by P. Then we have

$$\rho(\boldsymbol{q}, \boldsymbol{q}') = \exp(-\beta\mathcal{H})\frac{1}{N!}\sum_P (\pm 1)^P \delta(\boldsymbol{q}' - P\boldsymbol{q}) . \tag{2.4.37}$$

This can be derived by using the wave functions introduced in Sect. 3.1, but the procedure is very complicated. We may also calculate the density matrix by integrating the *Bloch equation* [2.18]

$$\frac{\partial}{\partial\beta}\rho = -\mathcal{H}\rho \tag{2.4.38}$$

under the initial condition that

$$\lim_{\beta\to 0} \rho = \frac{1}{N!}\sum (\pm 1)^P \delta(\boldsymbol{q}' - P\boldsymbol{q}) . \tag{2.4.39}$$

Let us consider the *classical limit*. Using the density matrix, we shall show a way of making the transition from quantum mechanics to classical mechanics. For simplicity, we shall discuss a one-dimensional problem, although extension to 3-dimensional systems and many-particle systems is straightforward. We use plane waves (periodic condition)

$$\varphi_n(x) = \frac{1}{\sqrt{L}}\exp(\mathrm{i}p_n x/\hbar) \tag{2.4.40}$$

normalized for the range L, as the complete set for calculating the representation of the density matrix. In (2.4.10), p_n is the eigenvalue of the momentum

$(\hbar/\mathrm{i})\partial/\partial x$:

$$p_n = \frac{2\pi\hbar}{L} n \quad (n = 0, \ \pm 1, \ \pm 2, \ldots) . \tag{2.4.41}$$

We assume a Hamiltonian of the form

$$\mathscr{H} = -\frac{\hbar^2}{2m}\frac{d^2}{dx^2} + U(x) . \tag{2.4.42}$$

To calculate

$$\exp(-\beta\mathscr{H})\exp(\mathrm{i}px/\hbar) = \sum_{s=0}^{\infty}\frac{1}{s!}(-\beta\mathscr{H})^s\exp(\mathrm{i}px/\hbar) ,$$

we first note

$$\mathscr{H}\exp(\mathrm{i}px/\hbar) = \left\{\frac{p^2}{2m} + U(x)\right\}\exp(\mathrm{i}px/\hbar) ,$$

or more generally,

$$\mathscr{H}\exp(\mathrm{i}px/\hbar)\phi(x) = \exp(\mathrm{i}px/\hbar)\left[\frac{1}{2m}\left(p + \frac{\hbar}{\mathrm{i}}\frac{d}{dx}\right)^2 + U(x)\right]\phi(x) .$$

Repeating operations, we are led to

$$\mathscr{H}^s\exp(\mathrm{i}px/\hbar) = \exp(\mathrm{i}px/\hbar)\left[\frac{1}{2m}\left(p + \frac{\hbar}{\mathrm{i}}\frac{d}{dx}\right)^2 + U(x)\right]^s ,$$

where it is promised that, if there is no function on the right of d/dx, the result vanishes. Thus we obtain

$$\begin{aligned}&\exp(-\beta\mathscr{H})\exp(\mathrm{i}px/\hbar)\\ &\quad = \exp(\mathrm{i}px/\hbar)\exp\left\{-\beta\left[\frac{1}{2m}\left(p + \frac{\hbar}{\mathrm{i}}\frac{d}{dx}\right)^2 + U(x)\right]\right\}\end{aligned}$$

and therefore,

$$\rho(x, x') = \frac{1}{L}\sum_n \exp[\mathrm{i}p_n(x - x')/\hbar]\exp\left\{-\beta\left[\frac{1}{2m}\left(p_n + \frac{\hbar}{\mathrm{i}}\frac{d}{dx}\right)^2 + U(x)\right]\right\} . \tag{2.4.43}$$

The sum over states is thus given as

$$\begin{aligned}Z &= \int_0^L \rho(x, x)\,dx\\ &= \frac{1}{L}\sum_n\int_0^L \exp\left\{-\beta\left[\frac{1}{2m}\left(p_n + \frac{\hbar}{\mathrm{i}}\frac{d}{dx}\right)^2 + U(x)\right]\right\}dx .\end{aligned} \tag{2.4.44}$$

So far, the calculation is rigorous. For sufficiently large L, the sum over p_n can be replaced by an integral:

$$\frac{1}{L}\sum_{n} \to \int \frac{dp}{2\pi\hbar} \,. \tag{2.4.45}$$

The classical limit can be derived by taking the formal limit as $\hbar \to 0$. That is, by writing

$$\mathscr{H}(x,p) = \frac{p^2}{2m} + U(x) \,, \tag{2.4.46}$$

we are led to the classical sum over states

$$Z = \frac{1}{2\pi\hbar}\iint dx\, dp \exp[-\beta\mathscr{H}(x,p)] \tag{2.4.47}$$

which is an integral in phase space (phase integral).

The classical sum over states for a system of N identical particles is given as a phase integral over different configurations of particles. If we integrate independently with respect to each particle, the result must be divided by $N!$, the number of permutations of particles (Sect. 2.1.4):

$$Z = \frac{1}{(2\pi\hbar)^f N!}\iint d^f\boldsymbol{q}\, d^f\boldsymbol{p} \exp[-\beta\mathscr{H}(\boldsymbol{q},\boldsymbol{p})] \,. \tag{2.4.47a}$$

Here f stands for the degree of freedom which is $f = 3N$ for N material points, and $\boldsymbol{q}$ and $\boldsymbol{p}$ are general coordinates and momenta. The classical approximation is obtained by taking the formal limit as the elementary phase volume $2\pi\hbar$ vanishes. This is valid when the density of particles is very small.

2.5 Subsystems with a Given Pressure

In the preceding section, we derived the canonical distribution $\exp(-\beta E_j)$ for a subsystem which can exchange energy with a large heat reservoir. If the subsystem is in equilibrium with the reservoir beyond a movable piston (Fig. 2.17), a discussion similar to that of Sect. 2.4.1 leads to a distribution probability which is proportional to

$$\exp[-\beta(E_j + PV)] \,, \tag{2.5.1}$$

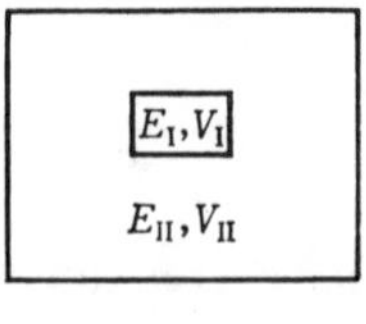

Fig. 2.17. Systems with a variable volume

where P is the pressure of the reservoir and V is the volume of the system under consideration. The factor $\exp(-\beta PV)$ is responsible for the decrease in the number of states of the reservoir when the volume V is shared to the system. It may be said to be the reaction of the reservoir. Though the above is an extension of the canonical distribution, it may be called the T–P distribution since the system is specified by a given T and P.

The sum over states for the T–P distribution is

$$Y(\beta, P) = \int_0^\infty dV \sum_j \exp[-\beta(E_j + PV)] \tag{2.5.2}$$

or

$$Y(\beta, P) = \exp(-\beta G) = \int_0^\infty dV \exp(-\beta PV) Z(\beta, V) . \tag{2.5.3}$$

This can be considered as the Laplace transform of the sum over states with respect to V. The average of the volume V for this distribution is given as

$$\begin{aligned} \langle V \rangle &= \int_0^\infty V dV \exp(-\beta PV) Z(\beta, V) / \int_0^\infty dV \exp(-\beta PV) Z(\beta, V) \\ &= -\frac{1}{\beta} \frac{\partial \ln Y(\beta, P)}{\partial P} . \end{aligned} \tag{2.5.4}$$

Since $Z(\beta, V)$ increases rapidly with V, $\exp(-\beta PV)Z(\beta, V)$ has a sharp maximum at some value $V = V^*$, which can be considered to coincide with $\langle V \rangle$.

We rewrite $\langle V \rangle$ as V, and the average of E_j as E. Then we have

$$E + PV = -\frac{\partial \ln Y(\beta, P)}{\partial \beta} \tag{2.5.5}$$

and so

$$V = \frac{\partial G}{\partial P}, \quad E + PV = \frac{\partial G}{\partial \beta} . \tag{2.5.6}$$

In thermodynamics, G is called the *Gibbs' free energy*, or *thermodynamic potential.*

We may write

$$\begin{aligned} Y(\beta, P) &= \exp(-\beta G) \\ &= \int_0^\infty dV \exp(-\beta PV) \int_0^\infty dE \exp(-\beta E) \Omega(E, V) . \end{aligned}$$

In general, the integrand, as a function of E and V, has a sharp maximum at some point $E = E^*$, $V = V^*$. If we write $\Omega(E^*, V^*) = \exp(S^*/k)$, then we may write

$$\exp(-\beta G) = \exp[-\beta(E^* - TS^* + PV^*)] .$$

Thus we have the relationship between the thermodynamic potential G and the free energy $F = -TS$, which is

$$G = F + PV\,. \tag{2.5.7}$$

As we have seen in Sect. 2.3, a system with a given pressure can be described as a system compressed by a piston with a weight. The total energy including the potential energy of the weight is $E + PV$, which appears in (2.5.7).

2.6 Subsystems with a Given Chemical Potential

2.6.1 Chemical Potential

In this section we consider the exchange of particles between two systems I and II through a boundary. Similar arguments to those used for the exchange of energy and volume, discussed in preceding sections, will apply. We assume that the boundary is fixed, but that particles can go through; since the volumes V_{I} and V_{II} of the two systems are fixed, they are omitted in Fig. 2.18. For simplicity, systems I and II are assumed to be composed of only one kind of molecule, whose numbers are, respectively, N_{I} and N_{II} (Fig. 2.18). The equilibrium state is characterized as a state of maximum $\Omega_{\mathrm{I}}(E_{\mathrm{I}}, N_{\mathrm{I}})\Omega_{\mathrm{II}}(E_{\mathrm{II}}, N_{\mathrm{II}})$, under the condition

$$N_{\mathrm{I}} + N_{\mathrm{II}} = N = \text{const.} \tag{2.6.1}$$

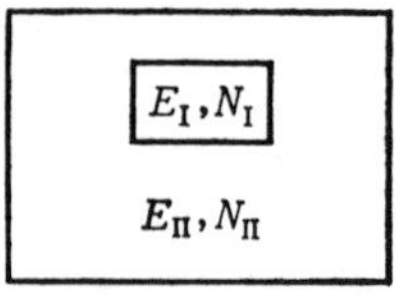

Fig. 2.18. Systems with a variable number of particles

(we also demand that $E_{\mathrm{I}} + E_{\mathrm{II}} = E = \text{const.}$). If we treat N_{I} and N_{II} as continuous variables, we may write the condition of maximum as

$$\frac{\partial \ln \Omega_{\mathrm{I}}(E_{\mathrm{I}}, N_{\mathrm{I}})}{\partial N_{\mathrm{I}}} = \frac{\partial \ln \Omega_{\mathrm{II}}(E_{\mathrm{II}}, N_{\mathrm{II}})}{\partial N_{\mathrm{II}}}\,. \tag{2.6.2}$$

Using the entropy $S = k \ln \Omega$, we define the *chemical potential* μ by

$$\left(\frac{\partial S}{\partial N}\right)_{E,V} = -\frac{\mu}{T}\,. \tag{2.6.3}$$

Then the condition of equilibrium with respect to the exchange of particles is written as

$$\mu_{\mathrm{I}}(E_{\mathrm{I}}, V_{\mathrm{I}}, N_{\mathrm{I}}) = \mu_{\mathrm{II}}(E_{\mathrm{II}}, V_{\mathrm{II}}, N_{\mathrm{II}})\,. \tag{2.6.4}$$

By using the relations $\partial S/\partial E = 1/T$, $\partial S/\partial V = P/T$, which also hold in the present case, we may write

$$dS = \frac{dE}{T} + \frac{P}{T}dV - \frac{\mu}{T}dN \, .$$

Rewriting, we have

$$dE = T\,dS - P\,dV + \mu\,dN$$

$$dF = -S\,dT - P\,dV + \mu\,dN \, .$$

Therefore, the chemical potential may be written as

$$\mu = \frac{\partial E(S, V, N)}{\partial N} = \frac{\partial F(T, V, N)}{\partial N} \, ,$$

i.e., as a function of the variables S, V, and N, or of T, V, and N. Similarly, since

$$dG = -S\,dT + V\,dP + \mu\,dN \, ,$$

we may write

$$\mu(T, P, N) = \frac{\partial G(T, P, N)}{\partial N} \, . \tag{2.6.5}$$

However, if T and P are kept constant, the size of the system increases with N. Therefore,

$$G(T, P, N) = N\mu(T, P) \tag{2.6.6}$$

must hold. Thus we see that the chemical potential, as a function of T and P, is independent of N.

When the system consists of several components, a chemical potential for each component can be similarly defined. If there are N_k molecules of the kth component, the thermodynamic potential can be written as

$$G(P, T, N_1, N_2, \ldots) = \sum_k N_k \mu_k \, , \tag{2.6.7}$$

since G is an extensive variable which is multiplied by a if $N_1, N_2, \ldots$ are simultaneously increased by the same factor a. That is, G is a homogeneous function of the first degree as defined by Euler. Therefore, each μ_k is a function of the intensive variables T, P and concentrations

$$\mu_k = \mu_k(T, P, c_1, c_2, \ldots) = \left(\frac{\partial G}{\partial N_k}\right)_{T, P, N_l(l \neq k)} \, , \tag{2.6.8}$$

where $c_k = N_k/\sum_l N_l$ is the concentration of the kth component.

2.6.2 Grand Partition Function

Suppose that a small system II is in contact with a large system I and they exchange particles. As in the case of energy exchange, we write

$$S_{\mathrm{I}}(E - E_{\mathrm{II}}, N - N_{\mathrm{II}}) = S_{\mathrm{I}}(E, N) - E_{\mathrm{II}}\frac{\partial S_{\mathrm{I}}}{\partial E} - N_{\mathrm{II}}\frac{\partial S_{\mathrm{I}}}{\partial N}$$

and higher terms can be neglected. $\partial S/\partial E = 1/T$ and $\partial S/\partial N = -\mu/T$ are intensive variables determined by the large system, and the probability for the partial system II to have the number of particles N and energy E_j is proportional to

$$\exp[-\beta(E_j - \mu N)] . \tag{2.6.9}$$

This is an extended canonical distribution (*T–μ distribution*) which is usually called the *grand canonical ensemble.*

The sum over states for a grand canonical ensemble is called the grand partition function, which is given as

$$\Xi(\beta, V, \mu) = \sum_{N=0}^{\infty} Z(\beta, V, N)\lambda^N, \quad \lambda = \exp(\beta\mu) . \tag{2.6.10}$$

The average number of particles of the partial system is

$$\langle N \rangle = \sum_{N=0}^{\infty} NZ\exp(\beta\mu N) \Big/ \sum_{N=0}^{\infty} Z\exp(\beta\mu N) = \frac{1}{\beta}\frac{\partial \ln \Xi}{\partial \mu} . \tag{2.6.11}$$

As in the case of $Y(\beta, P)$, we denote by N^* the value of N for which $Z\exp(\beta\mu N)$ takes the maximum value. Then we have

$$\Xi = \exp[-\beta(F^* - \mu N^*)] .$$

However, since μN is the thermodynamic potential $G = F + PV$, we see that

$$\Xi = \exp(\beta PV), \quad PV = \frac{1}{\beta}\ln \Xi . \tag{2.6.12}$$

The corresponding thermodynamic characteristic function leads to

$$d(\beta PV) = -E\,d\beta + \beta P\,dV + N\beta\,d\mu .$$

Sometimes we call

$$\lambda = \exp(\beta\mu) \tag{2.6.13}$$

the absolute activity. When there are many components, we have the grand partition function of the form

$$\Xi = \sum_{N_1}\sum_{N_2} \cdots Z(\beta, V, N_1, N_2, \ldots)\lambda_1^{N_1}\lambda_2^{N_2} \cdots . \tag{2.6.14}$$

2.7 Fluctuation and Correlation

There is a general relationship between the fluctuation of energy and the specific heat, which can be shown as follows.

The average of the energy can be written as

$$\langle E \rangle = \frac{\sum E\Omega(E)\exp(-\beta E)}{\sum \Omega(E)\exp(-\beta E)} = -\frac{Z'}{Z}, \tag{2.7.1}$$

where $Z = \sum \Omega(E)\exp(-\beta E)$ is the sum over states and $Z' = \partial Z/\partial\beta$. Similarly, we write $Z'' = \partial^2 Z/\partial\beta^2$. Then the average of the square of energy is given as

$$\langle E^2 \rangle = \frac{\sum E^2\Omega(E)\exp(-\beta E)}{\sum \Omega(E)\exp(-\beta E)} = \frac{Z''}{Z}. \tag{2.7.2}$$

Thus

$$\langle E^2 \rangle - \langle E \rangle^2 = \frac{ZZ'' - (Z')^2}{Z^2} = \frac{\partial}{\partial\beta}\frac{Z'}{Z}, \tag{2.7.3}$$

where the left-hand side is equal to the fluctuation $\langle (E - \langle E \rangle)^2 \rangle$. On the other hand, the specific heat is given as $C_v = \partial\langle E \rangle/\partial T$. Therefore, we have the relation

$$\langle (E - \langle E \rangle)^2 \rangle = kT^2 C_v. \tag{2.7.4}$$

The specific heat is thus proportional to the fluctuation of energy.

This result can be obtained in an alternative way. When the temperature T_0 and the volume is given, the average energy $\langle E \rangle$ is equal to E^* for which the free energy $F = E - T_0 S$ is a minimum. Expanding F in this neighborhood, we obtain

$$F(E) = F(\langle E \rangle) + \frac{(E - \langle E \rangle)^2}{2C_v T_0} + \dots, \tag{2.7.5}$$

where the coefficients are determined by using the relations

$$\frac{\partial S}{\partial E} = \frac{1}{T(E, V)}, \quad \frac{\partial^2 S}{\partial E^2} = \frac{\partial}{\partial E}\left(\frac{1}{T}\right) = -\frac{1}{C_v T^2}.$$

The probability that the system has energy E is given by

$$\Pr(E)dE \propto \exp[-F(E)/kT_0]\,dE. \tag{2.7.6}$$

Using this probability, we can calculate the average of $(E - \langle E \rangle)^2$, and the result agrees with (2.7.4) when T_0 is replaced by T.

Further, let $E(\boldsymbol{r})$ be the energy density at a position $\boldsymbol{r}$ in a volume of matter. The deviation of the energy from the average is $\Delta E(\boldsymbol{r}) = E(\boldsymbol{r}) - \langle E \rangle$, and the fluctuation of the energy in a unit volume is given by

$$\langle [\Delta E(\boldsymbol{r})]^2 \rangle = \langle (\int \Delta E(\boldsymbol{r})d\boldsymbol{r})^2 \rangle = \int \langle \Delta E(\boldsymbol{r})\Delta E(\boldsymbol{r}') \rangle\, d\boldsymbol{r}'. \tag{2.7.7}$$

Therefore, the specific heat per unit volume can be written as

$$C_v = \frac{1}{kT^2}\int \langle \Delta E(\boldsymbol{r}) \Delta E(\boldsymbol{r}') \rangle \, d\boldsymbol{r}' \, . \tag{2.7.8}$$

Similar arguments can be applied to other quantities. As an example, let us consider the magnetization M per unit volume. When the external magnetic field H is fixed, the magnetization is given by the condition that $F(M) - MH$ takes its minimum value, that is, by

$$F(M) - MH = \text{min.}, \quad \frac{\partial F}{\partial M} = H \, . \tag{2.7.9}$$

The free energy $F(M)$ is an even function of M and can be expanded as

$$F(M) = F_0 + \frac{(M - \langle M \rangle)^2}{2\chi} + \cdots , \tag{2.7.10}$$

where $\langle M \rangle$ is the spontaneous magnetization and χ the magnetic susceptibility. If we write $\Delta M = M - \langle M \rangle$, by a similar argument as in the case of ΔE, we obtain

$$\chi = \frac{1}{kT}\langle (\Delta M)^2 \rangle = \frac{1}{kT}\int \langle \Delta M(\boldsymbol{r}) \Delta M(\boldsymbol{r}') \rangle \, d\boldsymbol{r}' \, . \tag{2.7.11}$$

Near the critical point, or Curie point, the fluctuation becomes enormously large. In such a case, we have to take higher terms into account.

2.8 The Third Law of Thermodynamics, Nernst's Theorem

At absolute zero temperature, the system takes the lowest energy state. If the degeneracy, or the number of quantum states of the lowest energy, is g_0, the entropy of the system is

$$S_0 = k \ln g_0 \tag{2.8.1}$$

at the absolute zero of temperature.

For a pure substance, the presence of quantum states with the same energy is a rather exceptional case; small external fields or interactions, which are usually neglected, will remove any degeneracy. Thus, it should be assumed that the lowest state is nondegenerate, so that $S_0 = 0$. In any case, for a pure substance S_0 is not proportional to the size of the system, and thus is not an extensive variable. Therefore, we may assert

$$\lim_{T \to 0} S = 0 \tag{2.8.2}$$

for a pure substance. This fact is called the *third law of thermodynamics, or Nernst-Planck's theorem.*

H. Nernst examined many chemical reactions and proposed the hypothesis in 1906 that any change of condensed matter at the absolute zero of temperature is performed without change in entropy. *M Planck* supplemented it by saying that the entropy of any pure substance at $T = 0$ is finite, and therefore it can be taken as zero.

But in practice, there are some cases where this law is apparently violated. In particular, since the interaction between nuclear spins is so weak they are, even at 10^{-3} K, randomly oriented, and have entropy. In order to have the ordering of orientation of nuclear spins, one has to lower the temperature to below 10^{-6} K. It will be practical to disregard the entropy due to nuclear spins, which remains even at very low temperatures, in applying the third law.

There are also effects of nuclear spins on the rotational state of molecules. For example, a hydrogen molecule H_2 has two states distinguished by the orientation of nuclear (proton) spins: ortho-hydrogen when they are parallel and para-hydrogen when antiparallel. At high temperatures, hydrogen is a mixture of these with the ratio 3:1, and at lower temperatures the concentration of para-hydrogen increases. If it were in equilibrium, all the molecules should be transformed into para-hydrogen as absolute zero is approached and the entropy should vanish. However, if there is no magnetic catalyst, the rate of para-ortho conversion is very slow at low temperatures, and when the temperature is lowered, equilibrium cannot be set up. Thus, ortho-hydrogen remains even at the lowest temperatures, and we have the entropy due to nuclear spins of ortho-hydrogen and the random occupation of ortho- and para-hydrogen molecules on the crystal lattice.

In ice, we have randomness of the position of hydrogen even at very low temperatures. Cooled glass-like materials retain a finite entropy.

The entropy still remaining at very low temperatures is called the *residual entropy*. To estimate it, we usually start from the gaseous state, since we know theoretically the absolute value of entropy of an ideal gas including the entropy due to nuclear and electronic spins. From this reference we estimate the entropy change using experimental data on specific heat and latent heat of phase transition, and extrapolate the entropy down to the absolute zero of temperature to find the residual entropy.

If we use the third law of thermodynamics, many important results can be obtained. Since these are thermodynamic relations, we shall only point out some of them.

(i) Specific heat vanishes at the absolute zero:

$$\lim_{T\to 0} C_v = 0, \quad \lim_{T\to 0} C_p = 0 . \tag{2.8.3}$$

(ii) The coefficient of thermal expansion vanishes at absolute zero:

$$\lim_{T\to 0} \left(\frac{\partial V}{\partial p}\right)_T = 0, \quad \lim_{T\to 0} \left(\frac{\partial p}{\partial T}\right)_V = 0 . \tag{2.8.4}$$

(iii) With respect to the temperature change in magnetization, from $(\partial S/\partial H)_T = (\partial M/\partial T)_T$, where H denotes the magnetic field, we have

$$\lim_{T\to 0}\left(\frac{\partial M}{\partial T}\right)_H = 0\,. \tag{2.8.5}$$

(iv) In general, the latent heat of phase transition vanishes at the absolute zero:

$$\lim_{T\to 0} L = \lim_{T\to 0} T\Delta S = 0\,. \tag{2.8.6}$$

(v) When, like in solid helium, melting occurs with finite volume change at absolute zero, with respect to the temperature variation of melting pressure we have

$$\lim_{T\to 0}\frac{dP}{dT} = 0\,. \tag{2.8.7}$$

(vi) The derivative of the surface tension of liquid helium with respect to temperature vanishes at absolute zero:

$$\lim_{T\to 0}\frac{d\sigma}{dT} = 0\,. \tag{2.8.8}$$

We may approach indefinitely nearer to absolute zero if appropriate methods are employed. However, according to Nernst's theorem, even if some ideal method is used, we can not reach absolute zero by a finite number of operations.

2.8.1 Method of Lowering the Temperature

If we compress a gas, the temperature rises. Thus, let us cool it, and expand it adiabatically. This way we may lower the temperature. This is one of the currently used methods of cooling, which enables us to liquefy many kinds of gases. At very low temperatures, the method of cooling by expanding gas into a vacuum or into a low pressure region is particularly effective. This is called the *Joule-Thomson effect*, which is due to the expansion against the attractive force between molecules. Thus, we can reach liquid helium temperatures, i.e., about 4 K. Further coooling by rapid evacuation of liquid helium lowers the temperature to about 1 K, beyond which this method fails. To attain a temperature below 1 K, the method of *adiabatic demagnetization* is used.

This method uses paramagnetic substances, such as the paramagnetic salt $Mn(NH_4)_2(SO_4)_2\cdot 6H_2O$. Substances with weak interaction between ions are favored for this method, because then the electronic spins are freely oriented and have large entropy. Using rare-earth salts, about 0.005 K can be reached. Beyond this, by adiabatic demagnetization using nuclear spins (*nuclear demagnetization*) one can reach 10^{-6} K, the temperature of nuclear spins. In crystal lattices it is difficult to attain such a low temperature. The relaxation time of nuclear spins is so long that heat penetrates from the outside and raises the temperature before they reach a state of equilibrium with the electrons and the lattice.

The method of adiabatic demagnetization involves applying a magnetic field to the salt leading to a spin alignment. The generated heat is removed and then the magnetic field is rapidly removed. Then the spins become randomly oriented to yield low temperatures.

Thus the method is, in principle, similar to an adiabatic expansion of a gas, where the expanding gas exerts work on the piston and thereby loses energy, resulting in a cooling of the gas itself. In adiabatic demagnetization, when the magnetic field is decreased, the spins begin to lose their orientation. By means of spin flips they exert work on the magnetic field and the corresponding energy transfer cools them down.

In addition to these, an entirely different method of cooling a gas was developed since the 1980's. It is based on the fact that the speed of an atom is changed when it absorbs or is hit by a photon. Low temperatures below 10^{-6}K are attained by exposing a gas to a set of laser beams. As a result, the Bose condensation of a gas at about 10^{-8}K was realized for the first time by applying such a method.

3. Applications

In the preceding chapter, we described the general principles of statistical mechanics which can be applied to many-particle systems. In quantum mechanics identical particles are indistinguishable and particles are classified into two groups, Bose particles and Fermi particles, according to the symmetry character of their wave functions. Quantum states must fulfill the demand of symmetricity, which means that the number of quantum states depends on the symmetry. This circumstance is taken into account in the so-called quantum statistics. In this chapter, the method of quantum statistics and its application to quantum ideal gases are discussed. Classical statistics is discussed as a limit of quantum statistics and the condition of its validity is classified. Application of classical statistics to nonideal gases is also given.

3.1 Quantum Statistics

3.1.1 Many-Particle System

We consider a system of N identical particles. Extending the concept of an ideal gas, we assume that the Hamiltonian of the total system can be written as

$$\mathcal{H} = \mathcal{H}^{(1)} + \mathcal{H}^{(2)} + \cdots + \mathcal{H}^{(N)}, \tag{3.1.1}$$

where $\mathcal{H}^{(j)}$ means the Hamiltonian of the jth particle. The solution of the eigenvalue equation $\mathcal{H}\varphi = E\varphi$ can be written as

$$\varphi(1, 2, \ldots, N) = \varphi_{r_1}(1)\varphi_{r_2}(2) \cdots \varphi_{r_N}(N) \tag{3.1.2}$$

$$E = \varepsilon_{r_1} + \varepsilon_{r_2} + \cdots + \varepsilon_{r_N}, \tag{3.1.3}$$

where ε_r is an energy eigenvalue of the equation $\mathcal{H}^{(j)}\varphi(j) = \varepsilon_r \varphi(j)$.

The state $\varphi(1, 2, \ldots, N)$ means that the particle 1 is in level r_1, the particle 2 in r_2, and so forth, distinguishing these particles.

However, in quantum mechanics, such individuality of particles is prohibited. And, in general, a quantum mechanical particle has spin coordinates as well as space coordinates. As the state of motion is expressed by momentum, the spin state is expressed by angular momentum. The magnitude of spin angular momentum depends on the nature of the particle, and is some multiple, $0, \frac{1}{2}, 1, \ldots$, of the unit $\hbar$. For example, electron spin has the magnitude $\frac{1}{2}$ in this unit, and it has two states $+\frac{1}{2}$ and $-\frac{1}{2}$ specified by the z-component of the spin. For a spin with magnitude $s\hbar$, the degeneracy of the spin state is $2s+1$ if there is no magnetic field.

Irrespective of whether particles are independent or not, the state of the system is not altered when some particles are exchanged if it consists of identical particles. Let j stand for the coordinate x_j and the spin-coordinate s_j as a whole, and $\psi(1, 2, \ldots, N)$ be the wave function of the total system. If, for example, we exchange particles 1 and 2, the fact that the system is not altered is expressed in quantum mechanics as

$$\psi(2, 1, 3, \ldots, N) = c\psi(1, 2, 3, \ldots, N)\,, \tag{3.1.4}$$

where c is a constant. If we again exchange 1 and 2, it must come back to the original ψ, so that $c^2 = 1$. Therefore, we have two cases $c = 1$ and $c = -1$. When $c = 1$ the particles are called *Bose particles* (*bosons*), and when $c = -1$ they are called *Fermi particles* (*fermions*). If P expresses the operation of changing $(1, 2, \ldots, N)$ into $(j_1, j_2, \ldots, j_N)$, for Bose particles we have

$$P\psi(1, 2, \ldots, N) = \psi(1, 2, \ldots, N)\,, \tag{3.1.5}$$

and for Fermi particles

$$P\psi(1, 2, \ldots, N) = (-1)^{\delta(P)}\psi(1, 2, \ldots, N)\,, \tag{3.1.6}$$

where $\delta(P)$ is the number of permutations.

It is known that particles with integral spin ($s = 0, 1, 2, \ldots$) are Bose particles, and particles with half-integral spin ($s = \frac{1}{2}, \frac{3}{2}, \ldots$) are Fermi particles. Electrons, protons, and their anti-particles have spin $\frac{1}{2}$ and are Fermi particles. In general, atoms (or molecules) consisting of an odd number of Fermi particles are Fermi particles, and those consisting of an even number of Fermi particles are Bose particles. For example, ^{3}He atoms are Fermi particles and ^{4}He atoms are Bose particles.

For a system of independent identical particles, if orbital motion and spin are independent, the wave function $\psi(j)$ of a particle can be expressed as a product of functions $\varphi_k(x_j)$ of the coordinate x_j and $\theta_\sigma(s_j)$ of the spin-coordinate s_j, in such a way that

$$\psi_r(j) = \varphi_k(x_j)\theta_\sigma(s_j)\,. \tag{3.1.7}$$

For an electron spin, the spin-function θ_σ has two states: α-state with $\sigma = \frac{1}{2}$, and β-state with $\sigma = -\frac{1}{2}$. These two states $s_j = \pm 1$ can be expressed by the components $\alpha(1) = \beta(-1) = 1$, and $\alpha(-1) = \beta(1) = 0$. The subscript $r = r(k, \sigma)$ represents the combination of the state of orbital motion k and the spin-^{e}state σ.

In the case of Bose particles, many particles can be in the same state. If n_r denotes the number of particles in the state r, the values allowed to n_r are

$$n_r = 0, 1, 2, \ldots\,.$$

In this case, we may write for a normalized wave function of N independent particles

$$\psi_{r_1,r_2,\ldots,r_N}(1,2,\ldots,N) = \frac{1}{\sqrt{N!n_1!n_2!\ldots}} \sum_P P\psi_{r_1}(1)\psi_{r_2}(2)\cdots\psi_{r_N}(N)$$

$$= \frac{1}{\sqrt{N!n_1!n_2!\ldots}} \begin{vmatrix} \psi_{r_1}(1)\psi_{r_1}(2)\ldots\psi_{r_1}(N) \\ \psi_{r_2}(1)\psi_{r_2}(2)\ldots\psi_{r_2}(N) \\ \ldots\ldots\ldots\ldots \\ \psi_{r_N}(1)\psi_{r_N}(2)\ldots\psi_{r_N}(N) \end{vmatrix}_+ . \qquad (3.1.8)$$

Here $|\ |_+$ is called a permanent, in which minus signs of all the terms in the determinant are replaced by $+1$. The above wave function represents the state of the system specified by the set of states $(r_1, r_2, \ldots, r_N)$. States of a system with different sets of states are mutually orthogonal.

In the case of Fermi particles we can use

$$\psi_{r_1,r_2,\ldots r_N}(1,2,\ldots,N) = \frac{1}{\sqrt{N!}} \sum_P (-1)^{\delta(P)} P\psi_{r_1}(1)\psi_{r_2}(2)\cdots\psi_{r_N}(N)$$

$$= \frac{1}{\sqrt{N!}} \begin{vmatrix} \psi_{r_1}(1)\psi_{r_1}(2)\ldots\psi_{r_1}(N) \\ \psi_{r_2}(1)\psi_{r_2}(2)\ldots\psi_{r_2}(N) \\ \ldots\ldots\ldots\ldots \\ \psi_{r_N}(1)\psi_{r_N}(2)\ldots\psi_{r_N}(N) \end{vmatrix} . \qquad (3.1.9)$$

The last equation is called the *Slater determinant*. From the nature of a determinant, if the same state appeared more than once among $r_1, r_2, \ldots, r_N$, then ψ would vanish. Therefore, two particles cannot be accommodated in a state, that is

$$n_r = 0, 1 .$$

This is called *Pauli's (exclusion) principle.*

3.1.2 Oscillator Systems (Photons and Phonons)

An assembly of photons is the simplest example of a Bose systems of independent particles. Photons are quanta of the radiation field. As a similar system, we may consider an assembly of phonons, or quanta of oscillation of an ideal crystal lattice.

An electromagnetic field in a vacuum can be described as an assembly of electromagnetic waves. However, it can be shown that an electromagnetic wave with frequency ν is equivalent to an oscillator with frequency ν. The amplitude of the electromagnetic wave corresponds to the amplitude of the oscillator, and the nth excited level of the oscillator corresponds to the state of the electromagnetic wave with n photons, each of which has the energy $\varepsilon = h\nu$. This is the viewpoint of second quantization. Each level $\varepsilon_j = h\nu_j$ accommodates an indefinite number of photons. Since the number of photons n_j can be considered as the n_jth excitation, the probability is proportional to $-\beta n_j \varepsilon_j$, and the average of n_j is given by

$$\langle n_j \rangle = \sum_{n_j=0}^{\infty} n_j \exp(-\beta n_j \varepsilon_j) \Bigg/ \sum_{n_j=0}^{\infty} \exp(-\beta n_j \varepsilon_j) = \frac{1}{\exp(\beta \varepsilon_j) - 1} . \tag{3.1.10}$$

This is the Bose distribution for the reciprocal temperature β (Sect. 3.1.3).

When we calculate the number $g(\nu)d\nu$ of proper oscillations of the electromagnetic waves which are between ν and $\nu + d\nu$, we can show that

$$g(\nu) = 2\frac{4\pi V}{c^3}\nu^2 \tag{3.1.11}$$

[cf. (3.2.8) below], where V is the volume of the radiation field and c the light velocity. $g(\nu)$ is called the frequency spectrum. The factor 2 on the right-hand side of (3.1.11) is due to the two directions of polarization of electromagnetic waves which are transverse in character. Thus we see that the energy of a radiation field per unit volume, which is in the frequency range between ν and $\nu + d\nu$, is given as

$$u(\nu, T)\,d\nu = \frac{8\pi}{c^3}\frac{h\nu^3}{\exp(h\nu/kT) - 1}d\nu . \tag{3.1.12}$$

If we rewrite (3.1.12) for the wavelength λ, we have

$$u(\lambda, T)\,d\lambda = \frac{8\pi hc}{\lambda^5}\frac{1}{\exp(ch/k\lambda T) - 1}d\lambda . \tag{3.1.12a}$$

This is *Planck's law of radiation* (Fig. 3.1). The total energy is

$$u(T) = \int_0^{\infty} u(\nu, T)\,d\nu = \frac{8\pi^5 k^4}{15c^3h^3}T^4 \tag{3.1.13}$$

which is proportional to T^4. This is *Stefan-Boltzmann's law of radiation.* The specific heat of vacuum per unit volume is

$$c_v = \frac{du(T)}{dT} = \frac{32\pi^5 k^4}{15c^3h^3}T^3 \tag{3.1.14}$$

which is proportional to T^3.

Vibration of a crystal lattice can be discussed in a similar way to the radiation field. The quanta associated with the lattice waves are phonons, to whose levels $\varepsilon_j = h\nu_j$ we may apply the same Bose distribution as (3.1.10). When the temperature is low, only the waves with small frequency are excited. Since

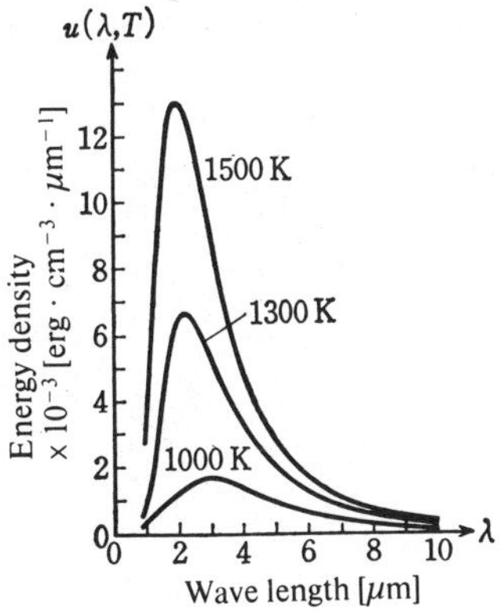

Fig. 3.1. Heat radiation $u(\lambda, T)$

the wavelength for these waves is much larger than the atomic spacing of the lattice, we may consider the lattice as a continuum. Then, the waves are quite analogous to those of the radiation field and the specific heat of solids should be proportional to T^3. This is verified by experiments and is known as *Debye's* T^3-*law*. The reason we get this law is due to the fact that the frequency spectrum is proportional to ν^2 for long wavelengths. However, this is not valid for short waves in crystal lattices. The exact form of the frequency spectrum of the lattice can be obtained only after elaborate calculation since it depends on the crystal form and the nature of interaction between atoms in the crystal. Calculations show that the frequency spectrum is in general quite complicated and differs from crystal to crystal. The frequency spectrum of a crystal lattice has a maximum frequency ν_m which is independent of the shape of crystals. In general, the frequency spectrum $g(\nu)$ has several peaks and vanishes rapidly near ν_m.

For small ν, $g(\nu)$ is proportional to ν^2. Since lattice waves consist of transverse waves with two polarizations and longitudinal waves, we have

$$g(\nu) = 4\pi V\left(\frac{2}{c_t^3} + \frac{1}{c_l^3}\right)\nu^2 , \tag{3.1.15}$$

where c_t is the velocity of transverse waves and c_l the velocity of longitudinal waves. *P. Debye* proposed the approximation where this spectrum is extended up to the maximum value ν_m. If N is the number of atoms, the total number of degrees of freedom, or the number of proper vibrations, is $3N$. Therefore, we have

$$\int_0^{\nu_m} g(\nu)\,d\nu = 3N . \tag{3.1.16}$$

In this approximation, the energy of the lattice vibration is given as

$$E = \int_0^{\nu_m} g(\nu)\frac{h\nu\,d\nu}{\exp(h\nu/kT) - 1} . \tag{3.1.17}$$

The specific heat is calculated by using this formula. The result can be written as

$$C_v = 3NkD\left(\frac{\Theta_D}{T}\right), \tag{3.1.18}$$

where D stands for the Debye function

$$D(x) = \frac{3}{x^3}\int_0^x \frac{\xi^4 \exp(\xi)\, d\xi}{[\exp(\xi) - 1]^2} \tag{3.1.19}$$

and Θ_D is the *Debye temperature* defined by

$$\Theta_D = \frac{h\nu_m}{k}. \tag{3.1.20}$$

The above is called *Debye's specific heat formula* (Fig. 3.2).

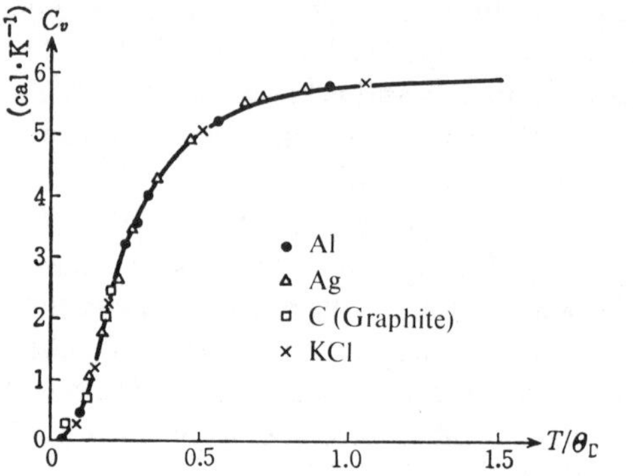

Fig. 3.2. Molar heat of solids

Before Debye, an approximation was proposed by *A. Einstein*, who assumed that atoms in a solid oscillate with the same frequency. *Einstein's specific heat formula* does not give the T^3-law, and fails in explaining experimental facts at low temperatures.

When the temperature is sufficiently high, all the modes of lattice vibration are excited and all $\langle n_j \rangle$ approach $kT/h\nu_j$. Thus, the total energy of lattice vibration becomes $3NkT$, and the specific heat takes the classical value

$$C_v \approx 3Nk \quad (kT \gg h\nu_m). \tag{3.1.21}$$

This is *Dulong–Petit's law.*

3.1.3 Bose Distribution and Fermi Distribution

We have so far considered Bose statistics for photons and phonons, which can be created and annihilated and their numbers are indefinite. If the number of particles is held constant as for usual material particles, we may think that there is a particle reservoir with which the system exchanges particles. If we denote by μ the chemical potential per particle of the particle reservoir, each level has the energy $\varepsilon_j - \mu$ referring to this standard. Therefore, in place of (3.1.10), we have the number of particles in each level given by

$$\langle n_j \rangle = \sum_{n_j=0}^{\infty} n_j \exp[-\beta n_j(\varepsilon_j - \mu)] \Big/ \sum_{n_j=0}^{\infty} \exp[-\beta n_j(\varepsilon_j - \mu)]$$

$$= \frac{1}{\exp[\beta(\varepsilon_j - \mu)] - 1} . \tag{3.1.22}$$

This is the *Bose distribution* (*Bose-Einstein distribution*). A similar argument can be carried out for Fermi particles.

We shall derive the same result using the grand partition function. For a system of independent particles, let n_r be the number of particles in the state with energy ε_r. Then the total number of particles is $N = \sum n_r$ and the total energy is

$$E = \sum \varepsilon_r n_r . \tag{3.1.23}$$

The sum over states is

$$Z_N = \sum_{\sum n_r = N} \exp\left(-\beta \sum_r \varepsilon_r n_r\right) . \tag{3.1.24}$$

Since we have the restriction $\sum n_r = N$, it is difficult to perform the summation in (3.1.24). Therefore, we turn to the grand partition function and find that

$$\Xi = \sum_{N=0}^{\infty} \lambda^N Z_N = \prod_r \sum_{n_r=0}^{\infty} [\lambda \exp(-\beta \varepsilon_r)]^{n_r} . \tag{3.1.25}$$

This is a product of the grand partition functions for states r which are considered as independent,

$$\xi_r = \sum_{n=0}^{\infty} \lambda^n \exp(-\beta \varepsilon_r n) . \tag{3.1.26}$$

If we write

$$\lambda = \exp(\beta\mu) , \tag{3.1.27}$$

then (3.1.26) means that particles independently enter each state with energy $\varepsilon_r - \mu$, measured from the chemical potential μ.

For Bose particles, $n_r = 0, 1, 2, \ldots$, so that

$$\xi_r = \frac{1}{1 - \lambda \exp(-\beta \varepsilon_r)} \tag{3.1.28}$$

$$\Xi = \prod \frac{1}{1 - \lambda \exp(-\beta \varepsilon_r)} . \tag{3.1.29}$$

The average value of n_r is

$$\langle n_r \rangle = \sum_n n\lambda^n \exp(-\beta\varepsilon_r n) \Big/ \sum_n \lambda^n \exp(-\beta\varepsilon_r n) = \frac{1}{\lambda^{-1}\exp(\beta\varepsilon_r) - 1}, \tag{3.1.30}$$

which is the Bose distribution.

For Fermi particles, $n_r = 0, 1$, so that

$$\xi_r = 1 + \lambda \exp(-\beta\varepsilon_r) \tag{3.1.31}$$

$$\Xi = \prod [1 + \lambda \exp(-\beta\varepsilon_r)] . \tag{3.1.32}$$

The average value of n_r is

$$\langle n_r \rangle = \frac{1}{\lambda^{-1}\exp(\beta\varepsilon_r) + 1}, \tag{3.1.33}$$

which is the *Fermi distribution* (*Fermi-Dirac distribution*).

λ, or the chemical potential μ, is determined by the condition for the total number of particles,

$$N = \sum_r \langle n_r \rangle = \sum_r \frac{1}{\lambda^{-1}\exp(\beta\varepsilon_r) \mp 1} . \tag{3.1.34}$$

In this equation and in the following discussion, the upper sign is for the Bose statistics and the lower sign is for the Fermi statistics.

When the particle density is small, $\langle n_r \rangle \ll 1$, then λ is very small, and we may neglect ± 1 in the denominator of (3.1.30) or (3.1.33) and obtain

$$\langle n_r \rangle = \lambda \exp(-\beta\varepsilon_r) ,$$

which is the *classical distribution* (*Maxwell-Boltzmann distribution*).

a) *Difference in the Degeneracy of Systems*

We shall now clarify some characteristics and differences between the Bose statistics and Fermi statistics. For a system of independent particles, when the total energy E is fixed, we have the degeneracy $\Omega(E)$ due to the ways of distributing particles among quantum states for each particle, under the condition that $E = \sum n_r \varepsilon_r$. For the Bose system we have no restriction for n_r, except that $N = \sum n_r$ is given. Let $\Omega_B(E)$ be the degeneracy for this system. On the other hand, for the Fermi system, only the values $n_r = 0, 1$ are allowed. If $\Omega_F(E)$ denotes the degeneracy for the Fermi system, it is clear that $\Omega_B(E) \geqq \Omega_F(E)$ (we are assuming that the total number of particles N is the same for both systems). In classical statistics, we distinguish particles, distribute them among states and then divide by $N!$, the number of ways of exchanging particles. It will be helpful to think of an infinite-dimensional space spanned by discrete coordinates $n_1, n_2, \ldots$ for the states $r = 1, 2, \ldots$, and the distribution of particles in this space. When we exchange two particles on n_j and n_k, we have another distribution if $n_j \neq n_k$. However, since particles are identical, these two distributions are equivalent. Therefore, there are $N!$ equivalent distributions if all the n_j's are different, and by dividing by $N!$ we have the right degeneracy of the system.

However, for distributions with some of the particles in the same state, division by $N!$ yields too small a value for the degeneracy. Thus the degeneracy Ω_{class} calculated by classical statistics yields an intermediate value

$$\Omega_{\text{B}}(E) \geqq \Omega_{\text{class}}(E) \geqq \Omega_{\text{F}}(E) ,$$

and for the sum over states $Z = \sum \Omega(E) \exp(-\beta E)$, we have

$$Z_{\text{B}} \geqq Z_{\text{class}} \geqq Z_{\text{F}} .$$

At low temperatures, we may say that the sum over states of a Fermi system is small because of the large zero-point energy. On the contrary, at high temperatures, the main contribution is from high energy states, so that distributions with more than two particles in the same state will have negligible contribution, and all the statistics give nearly the same sum over states, or free energy.

However, the above result does not mean that classical statistics gives properties between those of Bose and Fermi statistics. In general, thermodynamic properties are given as some logarithmic derivative of the sum over states, and the magnitude of such derivatives are not directly related to that of the sum over states.

Except at high temperatures, Bose statistics has the tendency to accommodate many particles in each state and Fermi statistics has the opposite tendency. The most remarkable differences appear in the lowest energy state or the zero-point energy (Sect. 3.2), and in the phenomena related to Bose condensation which take place for three-dimensional systems (Sect. 3.2).

b) *A Special Case*

There is a very special system for which Bose and Fermi statistics give the same result with respect to excitations from the lowest state [3.1]. This is the system of independent particles distributed among equidistant quantum levels, or more concretely, the system of particles in a one-dimensional parabolic potential. Alternatively, we may think of a two-dimensional ideal gas, since the system has constant level density when the system size is sufficiently large. Thus, the specific heat, for example, shows no difference between a two-dimensional Bose gas and a Fermi gas.

If we consider a parabolic potential and take the unit of energy in such a way that $\hbar\omega = 1$, we have the energy level $\varepsilon_n = n$ $(n = 0, 1, 2, \ldots)$ measured from the lowest energy.

As a Bose distribution, suppose that levels $n_1^{\text{B}}, n_2^{\text{B}}, \ldots, n_N^{\text{B}}$ are occupied from below, each by a particle. Since two particles may be in the same level, we have

$$0 \leqq n_1^{\text{B}} \leqq n_2^{\text{B}} \leqq \cdots \leqq n_N^{\text{B}} .$$

For a Fermi system, since no level can be occupied by more than two particles,

$$0 \leqq n_1^{\text{F}} < n_2^{\text{F}} < \cdots < n_N^{\text{F}} .$$

Therefore, by putting

$$n_j^{\text{B}} = n_j^{\text{F}} - j + 1 ,$$

we have a one-to-one correspondence between Bose and Fermi systems. The total energy is given, respectively, by

$$E^{\mathrm{B}} = n_1^{\mathrm{B}} + n_2^{\mathrm{B}} + \cdots + n_N^{\mathrm{B}}$$

$$E^{\mathrm{F}} = n_1^{\mathrm{F}} + n_2^{\mathrm{F}} + \cdots + n_N^{\mathrm{F}} = E^{\mathrm{B}} + E_0^{\mathrm{F}}(N)$$

with

$$E_0^{\mathrm{F}}(N) = 0 + 1 + 2 + \cdots + N - 1 = N(N-1)/2 \,.$$

It is seen that there is also a one-to-one correspondence between E^{B} and E^{F}, the difference being a constant, that is, $E_0^{\mathrm{F}}(N)$, or the zero-point energy of the Fermi system. Therefore, their degeneracies are the same, and the sum over states $Z_{\mathrm{F}}(N)$ of the Fermi system is given in terms of $Z_{\mathrm{B}}(N)$ of the Bose system as

$$Z_{\mathrm{F}}(N) = \exp[-\beta E_0^{\mathrm{F}}(N)] Z_{\mathrm{B}}(N) \,.$$

Moreover, we may show that

$$Z_{\mathrm{B}}(N) = \frac{1}{(1-x)(1-x^2)\cdots(1-x^N)}, \qquad x = \exp(-\beta\hbar\omega) \,.$$

We may discard the common potential field. If there is a harmonic potential proportional to $1/N$ between the particles, then the one-dimensional system is equivalent to that of independent particles in a common parabolic field. Such Bose and Fermi systems are equivalent.

If we treat levels $\varepsilon_n = n$ as continuous, we can easily show that

$$\lambda_{\mathrm{B}}^{-1} = \lambda_{\mathrm{F}}^{-1} + 1 = [1 - \exp(-\beta N)]^{-1}$$

holds for absolute activities λ_{B} and λ_{F}, respectively, of the Bose and Fermi statistics. Writing the jth level from below ($j \gg 1$) as ε_{B} and ε_{F} for the Bose and Fermi statistics, respectively, we can show that

$$\varepsilon_{\mathrm{B}} = \varepsilon_{\mathrm{F}} - j$$

$$j = \int_0^{\varepsilon_{\mathrm{B}}} \frac{d\varepsilon}{\lambda_{\mathrm{B}}^{-1}\exp(\beta\varepsilon) - 1} = \int_0^{\varepsilon_{\mathrm{F}}} \frac{d\varepsilon}{\lambda_{\mathrm{F}}^{-1}\exp(\beta\varepsilon) + 1} \,,$$

and using these relations, we have

$$\frac{d\varepsilon_{\mathrm{B}}}{\lambda_{\mathrm{B}}^{-1}\exp(\beta\varepsilon_{\mathrm{B}}) - 1} = \frac{d\varepsilon_{\mathrm{F}}}{\lambda_{\mathrm{F}}^{-1}\exp(\beta\varepsilon_{\mathrm{F}}) + 1} \,.$$

Thus, we recover the identity of such systems except for the zero-point energy.

3.1.4 Detailed Balancing and the Equilibrium Distribution

There seems to be no system of perfectly independent particles. Because of the small interaction between particles, or due to some impurities present, particles may be subject to transitions between levels. Under such circumstances, we may derive Bose and Fermi distributions. These must be stable to such perturbations, and thus we may consider the condition that entropy should have a stationary value for the equilibrium state. However, we shall derive the distribution by using the *transition probability*.

We assume n_r particles in the level r, and that the total system is specified by the set $\{n_r\}$. The transition probability from state i to f by the perturbation H_1 is given by quantum mechanics as

$$P_{fi} = \frac{2\pi}{\hbar} |\langle f | H_1 | i \rangle|^2 \, \delta(E_f - E_i) . \tag{3.1.35}$$

We shall assume that H_1 consists of interaction between particles, so that

$$H_1 = \sum_{k>l=1}^{N} v(k, l) . \tag{3.1.36}$$

For the transition $(r, r') \to (s, s')$, we have

$$\begin{aligned} |i\rangle &= |\ldots, n_r, \ldots, n_{r'}, \ldots, n_s, \ldots, n_{s'}, \ldots \rangle \\ |f\rangle &= |\ldots, n_r - 1, \ldots, n_{r'} - 1, \ldots, n_s + 1, \ldots, n_{s'} + 1, \ldots \rangle . \end{aligned} \tag{3.1.37}$$

Using (3.1.8) or (3.1.9) for the Bose or Fermi system, we can show that

$$P_{ss', rr'} = A_{ss', rr'}(1 \pm n_s)(1 \pm n_{s'}) n_r n_{r'} , \tag{3.1.38}$$

where + refers to the Bose system and − to the Fermi system. The coefficient $A_{ss', rr'}$ is proportional to the square of the element of v, or to $|v_{ss', rr'}|^2$, and is symmetric with respect to (ss') and (rr'):

$$A_{rr', ss'} = A_{ss, rr'} . \tag{3.1.39}$$

Now we consider an ensemble of such systems, and denote the average of n_r by $\langle n_r \rangle$. Further, we assume the average number of collisions such that the transition $(rr') \to (ss')$ is proportional to

$$(1 \pm \langle n_s \rangle)(1 \pm \langle n_{s'} \rangle)\langle n_r \rangle \langle n_{r'} \rangle . \tag{3.1.40}$$

The reverse transition is proportional to

$$(1 \pm \langle n_r \rangle)(1 \pm \langle n_{r'} \rangle)\langle n_s \rangle \langle n_{s'} \rangle . \tag{3.1.41}$$

In the state of equilibrium, these transitions are assumed to be balanced by *detailed balancing*. Then we should have

$$\frac{\langle n_s \rangle}{1 \pm \langle n_s \rangle} \frac{\langle n_{s'} \rangle}{1 \pm \langle n_{s'} \rangle} = \frac{\langle n_r \rangle}{1 \pm \langle n_r \rangle} \frac{\langle n_{r'} \rangle}{1 \pm \langle n_{r'} \rangle} . \tag{3.1.42}$$

Further, energy should be conserved by such a collision:

$$\varepsilon_s + \varepsilon_{s'} = \varepsilon_r + \varepsilon_{r'} . \tag{3.1.43}$$

Therefore, we should have

$$\frac{\langle n_r \rangle}{1 \pm \langle n_r \rangle} = \exp(-\beta \varepsilon_r - \alpha) , \tag{3.1.44}$$

where α and β are certain constants. Thus we are led to

$$\langle n_r \rangle = \frac{1}{\exp(\alpha + \beta\varepsilon_r) \mp 1} \tag{3.1.45}$$

which are the Bose and Fermi distributions already known. Such an argument can also be applied to a system consisting of different kinds of particles.

Einstein discussed radiation in a similar way. We assume some atoms of a certain kind to be fixed in the radiation field. For simplicity, we assume that an atom has only an excited state 1 above the ground state 0, and designate by N_1 and N_0 the number of atoms in these states. The number of transitions $P_{0\to 1}$ per unit time, in which the atom absorbs radiation and is raised from the ground state to the excited state, will be proportional to N_0 and the energy density of the radiation field $u = u(T, \nu)$, where $\varepsilon = h\nu$ is the energy of excitation of the atom. Therefore, we may write

$$P_{0\to 1} = N_0 B_{01} u(T, \nu) . \tag{3.1.46}$$

The number of inverse transitions $P_{1\to 0}$ per unit time will be given as

$$P_{1\to 0} = N_1 A_{10} + N_1 B_{10} u(T, \nu) , \tag{3.1.47}$$

where the coefficient A_{10} in the first term denotes the probability of spontaneous transition and the second term represents the induced transition. Detailed balancing implies $P_{0\to 1} = P_{1\to 0}$, so that

$$N_1(A_{10} + B_{10}u) = N_0 B_{01} u . \tag{3.1.48}$$

In addition, for the excitation probability of atoms we have

$$N_1/N_0 = \exp(-\beta\varepsilon) , \tag{3.1.49}$$

and quantum mechanics yields

$$B_{01} = B_{10} \tag{3.1.50}$$

$$A_{10} = \frac{8\pi h\nu^3}{c^3} B_{10} . \tag{3.1.51}$$

Thus we have

$$u = \frac{N_1 A_{10}}{(N_0 - N_1)B_{10}} = \frac{8\pi h\nu^3}{c^3} \frac{1}{\exp(\beta\varepsilon) - 1} \tag{3.1.52}$$

which is Planck's radiation formula (3.1.12).

3.1.5 Entropy and Fluctuations

The sum over states can be written from (2.6.10) as

$$Z_N = \frac{1}{2\pi i} \oint \frac{d\lambda}{\lambda^{N+1}} \Xi(\lambda) , \tag{3.1.53}$$

where the path of integration encircles the origin on the complex plane. We may write $\lambda = |\lambda|\exp(i\varphi)$ and integrate over φ from 0 to 2π. If we choose $|\lambda|$ in such a way that $\lambda^{-N}\Xi$ is minimum on the real axis, $\lambda^{-N}\Xi$ diminishes rapidly on both sides of the axis. Thus the integral can be evaluated by the value of the integrand at this saddle point as [3.2, 3]

$$Z_N \sim \lambda^{-N}\Xi(\lambda)\,, \tag{3.1.54}$$

where λ is the saddle point given by

$$\frac{\partial}{\partial\lambda}\ln[\lambda^{-N}\Xi(\lambda)] = 0 \tag{3.1.55}$$

or

$$N = \lambda\frac{\partial}{\partial\lambda}\ln\Xi(\lambda) = \sum_r \frac{1}{\lambda^{-1}\exp(\beta\varepsilon_r) \mp 1}\,. \tag{3.1.56}$$

Therefore, free energy is given as

$$\begin{aligned} F &= -kT\ln Z_N = NkT\ln\lambda - kT\ln\Xi \\ &= NkT\ln\lambda \mp kT\sum_r \ln[1 \mp \lambda\exp(-\varepsilon_r/kT)]\,, \end{aligned} \tag{3.1.57}$$

where λ is determined by (3.1.56) as a function of N and T.

Energy is given by the general formula $E = \partial(F/T)/\partial(1/T)$ or by $E = \sum \varepsilon_r\langle n_r\rangle$, valid for independent particles, as

$$E = \sum_r \frac{\varepsilon_r}{\lambda^{-1}\exp(\beta\varepsilon_r) \mp 1}\,. \tag{3.1.58}$$

Entropy S may be evaluated by $S = (E - F)/T$ or, using the number of particles in each state

$$\langle n_r\rangle = \frac{1}{\lambda^{-1}\exp(\beta\varepsilon_r) \mp 1}\,, \tag{3.1.59}$$

we may write

$$S = k\sum_r [\pm(1 \pm \langle n_r\rangle)\ln(1 \pm \langle n_r\rangle) - \langle n_r\rangle\ln\langle n_r\rangle]\,, \tag{3.1.60}$$

where the upper sign is for a Bose and the lower for a Fermi system. For a large system, levels are dense and are degenerate, in general. If some levels have nearly the same energy, we may group them together. A group of $g_{(r)}$ levels with energy $\varepsilon_{(r)}$ accommodates $g_{(r)}\langle n_r\rangle$ particles. Denoting

$$g_{(r)}\langle n_r\rangle = \langle N_{(r)}\rangle\,, \tag{3.1.61}$$

we have

$$S = k\sum_{(r)}\left[\pm g_{(r)}\ln\left(1 \pm \frac{\langle N_r\rangle}{g_{(r)}}\right) + \langle N_{(r)}\rangle\ln\left(\frac{g_{(r)}}{\langle N_{(r)}\rangle} \pm 1\right)\right], \tag{3.1.62}$$

where (r) represents the group with energy $\varepsilon_{(r)}$.

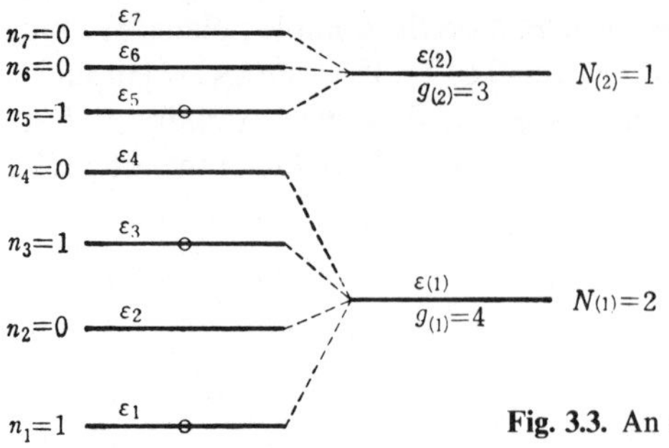

Fig. 3.3. An example of particles distributed among levels

We shall show that the above entropy is related to the number of ways W of distributing particles among levels by Boltzmann's relation $S = k \ln W$ (Fig. 3.3).

First we calculate the number of ways $W_{(r)}$ of distributing $N_{(r)}$ particles among $g_{(r)}$ levels, by regarding each level as a box which accommodates particles.

Let us first consider a Bose system. We begin by placing one of the $g_{(r)}$ boxes, and to the right we place in a row the particles to be put into this box. Then we place another box, and to the right we place particles to be put into this box, and so on (Fig. 3.4). As a result, we place $g_{(r)} + N_{(r)}$ boxes and particles. But we have to omit the first box, and thus place $g_{(r)} + N_{(r)} - 1$ objects (particles and boxes) so that the total number of ways is $(g_{(r)} + N_{(r)} - 1)!$. However, since we do not distinguish particles we divide it by $N_{(r)}!$, and since we do not distinguish boxes we divide it by $(g_{(r)} - 1)!$. Therefore,

$$W_{(q)} = \frac{(g_{(r)} + N_{(r)} - 1)!}{(g_{(r)} - 1)!\,N_{(r)}!} \,. \tag{3.1.63}$$

] ○ ○ □ □ ○ □ ○ ○ ○ □ ···

Fig. 3.4. $g_{(r)}$ boxes □ and $N_{(r)}$ particles ○ in a row

For a Fermi system, it is convenient to consider putting $N_{(r)}$ particles one by one into $g_{(r)}$ boxes. The first particle is put into one of the $g_{(r)}$ boxes. The second one is put into one of the remaining $g_{(r)} - 1$ boxes, and so on. The last particle is put into one of the $g_{(r)} - N_{(r)} + 1$ boxes. Thus the number of ways of putting particles is $g_{(r)}(g_{(r)} - 1) \cdots (g_{(r)} - N_{(r)} + 1) = g_{(r)}!/(g_{(r)} - N_{(r)})!$. However, since we do not distinguish particles, we divide it by $N_{(r)}!$ to obtain

$$W_{(r)} = \frac{g_{(r)}!}{(g_{(r)} - N_{(r)})!\,N_{(r)}!} \,. \tag{3.1.64}$$

In both cases, the number of ways of distributing all the particles is given by

$$W = W(N_{(1)}, N_{(2)}, \ldots) = \prod_{(r)} W_{(r)} \,. \tag{3.1.65}$$

If we use Stirling's formula $\ln n! = n(\ln n - 1)(n \gg 1)$ for $g_{(r)}$ and $N_{(r)}$, assuming they are much larger than unity, we obtain

$$\ln W = \sum_{(r)} \left[\pm g_{(r)} \ln\left(1 \pm \frac{N_{(r)}}{g_{(r)}}\right) + N_{(r)} \ln\left(\frac{g_{(r)}}{N_{(r)}} \pm 1\right)\right] . \tag{3.1.66}$$

The most probable value of $N_{(r)}$ is given by maximizing $\ln W$ under the condition that E and N are kept constant. Then we have

$$\frac{\partial}{\partial N_{(r)}}(\ln W - \beta E - \alpha N) = 0 , \tag{3.1.67}$$

where α and β are indeterminate multipliers. Thus we are led to the most probable value of $N_{(r)}$, which coincides with the result we already know for $\langle N_{(r)} \rangle [\lambda = \exp(-\alpha)]$, and $k \ln W$ agrees with the entropy formula (3.1.62).

Around $\langle N_{(r)} \rangle$, the distribution W can be approximated by a Gaussian distribution and the fluctuation can be expressed in terms of the second derivative of $\ln W$:

$$\left(\frac{\partial^2}{\partial N_{(r)}^2} \ln W\right)_{N_{(r)} = \langle N_{(r)} \rangle} = -\frac{1}{\langle N_{(r)} \rangle \left(1 \pm \dfrac{\langle N_{(r)} \rangle}{g_{(r)}}\right)} . \tag{3.1.68}$$

Thus we have the fluctuation

$$\begin{aligned} \langle (N_{(r)} - \langle N_{(r)} \rangle)^2 \rangle &= \frac{\sum (N_{(r)} - \langle N_{(r)} \rangle)^2 W(N_{(1)}, N_{(2)}, \dots)}{\sum W(N_{(1)}, N_{(2)}, \dots)} \\ &\approx \langle N_{(r)} \rangle \left(1 \pm \frac{\langle N_{(r)} \rangle}{g_{(r)}}\right) \\ &= g_{(r)} \frac{\exp(\alpha + \beta \varepsilon_{(r)})}{[\exp(\alpha + \beta \varepsilon_{(r)}) \pm 1]^2} . \end{aligned} \tag{3.1.69}$$

We may derive the fluctuation directly from the grand partition function

$$\Xi = \sum_N \exp(-\alpha N) \sum_{\substack{n_r \\ \sum n_r = N}} \exp\left(-\beta \sum_r n_r \varepsilon_r\right) . \tag{3.1.70}$$

Then we have

$$\langle n_r \rangle = -\frac{1}{\beta} \frac{\partial}{\partial \varepsilon_r} \ln \Xi = \frac{1}{\exp(\alpha + \beta \varepsilon_r) \mp 1} \tag{3.1.71}$$

$$\langle (n_r - \langle n_r \rangle)^2 \rangle = \frac{1}{\beta^2} \frac{\partial^2}{\partial \varepsilon_r^2} \ln \Xi = \langle n_r \rangle \pm \langle n_r \rangle^2 . \tag{3.1.72}$$

Since the numbers of particles in different levels are independent, adding the same formula for $g_{(r)}$ levels in a group we recover (3.1.69). Adding for all the levels, we get

$$\langle (N - \langle N \rangle)^2 \rangle = \sum_r \frac{\exp(\alpha + \beta\varepsilon_r)}{[\exp(\alpha + \beta\varepsilon_r) \mp 1]^2} \tag{3.1.73}$$

which may be directly derived by calculating $\partial^2 \ln \Xi / \partial\alpha^2$.

Further, for a Fermi system, n_r is limited to 0 and 1 so that $\langle n_r^2 \rangle = \langle n_r \rangle$ holds. For a Bose system we may note

$$\frac{\langle (n_r - \langle n_r \rangle)^2 \rangle}{\langle n_r \rangle^2} = \frac{1}{\langle n_r \rangle} + 1 . \tag{3.1.74}$$

If we refer to Sect. 1.2, we see that the first term on the right-hand side represents the property of classical particles and the second, the nature of waves.

3.2 Ideal Gases

3.2.1 Level Density of a Free Particle

In order to discuss an ideal gas, we must examine the eigenstates of a free particle. We shall here omit spin states. The Schrödinger equation for an energy eigenvalue E_n is

$$-\frac{\hbar^2}{2m}\nabla^2 \varphi_n = E_n \varphi_n . \tag{3.2.1}$$

For simplicity, we adopt a periodic boundary condition for a cubic box of sides L. The eigenfunction is

$$\varphi_n(x, y, z) = \frac{1}{\sqrt{V}} \exp\left[\frac{\mathrm{i}}{\hbar}(p_x x + p_y y + p_z z)\right], \tag{3.2.2}$$

where $V = L^3$ is the volume, and

$$\begin{gathered} p_x = \frac{n_1}{L}h, \quad p_y = \frac{n_2}{L}h, \quad p_z = \frac{n_3}{L}h , \\ n_1, n_2, n_3 = 0, \ \pm 1, \ \pm 2, \ \pm 3, \dots , \end{gathered} \tag{3.2.3}$$

where $h = 2\pi\hbar$, p_x, p_y, and p_z are components of momentum of the particle, and n_1, n_2, and n_3 are quantum numbers. The energy eigenvalue is given by

$$E_n = \frac{1}{2m}(p_x^2 + p_y^2 + p_z^2) . \tag{3.2.4}$$

When L is sufficiently large, momentum is quasicontinuous. We consider the momentum space (p_x, p_y, p_z) in which, corresponding to integers n_1, n_2, and n_3, we have eigenstates with the intervals $\Delta p_x = \Delta p_y = \Delta p_z = h/L$ (Fig. 3.5). Therefore, we see that the number of eigenstates in the small region $dp_x dp_y dp_z$ of the momentum space is (omitting spin)

$$\frac{V dp_x dp_y dp_z}{h^3} . \tag{3.2.5}$$

The number of states with energy below E is given as

$$j(E) = \frac{V}{h^3} 4\pi \int_0^{\sqrt{2mE}} p^2 dp = \frac{V}{h^3} \frac{4\pi}{3} (2mE)^{3/2}$$

and the state density $g(E) = dj/dE$ is

$$g(E) = V \cdot 2\pi \left(\frac{2m}{h^2} \right)^{3/2} \sqrt{E} . \tag{3.2.6}$$

Thus, the state density is proportional to V and $\sqrt{E}$ (Fig. 3.6).

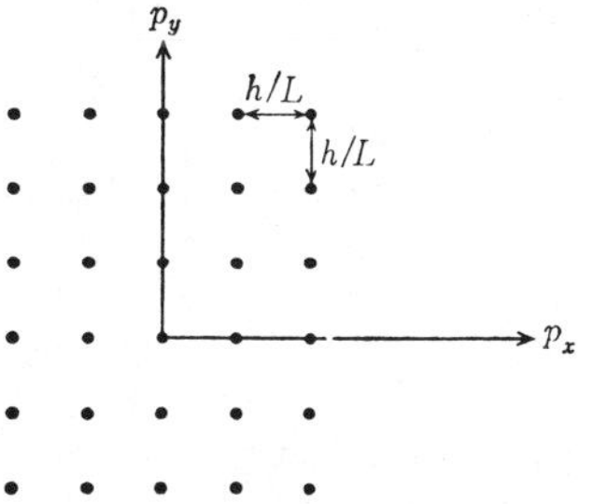

Fig. 3.5. Eigenstates for a free particle

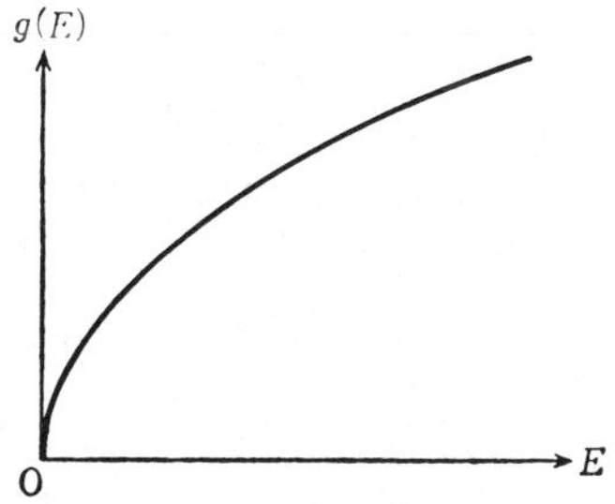

Fig. 3.6. State density for a free particle

For a relativistic particle, the state density is of course different from that of a classical particle. However, the energy E of the particle is also specified by the value of momentum p by the formula

$$E^2 = c^2 p^2 + m^2 c^4 , \tag{3.2.7}$$

where c is the light velocity, m the mass. The eigenvalues for each component of momentum is also given, for example, as $p_x = n_1 h/L$, and we have the same formula as (3.2.5) for the number of eigenstates in an elementary volume of the momentum space. The case of photons is given as the limit $m \to 0$, and $E = cp (= h\nu)$. The number of eigenstates for photons is, therefore, given as

$$2V \frac{d\boldsymbol{p}}{h^3} = 2V \frac{4\pi p^2 dp}{h^3} = 2V \frac{4\pi \nu^2}{c^3} d\nu , \tag{3.2.8}$$

where the coefficient 2 is attached since each electromagnetic wave has two states of polarization.

3.2.2 An Ideal Gas

As we have seen, the level density for a particle of an ideal gas is given as, cf. (3.2.6),

$$g(\varepsilon) = AV\sqrt{\varepsilon} \ , \tag{3.2.9}$$

where ε is the energy of the particle and V the volume of the ideal gas. If we count the spin weight g_s, we have

$$A = g_s \cdot 2\pi \frac{(2m)^{3/2}}{h^3} \ . \tag{3.2.9a}$$

In the preceding section we discussed systems of independent particles and obtained formulas for free energy F, energy E, and other quantities of systems of independent particles. If we replace the sums $\sum_r$ there by integrals in such a way that

$$\sum_r f_r \rightarrow \int g(\varepsilon) d\varepsilon f(\varepsilon) \ , \tag{3.2.9b}$$

we obtain formulas for thermodynamic quantities of ideal gases.

The equation of state may be calculated by the pressure equation $P = -(\partial F/\partial V)_{T,N}$, or by

$$PV = kT \ln \Xi \ . \tag{3.2.10}$$

In what follows, the upper sign is for a Bose system and the lower sign for a Fermi system. Pressure is given by [cf. (3.1.29, 33); $\lambda = \exp(-\alpha)$]

$$\frac{P}{kT} = \mp A \int_0^\infty \ln[1 \mp \exp(-\alpha - \varepsilon/kT)]\sqrt{\varepsilon}\, d\varepsilon \ . \tag{3.2.11}$$

If we perform partial integration in which we integrate $\sqrt{\varepsilon}$ and differentiate the logarithmic function, we obtain the relation

$$PV = \frac{2}{3} E \ . \tag{3.2.12}$$

This equation (Bernoulli's equation) was already derived by using the virial theorem, cf. (1.2.18), and is valid for all nonrelativistic ideal gases. In (3.2.12), E stands for energy

$$E = VA \int_0^\infty \frac{\varepsilon^{3/2} d\varepsilon}{\exp(\alpha + \varepsilon/kT) \mp 1} \ . \tag{3.2.13}$$

The constant α is determined by the condition of the total particle number:

$$N = VA \int_0^\infty \frac{\sqrt{\varepsilon}\, d\varepsilon}{\exp(\alpha + \varepsilon/kT) \mp 1} \ . \tag{3.2.13a}$$

a) *Adiabatic Change*

Since $\alpha = -\mu/kT$, we may write

$$PV = kT \ln \Xi = VT^{5/2} f\left(\frac{\mu}{T}\right) \tag{3.2.14}$$

for an ideal gas. Here f is a function of μ/T only and independent of V. Therefore, we may further write

$$S = \left(\frac{\partial PV}{\partial T}\right)_{V,\mu} = VT^{3/2} f_S\left(\frac{\mu}{T}\right) \tag{3.2.14a}$$

$$N = \left(\frac{\partial PV}{\partial \mu}\right)_{T,V} = VT^{3/2} f_N\left(\frac{\mu}{T}\right), \tag{3.2.14b}$$

where f_S, f_N are functions of μ/T and independent of V. Thus, $S/N = \varphi(\mu/T)$ is independent of V. It means that if we keep entropy constant and change the volume, namely, for adiabatic change, μ/T is kept constant. Therefore, for adiabatic change, we have

$$VT^{3/2} = \text{const.}, \quad \frac{P}{T^{5/2}} = \text{const.}, \quad PV^{5/3} = \text{const.} \tag{3.2.15}$$

These relations look similar to Poisson's adiabatic equation of state, $PV^{\gamma} =$ const. ($\gamma = C_p/C_v$ is the ratio of specific heats). However, for a degenerate quantum gas, the ratio of specific heats is not 5/3; the exponent 5/3 of its adiabatic equation of state, $PV^{5/3} =$ const., is due to the assumption that the particles are monatomic.

b) *High Temperature Expansion*

If $\alpha > 0$ in (3.2.11, 13) we may expand the integrands in powers of $\exp(-\alpha - \varepsilon/kT)$ and integrate each term to have

$$\frac{P}{kT} = g_s \frac{(2\pi mkT)^{3/2}}{h^3} \sum_{n=1}^{\infty} \frac{(\pm 1)^{n-1}}{n^{5/2}} \exp(-n\alpha) \tag{3.2.16}$$

$$\frac{N}{V} = g_s \frac{(2\pi mkT)^{3/2}}{h^3} \sum_{n=1}^{\infty} \frac{(\pm 1)^{n-1}}{n^{3/2}} \exp(-n\alpha)\,. \tag{3.2.16a}$$

For a Bose gas, since the number in the lowest level $\langle n_0 \rangle = 1/[\exp(\alpha) - 1]$ should be positive finite, α is always positive and we can use the above expansion. However, for a Fermi gas, α is negative when the density is high, as we shall see later, and the above expansion applies only when the density is sufficiently low.

In either case, when N/V is small and T is large, $\exp(-\alpha)$ will be sufficiently small and we may eliminate it by successive approximation. Thus we have

$$\frac{PV}{NkT} = 1 \mp \frac{1}{2^{5/2} g_s} \frac{h^3}{(2\pi mkT)^{3/2}} \frac{N}{V} + \cdots, \tag{3.2.17}$$

where the first term represents the Boyle-Charles law and higher terms are quantum corrections. To examine the effect of the second term, we assume $g_s = 1$, the molecular weight M of the gas ($m = M/N_A$, where N_A is Avogadro's number), and express the pressure P in units of atmospheric pressure to obtain

$$\frac{h^3}{(2\pi mkT)^{3/2}} \frac{N}{V} \approx 38 \frac{P}{M^{3/2} T^{5/2}} .$$

As an example, for helium under 1 atm. and at 4.7 K, the value of the right-hand side is about 0.13. Thus, the second term of (3.2.17) is not large even for light molecules and low temperature, so that the quantum effect can usually be neglected for gases heavier than helium. However, for the electron gas in metals, m is so small and the density so high that the quantum effect is important and the above expansion cannot be used. But if the electron density is low, as in the case of carriers in semi-conductors or electrons outside metals, the quantum effect can be neglected.

c) *Density Fluctuation*

To calculate the density fluctuation of an ideal gas, we may either use formulas already obtained for independent particles, or use the general formula (1.2.12). As we see from (1.2.12, 13), density fluctuation is related to the radial distribution function $g(R)$ [$g(R)$ in Chap. 1]. We can show that

$$\langle [n(\boldsymbol{r}_1) - \langle n \rangle][n(\boldsymbol{r}_2) - \langle n \rangle] \rangle = \langle n \rangle \delta(\boldsymbol{r}_2 - \boldsymbol{r}_1) + \langle n \rangle^2 [g(\boldsymbol{r}_2 - \boldsymbol{r}_1) - 1] , \qquad (3.2.18)$$

where $n(\boldsymbol{r})$ is the number density of particles at $\boldsymbol{r}$, and that

$$\langle n \rangle^2 [g(R) - 1] = \frac{\pm g_s}{h^6} \left| \int \frac{\exp(i\boldsymbol{p} \cdot \boldsymbol{R}/\hbar)}{\exp(\alpha + \beta\varepsilon) \mp 1} d\boldsymbol{p} \right|^2 . \qquad (3.2.19)$$

For a Bose gas we see that $g(R) - 1 > 0$, and for a Fermi gas $g(R) - 1 < 0$. That is, in a Bose gas, the probability of finding a particle is increased by the presence of a nearby particle of the same kind; the correlation looks as if there is an attraction force between particles. On the contrary, in a Fermi gas, repulsion acts between particles of the same kind which is a result of Pauli's exclusion principle.

3.2.3 Bose Gas

Suppose we have a Bose gas. The number of particles which fall into the lowest state $\varepsilon_{(0)} = 0$ is

$$\langle N_{(0)} \rangle = \frac{g_{(0)}}{\exp \alpha - 1} . \qquad (3.2.20)$$

Therefore, α is always positive. When the density is low we see that $\alpha \gg 1$ and $N_{(r)} \ll 1$ for each group of levels. However, when the density becomes high, α will approach 0. Equation (3.2.16a) determines α as a function of N/V and T. It can be seen that the right-hand side of (3.2.16a) has the limiting value

$$\sum_{n=1}^{\infty} \frac{1}{n^{3/2}} = 2.612$$

for $\alpha \to 0$. Therefore, if V or T is so small that

$$\frac{N}{V} \Bigg/ g_s \left(\frac{2\pi mkT}{h^2} \right)^{3/2} > 2.612 ,$$

(3.2.16a) may look violated. However, this failure comes from the fact that we have discarded the weight of the lowest level $\varepsilon = 0$ in replacing the sum in (3.2.9a) by the integration. When the density N/V becomes so high, or when the temperature so low, that 2.612 is exceeded then α will approach zero in such a way that $\alpha = O(1/N) \approx 0$, and the number $\langle N_{(0)} \rangle$ in the lowest level becomes appreciable compared with N. Then the remaining $N - \langle N_{(0)} \rangle$ particles will be distributed among the higher levels $\varepsilon > 0$. Thus, the number of excited particles is given as

$$N^* = N - \langle N_{(0)} \rangle = V \frac{(2\pi mkT)^{3/2}}{h^3} \sum_{n=1}^{\infty} \frac{\exp(-n\alpha)}{n^{3/2}} . \tag{3.2.21}$$

Suppose, for example, we cool down a Bose gas. Then there is a critical temperature T_0 given by

$$\frac{N}{V} = 2.612 \frac{(2\pi mkT_0)^{3/2}}{h^3} . \tag{3.2.22}$$

Above T_0, we have $N^* \gg N_{(0)} \approx 0$; when V is sufficiently large, we have $N_{(0)}/N = 0$ if $T > T_0$. Below T_0, we have $\alpha \approx 0$.

When we compress a Bose gas keeping the temperature constant, there is a critical volume V_0 given by

$$\frac{N}{V_0} = 2.612 \frac{(2\pi mkT)^{3/2}}{h^3} . \tag{3.2.23}$$

Above V_0, the pressure of the gas changes as

$$P = kT \frac{(2\pi mkT)^{3/2}}{h^3} \sum_{n=1}^{\infty} \frac{\exp(-n\alpha)}{n^{5/2}} . \tag{3.2.24}$$

Below V_0 we have $\alpha \approx 0$, so that the pressure is constant (Fig. 3.7a) which is given by the sum

$$\sum_{n=1}^{\infty} \frac{1}{n^{5/2}} = 1.341 .$$

It can be shown that dP/dV is continuous at V_0.

Since below the critical point ($V < V_0$ or $T < T_0$) an appreciable number of particles fall down to the lowest level, this may be called condensation in

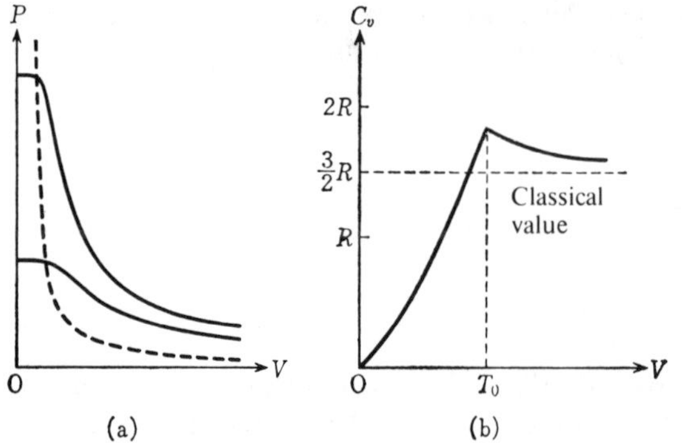

Fig. 3.7 a, b. Bose condensation

momentum space, or *Bose condensation.* At the transition temperature T_0, the curve for the specific heat C_v has a kink, and below T_0, we have $C_v \propto T^{3/2}$ (Fig. 3.7b).

Real gases have no such transition because they all turn into liquids or solids under the conditions required for Bose condensation to occur. However, liquid helium (^{4}He) has two phases called He I and He II, and He II has anomalous thermal and mechanical properties. Since ^{4}He obeys Bose statistics, we may calculate the transition temperature which turns out to be $T_0 = 3.14$ K for $N/V = 2.3 \times 10^{22}\ \text{cm}^{-3}$, the density of liquid helium. The transition temperature between He I and He II is 2.19 K, where the specific heat curve has a sharp maximum, and it is called the *λ-transition.* Therefore, He II is considered to be a liquid which has undergone Bose condensation. ^{3}He behaves quite differently, showing the difference in statistics.

3.2.4 Fermi Gas

Next, we consider a Fermi gas. Putting $\alpha = -\mu/kT$, we have the Fermi distribution

$$f(\varepsilon) = \frac{1}{\exp[(\varepsilon - \mu)/kT] + 1}\,, \tag{3.2.25}$$

where μ is the chemical potential which is a function of the temperature and the particle density. At low temperatures, $f \approx 1$ for $\varepsilon < \mu$ and $f \approx 0$ for $\varepsilon > \mu$ (Fig. 3.8). This means that low-lying levels are all occupied according to the Pauli principle. μ is called the *Fermi energy* or the *Fermi level.* If ε_F and E_0 denote, respectively, the Fermi energy and the total energy at $T = 0$, we have

$$N = \int_0^{\varepsilon_F} g(\varepsilon)\,d\varepsilon, \quad E_0 = \int_0^{\varepsilon_F} \varepsilon g(\varepsilon)\,d\varepsilon\,.$$

E_0 is called the *zero-point energy* of the Fermi gas. For an electron gas, the spin

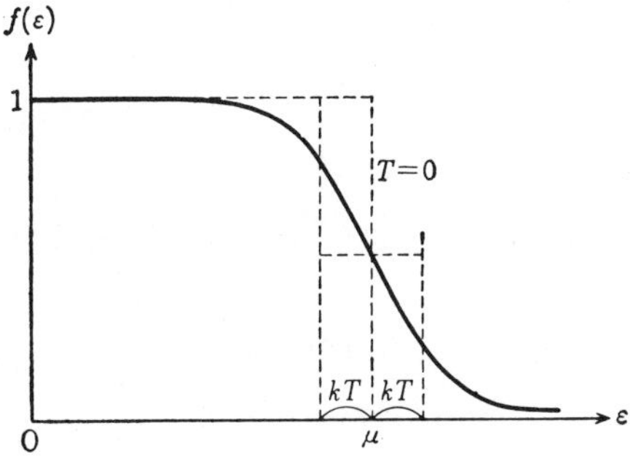

Fig. 3.8. Fermi distribution

weight is $g_s = 2$, so that

$$\varepsilon_F = \frac{\hbar^2 \pi^2}{2m}\left(\frac{3N}{\pi V}\right)^{2/3} \tag{3.2.26}$$

and

$$E_0 = \frac{3}{5} N \varepsilon_F \,. \tag{3.2.27}$$

At temperatures sufficiently lower than the *Fermi temperature*

$$T_F = \frac{\varepsilon_F}{k} \,,$$

the distribution is said to be *degenerate*. For a free-electron gas in a metal, T_F is of the order of tens of thousands of degrees, so that it is nearly completely degenerate in this sense, even at room temperature.

The Fermi distribution $f(\varepsilon)$ changes from 1 to 0 at about $\varepsilon = \mu$ in the range $\varepsilon = \mu \pm kT$. This means that electrons in the energy interval kT have had their energy increased and the energy of excitation is of the order kT, so that the total energy at T can be written as

$$E = E_0 + \frac{\gamma}{2} T^2 \tag{3.2.28}$$

when $T \ll T_F$. Therefore, the *electronic specific heat* is proportional to T:

$$C_e = \gamma T \,, \tag{3.2.29}$$

where the coefficient γ is called the *Sommerfeld constant*. As is seen from the above argument, γ is of the order of $g(\varepsilon_F)k^2$, where $g(\varepsilon_F)$ is the level density at the Fermi energy, so that $\varepsilon_F f(\varepsilon_F) \approx N$.

If we want to have an exact calculation of γ, we have to deal with

$$N = \int_0^\infty \frac{g(\varepsilon)d\varepsilon}{\exp[(\varepsilon - \mu)/kT] + 1} \,, \quad E = \int_0^\infty \frac{\varepsilon g(\varepsilon)d\varepsilon}{\exp[(\varepsilon - \mu)/kT] + 1} \,.$$

Consider the integral

$$I = \int_0^\infty \frac{F(\varepsilon)d\varepsilon}{\exp[(\varepsilon - \mu)/kT] + 1}, \tag{3.2.30}$$

where $F(\varepsilon)$ is some function which make the integral convergent. Simple calculation yields

$$I = \int_0^\mu F(\varepsilon)d\varepsilon - kT \int_0^{\mu/kT} \frac{F(\mu - kTz)dz}{\exp(z) + 1} + kT \int_0^\infty \frac{F(\mu + kTz)dz}{\exp(z) + 1}.$$

When $\mu/kT \gg 1$, we may replace the upper limit on the integral of the second term by ∞. We are neglecting terms of the order $\exp(-\mu/kT)$ and are led to an asymptotic series. Thus we have

$$I = \int_0^\mu F(\varepsilon)d\varepsilon + kT \int_0^\infty \frac{F(\mu + kTz) - F(\mu - kTz)}{\exp(z) + 1} dz$$

so that a Taylor series expansion of the integrand in the second term gives, after integration,

$$I = \int_0^\mu F(\varepsilon)d\varepsilon + \frac{\pi^2}{6}(kT)^2 F'(\mu) + \frac{7\pi^4}{360}(kT)^4 F'''(\mu) + \cdots. \tag{3.2.31}$$

Therefore, we have

$$N = \int_0^\mu g(\varepsilon)d\varepsilon + \frac{\pi^2}{6}(kT)^2 g'(\mu) + \cdots$$

which gives the dependence of μ on temperature. Since at absolute zero $N = \int_0^{\varepsilon_F} g(\varepsilon)d\varepsilon$, we have

$$\mu = \varepsilon_F - \frac{\pi^2}{6}\frac{g'(\varepsilon_F)}{g(\varepsilon_F)}(kT)^2 + \cdots.$$

Equation (3.2.31) leads to the energy

$$E = \int_0^\mu \varepsilon g(\varepsilon)d\varepsilon + \frac{\pi^2}{6}(kT)^2[g(\mu) + \mu g'(\mu)] + \cdots,$$

where the first term

$$\int_0^\mu \varepsilon g(\varepsilon)d\varepsilon = \int_0^{\varepsilon_F} \varepsilon g(\varepsilon)d\varepsilon + (\mu - \varepsilon_F)\varepsilon_F g(\varepsilon_F) + \cdots$$

can be modified by using the above dependence of μ on T. As a result, the term including $g'(\mu)$ is cancelled up to $(kT)^2$, leading to

$$E = E_0 + \frac{\pi^2}{6}(kT)^2 g(\varepsilon_F) + \cdots, \tag{3.2.32}$$

where $E_0 = \int_0^{\varepsilon_F} \varepsilon g(\varepsilon) d\varepsilon$ is the energy at absolute zero. Thus we have

$$C_e = \frac{\pi^2}{3} g(\varepsilon_F) k^2 T \,. \tag{3.2.33}$$

For an ideal gas, using (3.2.9) we have

$$C_e = Nk \frac{\pi^2}{2} \frac{kT}{\varepsilon_F} \,. \tag{3.2.33a}$$

Since $kT \ll \varepsilon_F$ for metallic electrons, C_e is very small compared with the specific heat $(3/2)Nk$ of a classical gas. At very low temperatures the specific heat of a metal is given as a sum of two terms: one is due to the vibration of a crystal lattice which is proportional to T^3, and the other is the electronic specific heat which is proportional to T. These two can be separated from experimental data. The electronic specific heat thus obtained is of the order of magnitude as predicted by the above theory (Fig. 3.9).

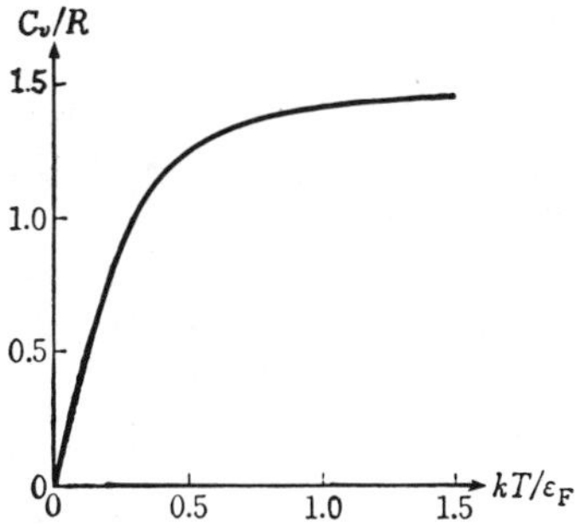

Fig. 3.9. Electronic specific heat

3.2.5 Relativistic Gas

For a relativistic quantum gas, most of the above argument stays valid, except that the energy is to be replaced by (3.2.7). The level density is $Vg_s \cdot 4\pi p^2 dp/h^3$. If we put $g_s = 1$, we have

$$N = \frac{4\pi}{h^3} V \int_0^\infty \frac{p^2\, dp}{\exp(\alpha + \beta\varepsilon) \mp 1}, \quad E = \frac{4\pi}{h^3} V \int_0^\infty \frac{\varepsilon p^2\, dp}{\exp(\alpha + \beta\varepsilon) \mp 1}, \tag{3.2.34}$$

where

$$\varepsilon = c\sqrt{p^2 + m^2 c^2}, \quad \beta = \frac{1}{kT} \,. \tag{3.2.35}$$

The grand partition function Ξ is given as

$$\frac{PV}{KT} = \ln \Xi = \mp \frac{4\pi}{h^3} V \int_0^\infty \ln[1 \mp \exp(-\alpha - \beta\varepsilon)] p^2 \, dp$$

$$= \mp \frac{4\pi}{h^3 c^3} V \int_{mc^2}^\infty \varepsilon(\varepsilon^2 - m^2c^4)^{1/2} \ln[1 \mp \exp(-\alpha - \beta\varepsilon)] \, d\varepsilon \,, \tag{3.2.36}$$

where we have used the fact that $p^2 dp = \varepsilon(\varepsilon^2 - m^2c^4)^{1/2} d\varepsilon/c^3$.

When kT is much larger than the rest energy mc^2 (highly relativistic case), we may put $\varepsilon = cp$ so that

$$PV = E/3 \,. \tag{3.2.37}$$

Since, in this case, we may write

$$PV = VT^4 f(\mu/T) \,, \tag{3.2.38}$$

similar arguments as we had for (3.2.14) lead to the relations

$$VT^3 = \text{const.}, \quad P/T^4 = \text{const.}, \quad PV^{4/3} = \text{const.} \tag{3.2.39}$$

for adiabatic changes of a highly relativistic gas.

a) *Photon Gas*

Consider a system of photons ($m = 0$) as an example of a highly relativistic Bose gas. Since the number of photons is indefinite, α in (3.1.67) is zero. Taking the factor 2 for the polarization into account, we have, cf. (3.1.13),

$$u = \frac{E}{V} = \frac{8\pi}{h^3c^3} \int_0^\infty \frac{\varepsilon^3 \, d\varepsilon}{\exp(\varepsilon/kT) - 1} = \frac{8\pi^5}{15h^3c^3} (kT)^4 \,. \tag{3.2.40}$$

The pressure equation for photons reduces to

$$P = -\frac{8\pi kT}{h^3c^3} \int_0^\infty \varepsilon^2 \ln[1 - \exp(-\alpha - \beta\varepsilon)] \, d\varepsilon$$

$$= \frac{8\pi}{3h^3c^3} \int_0^\infty \frac{\varepsilon^3 \, d\varepsilon}{\exp(\varepsilon/kT) - 1} = \frac{E}{3V} \,. \tag{3.2.41}$$

b) *Fermi Gas*

As an example of a highly relativistic Fermi gas with $m = 0$, consider a perfectly degenerate case. If we write the highest energy as $\varepsilon_0 = cp_0$, we have

$$N = g_s \frac{V}{h^3} \int_0^{p_0} 4\pi p^2 \, dp, \quad E = g_s \frac{V}{h^3} \int_0^{p_0} 4\pi c p^3 \, dp \,.$$

Since $PV = E/3$ holds, the pressure is given as

$$P = \frac{E}{3V} = \frac{1}{4} \left(\frac{3}{4\pi g_s} \right)^{1/3} hc \left(\frac{N}{V} \right)^{4/3} \,. \tag{3.2.42}$$

If it is not highly relativistic, $PV \neq E/3$. For a perfectly degenerate Fermi gas with $m \neq 0$, we can easily show that

$$\frac{E}{V} = \frac{g_s \pi}{8h^3} m^4 c^5 (\sinh \xi - \xi)$$
$$P = \frac{g_s \pi}{8h^3} m^4 c^5 \left(\frac{1}{3} \sinh \xi - \frac{8}{3} \sinh \frac{\xi}{2} + \xi \right), \tag{3.2.43}$$

where $\xi = 4 \sinh^{-1}(p_0/mc)$.

c) *Classical Gas*

The nondegenerate case or the classical case can be obtained by assuming $\exp \alpha \gg 1$. Thus, for a dilute relativistic gas

$$N = \frac{4\pi}{h^3 c^3} V \exp(-\alpha) \int_{mc^2}^{\infty} \exp(-\beta\varepsilon)\varepsilon \sqrt{\varepsilon^2 - m^2 c^4}\, d\varepsilon \tag{3.2.44}$$

$$E = \frac{4\pi}{h^3 c^3} V \exp(-\alpha) \int_{mc^2}^{\infty} \exp(-\beta\varepsilon)\varepsilon^2 \sqrt{\varepsilon^2 - m^2 c^4}\, d\varepsilon \tag{3.2.44a}$$

$$\frac{PV}{kT} = \frac{4\pi}{h^3 c^3} V \exp(-\alpha) \int_{mc^2}^{\infty} \exp(-\beta\varepsilon)\varepsilon \sqrt{\varepsilon^2 - m^2 c^4}\, d\varepsilon\ . \tag{3.2.44b}$$

As we see, for a relativistic classical gas

$$PV = NkT \tag{3.2.45}$$

holds, though the energy E is not equal to $\frac{3}{2}NkT$.

3.3 Classical Systems

3.3.1 Quantum Effects and Classical Statistics

Molecules (atoms) such as helium and argon are spherical. We shall consider substances, liquids for example, consisting of such simple molecules. Molecules interact through repulsive forces acting at short distances and attractive forces at relatively large distances. The attractive force between two electrically neutral molecules has a potential energy inversely proportional to the sixth power of the distance r, and is called the *van der Waals attraction.* The repulsive force comes from the overlapping of the wave functions of external electrons in molecules and is given roughly as an exponential function of r. However, for convenience, the repulsive force is also expressed as an inverse power of r. For example, the interaction potential between two molecules is expressed as

$$\phi(r) = \frac{\lambda}{r^{12}} - \frac{\lambda}{r^6} \quad (\lambda, \mu > 0)\ . \tag{3.3.1}$$

This is called the *Lennard-Jones potential.* Assuming this form of the potential, the coefficients λ of the repulsive and μ of the attractive forces are determined by the experimental data of the *second virial coefficient*, which is a measure of the derivation of the equation of state for a gas from the ideal gas law [3.4].

We assume the additivity of the interaction potential, so that the total potential energy of the system is the sum of potential energy between molecular pairs. Though this assumption may, in some circumstances, give rise to an error of a few percent, we shall neglect it. Then the total energy of a system consisting of identical molecules is given as an eigenvalue of the Schrödinger equation,

$$\left[-\frac{\hbar^2}{2m}\sum_{j=1}^{N}\nabla_j^2+\sum_{j>k}\phi(r_{jk})\right]\psi_n=E_n\psi_n\,. \tag{3.3.2}$$

If the potential energy between molecules is specified by two parameters as in (3.3.1), then we will have certain similarities between different substances because the interaction is similar. We may use, instead of λ and μ, the distance σ where $\phi(r)$ takes the minimum value and the minimum value $-\phi_0$ itself (Fig. 3.10). We write

$$\phi(r)=\phi_0\varphi\left(\frac{r}{\sigma}\right) \tag{3.3.3}$$

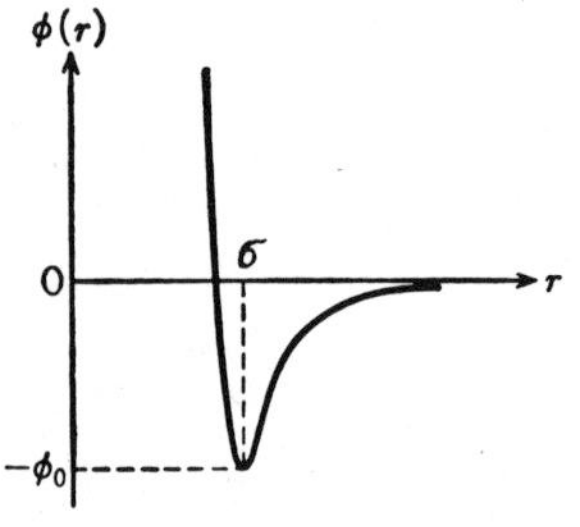

Fig. 3.10. Interaction potential between molecules

and assume that the function φ is common to various kinds of substances. If we measure the distance in units of σ and energy in units of φ_0, and express them by superscripts *, then the eigenvalue equation becomes [3.5]

$$\left[-\frac{\Lambda^2}{8\pi^2}\sum_j\nabla_j^{*2}+\sum_{j>k}\varphi(r_{jk}^*)-NE_n^*\right]\psi(r_1^*,r_2^*,\dots,r_N^*)=0\,, \tag{3.3.4}$$

where

$$\Lambda=\frac{h}{\sigma\sqrt{m\phi_0}} \tag{3.3.5}$$

$$E_n^*=\frac{E_n}{N\phi_0},\quad V^*=\frac{V}{N\sigma^3}\,. \tag{3.3.6}$$

Thus, the sum over states is a function of the reduced temperature

$$T^* = \frac{kT}{\phi_0} \tag{3.3.6a}$$

and V^*. It is also a function of the de Boer parameter Λ, since each energy eigenvalue E_n^* is a function of V^* and Λ. The de Boer parameter Λ is the ratio of the de Broglie wavelength for a molecule with energy ϕ_0 to σ. If Λ is large, it means that the quantum effect is serious. When we compare the values of thermal properties by reducing them in terms of σ and ϕ_0, they are expressed as functions of Λ for similar substances.

For example, the critical temperature, melting temperature, vapor pressure and so forth can be compared in this way (*de Boer's corresponding state*). Figure 3.11 shows the relationship between the reduced *critical temperature* $T_c^* = kT_c/\phi_0$ and Λ.

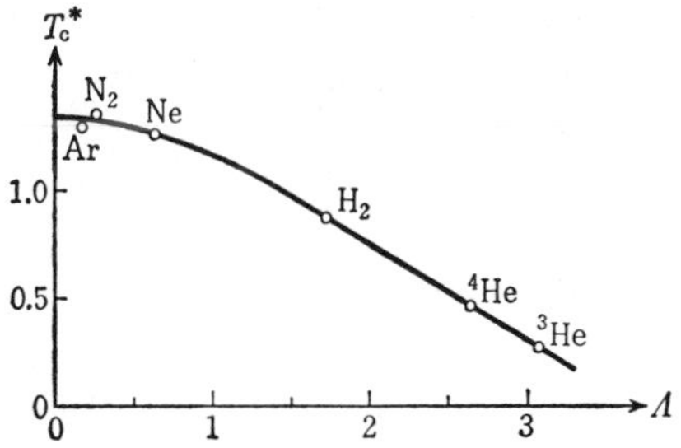

Fig. 3.11. De Boer's correspondence principle for the critical temperatures of several gases

In such similar properties differences in statistics do not seem to appear. However, since liquid helium shows the λ-transition which is a clear evidence of statistics, we have to think that the effect of statistics should appear at sufficiently low temperatures. This will mean that, though the effect of statistics is remarkable for low energy states, for higher energy the wave-functions are well localized, so that the level density is not affected by the difference in statistics so much. Near and above the melting temperature, thermal properties will be determined mainly by such high energy states.

It is likely that the quantum effect is manifest to a small extent in neon ($\Lambda = 0.591$), but for argon ($\Lambda = 0.187$) no quantum effect seems to appear. Classical statistics is valid when $\Lambda \ll 1$, so that the classical correspondence principle, in the limit $\Lambda = 0$, does not include Λ.

a) *Classical Statistics*

If molecules can be considered as particles without any internal degree of freedom, the partition function in classical statistics is a product

$$\begin{aligned} Z &= \frac{1}{N!h^{3N}} \iint d^N\boldsymbol{p}\, d^N\boldsymbol{q} \exp[-\mathscr{H}(\boldsymbol{p}, \boldsymbol{q})/kT] \\ &= J(T)Q(T, V)\,, \end{aligned} \tag{3.3.7}$$

where J is the part of the partition function related to momentum p and Q is that of the coordinate space:

$$J(T)=\frac{1}{h^{3N}}\left[\int_{-\infty}^{\infty}\exp(-p^2/2mkT)\,dp\right]^{3N}=\left(\frac{2\pi mkT}{h^2}\right)^{3N/2} \tag{3.3.7a}$$

$$Q(T,V)=\frac{1}{N!}\int_{(V)}\cdots\int\exp(-U/kT)dx_1dy_1dz_1\cdots dz_N\,. \tag{3.3.7b}$$

We have written

$$\mathscr{H}(p,q)=\frac{1}{2m}\sum(p_{x_j}^2+p_{y_j}^2+p_{z_j}^2)+U(q)\,, \tag{3.3.7c}$$

where U stands for the potential energy which is a function of $\boldsymbol{q}=(x_1,y_1,z_1,\dots,z_N)$, the molecular configuration. The factor $1/N!$ implies that we have divided by the number of permutations of molecules, so that in the spatial part Q, the integration is taken independently of all the molecules . This factor is derived when we consider the classical limit of the quantum statistics as we saw in Sect. 2.1. In classical mechanics, the sum over states, or the partition function, is sometimes called the phase integral.

The average of a quantity $A(p,q)$ is given in classical statistics by

$$\langle A\rangle=\frac{\iint A(p,q)\exp[-\mathscr{H}(p,q)/kT]dpdq}{\iint\exp[-\mathscr{H}(p,q)/kT]dpdq}\,, \tag{3.3.8}$$

where $dpdq$ is the elementary volume in phase space.

b) *Law of Equipartition of Energy*

By simple calculation we have

$$\left\langle\frac{1}{2m}p_{x_j}^2\right\rangle=\left\langle\frac{1}{2m}p_{y_j}^2\right\rangle=\left\langle\frac{1}{2m}p_{z_j}^2\right\rangle=\frac{kT}{2}\,. \tag{3.3.9}$$

If the coordinate q_l is included in U in the form cq_l^2 ($c=$ const.), we have similarly

$$\langle cq_l^2\rangle=\frac{kT}{2}\,. \tag{3.3.10}$$

In general, if the Hamiltonian includes p_n and q_l, we have

$$\left\langle p_n\frac{\partial\mathscr{H}}{\partial p_n}\right\rangle=\left\langle q_l\frac{\partial\mathscr{H}}{\partial q_l}\right\rangle=kT\,. \tag{3.3.11}$$

This is called the *law of equipartition of energy*, which is one of the most important results in classical mechanics. In quantum mechanics this law does not hold.

3.3.2 Pressure

Since the distribution density in phase space of an ensemble is proportional to $\exp[-\mathcal{H}(p,q)/kT]$, the probability of a given configuration is proportional to $\exp(-U/kT)$. This is called the *Boltzmann factor*.

In particular, the probability that two molecules are at $\boldsymbol{r}$ and $\boldsymbol{r}'$ is given by the two-body distribution function

$$n^{(2)}(\boldsymbol{r},\boldsymbol{r}') = \frac{N(N-1)\int\int\cdots\int \exp[-U(\boldsymbol{r},\boldsymbol{r}',\boldsymbol{r}_3,\ldots,\boldsymbol{r}_N)/kT]d\boldsymbol{r}_3\cdots d\boldsymbol{r}_N}{\int\int\cdots\int \exp[-U(\boldsymbol{r}_1,\boldsymbol{r}_2,\boldsymbol{r}_3,\ldots,\boldsymbol{r}_N)/kT]d\boldsymbol{r}_1 d\boldsymbol{r}_2 d\boldsymbol{r}_3\cdots d\boldsymbol{r}_N}\,. \tag{3.3.12}$$

For a gas or a liquid the distribution of molecules is the same at any position. In such a case, writing $n = N/V$, we may write $n^{(2)}(\boldsymbol{r},\boldsymbol{r}') = n^2 g(|\boldsymbol{r}-\boldsymbol{r}'|)$, where $g(R)$ is the radial distribution function.

We assume that the potential energy due to molecular interaction is given as a sum of the interaction energy $\phi(R)$ between molecular pairs (additivity), so that

$$U = \sum_{j<k} \phi(r_{jk}), \quad (r_{jk} = |\boldsymbol{r}_j - \boldsymbol{r}_k|)\,. \tag{3.3.13}$$

The total energy of the system is then given as

$$E = -\frac{\partial \ln Z}{\partial(1/kT)} = \frac{3}{2}NkT + \frac{2\pi N^2}{V}\int_0^\infty \phi(r)g(r)r^2\,dr\,, \tag{3.3.14}$$

where the first term is the kinetic energy and the second the energy of interaction.

The equation of state is given as $P = kT\partial \ln Q/\partial V$. To perform differentiation with respect to volume, it is convenient to assume the system to be in a cubic box of sides l, namely $V = l^3$. We change variables from $x_1, y_1, z_1, \ldots, x_N, y_N, z_N$ to[1]

$$\xi_1 = \frac{x_1}{l},\quad \eta_1 = \frac{y_1}{l},\quad \zeta_1 = \frac{z_1}{l},\quad \ldots,\quad \xi_N = \frac{x_N}{l},$$

$$\eta_N = \frac{y_N}{l},\quad \zeta_N = \frac{z_N}{l}, \tag{3.3.15}$$

and at the same time we use

$$\rho_{jk} = \frac{r_{jk}}{l} \tag{3.3.15a}$$

[1] The method used here has previously only been presented in Japanese [3.6].

in place of r_{jk}. By this transformation, we obtain

$$Q(T,V)=\frac{l^{3N}}{N!}\int_0^1\cdots\int_0^1\exp\left[-\frac{1}{kT}\sum\phi(l\rho_{jk})\right]d\xi_1 d\eta_1\cdots d\zeta_N\,. \tag{3.3.16}$$

Thus we have

$$l\frac{\partial\ln Q}{\partial l}$$

$$=3N-\frac{1}{kT}\sum_{j<k}\frac{\int\int\cdots\int l\rho_{jk}\phi'(l\rho_{jk})\exp\left[-\frac{1}{kT}\sum\phi(l\rho_{jk})\right]d\xi_1 d\eta_1\cdots d\zeta_N}{\int\int\cdots\int\exp\left[-\frac{1}{kT}\sum\phi(l\rho_{jk})\right]d\xi_1 d\eta_1\cdots d\zeta_N}$$

$$=3N-\frac{N(N-1)}{2kT}\frac{\int\int\cdots\int r_{12}\phi'(r_{12})\exp(-U/kT)\,dx_1 dy_1\cdots dz_N}{\int\int\cdots\int\exp(-U/kT)dx_1 dy_1\cdots dz_N}\,.$$

By virtue of $\partial/\partial V=(1/3V)l\partial/\partial l$ and the definition of $n^2=n^2g(r)$, cf. (1.2.5), we are led to an equation of state of the form

$$PV=NkT-\frac{2\pi N^2}{3V}\int_0^\infty r^3\phi'(r)g(r)\,dr \tag{3.3.17}$$

which implies the virial theorem (1.2.18) already stated in Chap. 1 except that, in the present case, the kinetic energy is $\frac{3}{2}NkT$ because of classical statistics.

Though pressure is given thus as the volume derivative of ln Q, it is more direct to consider the force between the container wall and the system. To calculate this force, we consider the potential energy $\sum_j u(l-x_j)$ beteween molecules and a wall at $x=l$, perpendicular to the x-axis. The force exerted on the wall by the molecule $j=1$ is $du(l-x_1)/dx_1$. If the potential u due to the wall is included in U, we thus obtain

$$P=N\frac{\frac{1}{l^2}\int\int\cdots\int\frac{du(l-x_1)}{dx_1}\exp(-U/kT)\,dx_1 dy_1\cdots dz_N}{\int\int\cdots\int\exp(-U/kT)\,dx_1 dy_1\cdots dz_N}\,.$$

Assuming a cubic container of volume $V=l^3$, we integrate partially to obtain

$$\int\int\cdots\int\exp(-U/kT)\,dx_1 dy_1\cdots dz_N$$

$$=\int\cdots\int dx_2\cdots dz_N\left[\int\int dy_1 dz_1 x_1\exp(-U/kT)\right]_{x_1=0}^{l}$$

$$+\frac{1}{kT}\int\cdots\int dx_2\cdots dz_N\int\int\int dx_1 dy_1 dz_1$$

$$\times x_1\left[\sum_j\frac{\partial\phi(r_{1j})}{\partial x_1}+\frac{du(1-x_1)}{dx_1}\right]\exp(-U/kT)\,.$$

In the first term on the right-hand side, the lower limit $x_1=0$ has a vanishing contribution because of x_1 in the integrand, and the upper limit $x_1 = l$ also yields nothing since U is ∞ at $x_1 = l$ because of the potential u of the wall. In the second term, $x_1 \partial\phi(r_{1j})\partial x_1$ gives rise to the virial of the molecular forces. Further, since $[du(l - x_1)/dx_1] \exp[-u(l - x_1)/kT]$ has a sharp peak around $x_1 = l$, we can replace x_1 by l in the integrand, and this term gives the pressure. Therefore, the above equation can be written as

$$\int\int \cdots \int \exp(-U/kT)\, dx_1 dy_1 \cdots dz_N$$

$$= \frac{N-1}{6kT} \int\int \cdots \int r_{12} \frac{d\phi_{12}}{dr_{12}} \exp(-U/kT)\, dx_1 dy_1 \cdots dz_N$$

$$+ \frac{PV}{NkT} \int\int \cdots \int \exp(-U/kT)\, dx_1 dy_1 \cdots dz_N$$

which agrees with the equation of state (3.3.17) we have already obtained. Thus, it has been shown directly that the pressure is the average of the force on the wall exerted by the molecules in the system.

3.3.3 Surface Tension

Surface tension of a liquid is related to the excess free energy per unit surface area. The corresponding formula in statistical mechanics can be obtained by changing the area of the interface between gas and liquid, maintaining the volumes of these phases so that no transfer of molecules occurs between them. We take the z-axis perpendicular to the surface assuming it to be flat. Though the state of aggregation of molecules changes through several molecular layers, the following formulas do not depend on the position of the origin of coordinates. Let $n_2 g(z_1, \boldsymbol{r}_{12})$ be the number density of molecules at $\boldsymbol{r}_{12}$ from a molecule situated at z_1, where n_2 is the average number density at the point 2. If the average number density at 1 is denoted by n_1, the exact formula for surface tension can be written as [3.7, 8]

$$\gamma = \frac{1}{2} \int\int \frac{x_{12}^2 - z_{12}^2}{2r_{12}} \frac{d\phi}{dr_{12}} n_1 n_2 g(z_1, \boldsymbol{r}_{12}) d\boldsymbol{r}_1 d\boldsymbol{r}_2 \,, \tag{3.3.18}$$

where integration is to be carried out over the unit area on the plane (x, y) around the origin. For liquid helium and liquid hydrogen, there is a quantum effect involved in surface tension and we have to add [3.9, 10]

$$\gamma' = \frac{1}{A} \operatorname{tr} \left\{ -\frac{\hbar^2}{2m} \sum_{j=1}^{N} \left(\frac{\partial^2}{\partial z_j^2} - \frac{\partial^2}{\partial x_j^2} \right) \rho \right\} \Big/ \operatorname{tr}\{\rho\} \tag{3.3.19}$$

to (3.3.18), where ρ is the density matrix, the summation is to be taken over all the molecules $j = 1, 2, \ldots, N$, and A is the surface area. γ' is equal to the difference between the kinetic energy along the z- and x-directions, and of course vanishes in classical statistical mechanics.

If we approximate the interface by a geometrical surface, (3.3.18) reduces to

$$\gamma = \frac{\pi}{8} n^2 \int_0^\infty r^4 \frac{d\phi(r)}{dr} g(r)dr \, .$$

The surface energy u, the excess energy per unit area, is related to surface tension by the formula $u = \gamma - Td\gamma/dT$, and can be obtained by measuring the surface tension, which is the work necessary to isothermally extend the area of the liquid surface.

3.3.4 Imperfect Gas

The partition function of a classical imperfect gas [3.3, 11–13] can be obtained when we integrate $\exp(-\beta U)$ over the configuration space, that is, by calculating the space part Q of the partition function. If we denote the interaction between molecules j and k by ϕ_{jk}, we have

$$\exp(-\beta U) = \prod_{i>j} \exp(-\beta\phi_{ij}) = \prod_{i>j} (1 + f_{ij}) \, , \tag{3.3.20}$$

where $\beta = 1/k_B T$ (k_B denotes Boltzmann's constant in this section) and

$$f_{ij} = \exp(-\beta\phi_{ij}) - 1 \tag{3.3.21}$$

is the so-called *Mayer's function*, which vanishes rapidly for large r_{ij} because $\phi \to 0$ at large distance (Fig. 3.12). f is greater than zero in the attractive region where $\phi < 0$.

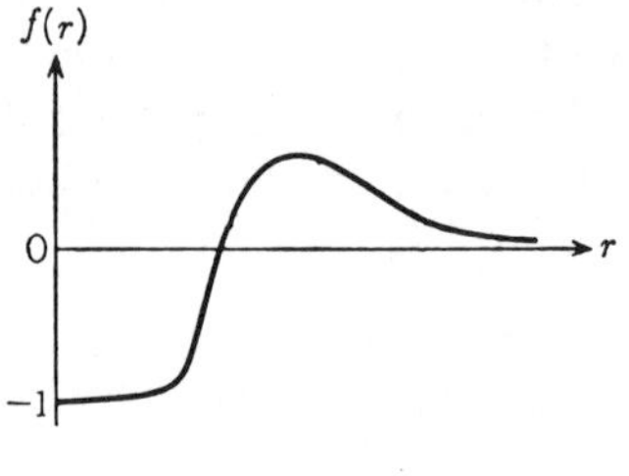

Fig. 3.12. Mayer's function (3.3.21)

What is important in statistical mechanics is the logarithm of the partition function, or the function W defined by

$$\exp(NW) = \left\langle \prod_{i>j} (1 + f_{ij}) \right\rangle = \frac{1}{V^N} \int\int \cdots \prod_{i>j} (1 + f_{ij}) d\boldsymbol{r}_1 d\boldsymbol{r}_2 \cdots d\boldsymbol{r}_N \, , \tag{3.3.22}$$

where $\langle \ \rangle$ means the average in configuration space. Such a logarithm of an average can be calculated by means of the cumulant expansion. However, we shall here indicate some results in a more elementary way.

If we expand the product in (3.3.20), we have

$$\prod_{N \geqq i > j \geqq 1} (1 + f_{ij}) = 1 + \sum_{i>j} f_{ij} + \sum\sum f_{ij} f_{kl} + \cdots .$$

We write the integrals for the aggregates of molecules connected by the subscripts ij as

$$\begin{aligned} b_1 &= \frac{1}{V} \int d\boldsymbol{r}_1 = 1 \\ b_2 &= \frac{1}{2V} \int\int f(r_{12}) d\boldsymbol{r}_1 d\boldsymbol{r}_2 = \frac{1}{2} \int_0^\infty 4\pi r^2 f(r) dr \\ b_3 &= \frac{1}{3!V} \int\int\int (f_{31} f_{21} + f_{32} f_{31} + f_{32} f_{21} + f_{32} f_{31} f_{21}) d\boldsymbol{r}_1 d\boldsymbol{r}_2 d\boldsymbol{r}_3 . \end{aligned} \tag{3.3.23}$$

In general, b_l is an integral with respect to the diagrams in which l molecules are directly or indirectly connected and is called a *cluster integral*, which is defined by

$$b_l = \frac{1}{l!V} \int\int \cdots \int \sum \prod_{l \geqq i > j \geqq 1} f_{ij} d\boldsymbol{r}_1 d\boldsymbol{r}_2 \cdots d\boldsymbol{r}_l , \tag{3.3.24}$$

where $\sum$ is the sum over the diagrams. If we want to treat a liquid filling volume V, then b_l for large l will become functions of V. However, for an imperfect gas, we may first fix the position of the molecule $j = 1$ and integrate with respect to the molecules $j = 2, 3, \ldots, l$, with a result independent of the position of $j = 1$. Thus, b_l is independent of the volume V for an imperfect gas.

The number of ways of dividing N molecules into m_1 isolated molecules, m_2 molecular pairs, m_3 sets of three molecules, $\ldots$, m_l sets of l molecules, etc., is given as

$$N! \bigg/ \prod_{l=1}^{N} (l!)^{m_l} m_l! . \tag{3.3.25}$$

This is made clear when we prepare $m_1, m_2, \ldots, m_l, \ldots$ boxes. If we first distinguish the molecules, we have $N!$ ways to put N molecules in these boxes, by placing a molecule in each of m_1 boxes, two molecules in each of m_2 boxes, $\ldots$, l molecules in each of m_l boxes, etc. But since all the molecules are indistinguishable, this number must be divided by $l!$, the number of permutations of molecules in each box, or for m_l boxes, by the product of $(l!)^{m_l}$. Furthermore, since the m_l boxes are identical, it must be also divided by $m_l!$, the ways of permuting the boxes. This holds true for $l = 1$ up to $l = N$. Thus we have

$$\begin{aligned} Q_N(T, V) &= \frac{1}{N!} \sum_{\substack{m_l \\ \sum l m_l = N}} \frac{N!}{\prod_l (l!)^{m_l} m_l!} \prod_l (l! V b_l)^m \\ &= \sum_{\substack{m_l \\ \sum l m_l = N}} \prod_l \frac{(V b_l)^{m_l}}{m_l} . \end{aligned} \tag{3.3.26}$$

If we multiply by $\xi^N = \prod(\xi^l)^{m_l}$ on both sides of (3.3.26) and add with respect to N, we have the grand partition function

$$\Xi(T, V, \xi) = \sum_{N=0}^{\infty} \xi^N Q_N = \exp\left(\sum_{l=1}^{\infty} Vb_l\xi^l\right) \tag{3.3.27}$$

and therefore, pressure P is given as

$$P = k_B T \sum_{l=1}^{\infty} b_l\xi^l . \tag{3.3.28}$$

If the density of the gas is lowered, ξ approaches the average molecular density $n = N/V$, as is seen by comparison with the case of an ideal gas. ξ is called the *fugacity*. Its relation to the number of molecules is given by

$$N = V \sum_{l=1}^{\infty} lb_l\xi^l . \tag{3.3.29}$$

Except for b_1, b_2 and b_3, because of the difficulty of integration, we have no detailed calculation of cluster integrals. Since b_l is an integral with respect to l molecules, we may have an approximation

$$b_l \propto \exp(-\alpha_s l^{2/3} - \alpha_1 l) , \tag{3.3.30}$$

since it will be related to the surface area proportional to $l^{3/2}$. The coefficient α_s is roughly equal to the surface tension (times the surface area of a molecule) divided by $k_B T$ and α_1 is the chemical potential of the liquid divided by $k_B T$, since the cluster is approximated by a liquid drop of l molecules. On the other hand, if we put $\xi = e^\alpha$, then $\mu = \alpha k_B T$ is the chemical potential of the gas. Now, suppose that we increase ξ without changing the temperature. Then V diminishes because $N/V = \sum lb_l\xi^l$ (we are assuming that the temperature is not very high, and therefore b_l's are positive). If $\exp(-\alpha_1)\xi = \exp(\alpha - \alpha_1)$ is thus increased and approaches 1, then $\sum lb_l\xi^l \approx \sum l \exp[(\alpha - \alpha_l)l] \exp(-\alpha_s l^{2/3})$ will diverge, so that V also diminishes very rapidly, but $P \propto \sum \exp[(\alpha - \alpha_1)l] \exp(-\alpha_s l^{2/3})$ will stay nearly constant. This corresponds to the liquefaction of a gas by compression. Therefore, during condensation we have $\alpha = \alpha_1$, that is, the chemical potentials of liquid and gas phases are equal.

Now, the integral b_l includes many diagrams connecting molecules in various ways. For example, though b_4 includes an integral of $f_{21}f_{31}f_{41}$, integration can be carried out independently over 2, 3, and 4, fixing 1, and it reduces to Vb_2^3. Clusters which can be thus decomposed are called *reducible clusters*. On the contrary, an integral of $f_{21}f_{32}f_{31}$, for example, cannot be reduced to b_l with smaller l. Such clusters are called irreducible clusters, and we write

$$\beta_k = \frac{1}{k!V}\int\int\cdots\int \sum^{(\mathrm{irr})} \prod_{k+1 \geq j > k \geq 1} f_{jk}\, dr_1\, dr_2 \cdots dr_{k+1} \tag{3.3.31}$$

for an *irreducible integral* of $k + 1$ molecules, where $\sum^{(\mathrm{irr})}$ means the summation

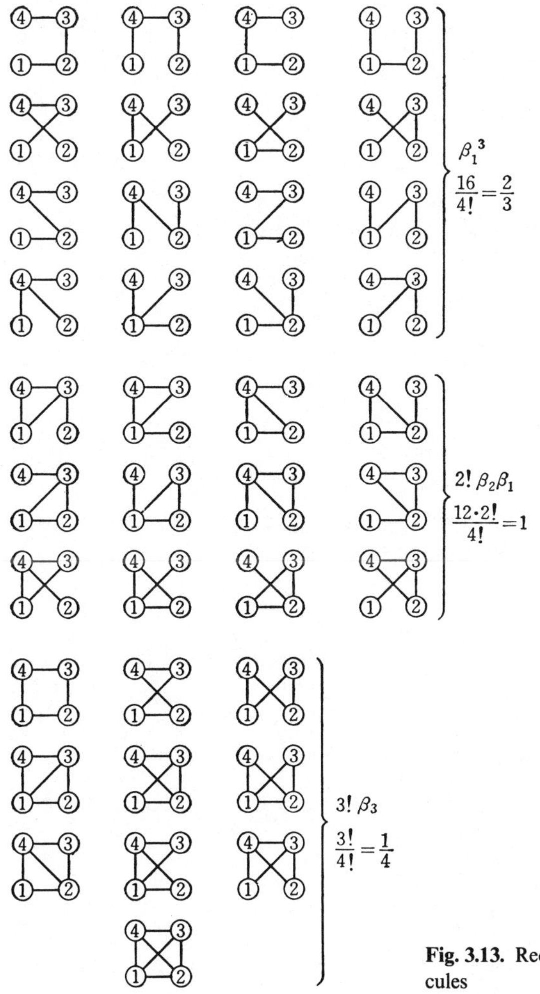

Fig. 3.13. Reducing reducible clusters of four molecules

over irreducible clusters. If we rewrite b_l in terms of β_k, we have (Fig. 3.13)

$$b_1 = \frac{1}{V}\int d\boldsymbol{r}_1 = \beta_0 \tag{3.3.32}$$

$$b_2 = \frac{1}{2V}\int\int f_{12}\, d\boldsymbol{r}_1 d\boldsymbol{r}_2 = \frac{1}{2}\beta_1 \tag{3.3.32a}$$

$$b_3 = \frac{1}{6V}\left(3\int\int\int f_{12}f_{23}\, d\boldsymbol{r}_1 d\boldsymbol{r}_2 d\boldsymbol{r}_3 + \int\int\int f_{12}f_{23}f_{13}\, d\boldsymbol{r}_1 d\boldsymbol{r}_2 d\boldsymbol{r}_3\right)$$

$$= \frac{1}{2}\beta_1^2 + \frac{1}{3}\beta_2 \tag{3.3.32b}$$

$$b_4 = \frac{2}{3}\beta_1^3 + \beta_1\beta_2 + \frac{1}{4}\beta_3 \tag{3.3.32c}$$

and so forth. In general, we can prove

$$l^2 b_l = \sum_{m_k}\prod_k \frac{(l\beta_k)^{m_k}}{m_k!}, \tag{3.3.33}$$

where $\sum_{m_k}$ means summation over all the sets of m_k satisfying $\sum_k k m_k = l - 1$.

ξ is determined by

$$v \sum_{l \geqq 1} l b_l \xi^l = 1 \tag{3.3.34}$$

with $v = V/N$. If we write

$$\xi = \frac{a_1}{v} + \frac{a_2}{v^2} + \frac{a_3}{v^3} + \frac{a_4}{v^4} + \cdots \tag{3.3.35}$$

and insert into (3.3.34), we have relations expressing a_1, a_2, a_3, and a_4 in terms of b_1, b_2, b_3 and b_4. Using (3.3.32), we rewrite them to obtain

$$a_1 = 1 \tag{3.3.36}$$

$$a_2 = -2b_2 = -\beta_1 \tag{3.3.36a}$$

$$a_3 = 8b_2^2 - 3b_3 = -\left(\beta_2 - \frac{1}{2}\beta_1^2\right) \tag{3.3.36b}$$

$$a_4 = -40b_2^3 + 30b_2b_3 - 4b_4 = -\left(\beta_3 - \beta_1\beta_2 + \frac{1}{6}\beta_1^3\right). \tag{3.3.36c}$$

Then from (3.3.3, 32 and 35) we obtain

$$\frac{P}{k_B T} = \sum b_l \xi^l = \frac{1}{v}\left(1 - \frac{\beta_1}{2v} - \frac{2}{3}\frac{\beta_2}{v^2} - \frac{3}{4}\frac{\beta_3}{v^3}\right). \tag{3.3.37}$$

It is known that

$$PV = Nk_B T\left[1 - \sum_{k=1}^{\infty}\frac{k}{k+1}\beta_k\left(\frac{N}{V}\right)^k\right]. \tag{3.3.38}$$

This is the *virial expansion* of the equation of state for an imperfect gas, and the coefficient of V^{-k}, $B_k(T) = k(k+1)^{-1}\beta_k N^k$, is called the kth *virial coefficient*.

In terms of W defined by (3.3.22), we may write

$$\frac{PV}{Nk_B T} = 1 - n\frac{\partial W}{\partial n} \quad \left(n = \frac{N}{V}\right). \tag{3.3.39}$$

Therefore, from the above equation of state,

$$W = \sum_{k=1}^{\infty}\frac{\beta_k}{k+1}n^k. \tag{3.3.40}$$

If we use the formal expansion

$$f = \exp(-\beta\phi) - 1 = \sum_{\nu=1}^{\infty} \frac{1}{\nu!}(-\beta\phi)^{\nu}, \tag{3.3.41}$$

we have

$$W = \sum_{k=1}^{\infty} \frac{n^k}{(k+1)!} \sum^{(\text{irr})} \int\int \cdots \int \prod_{k+1 \geqq i > j \geqq 1} \sum_{\nu_{ij}=1}^{\infty} \frac{(-\beta\phi_{ij})^{\nu_{ij}}}{\nu_{ij}!} d\boldsymbol{r}_2 d\boldsymbol{r}_3 \cdots d\boldsymbol{r}_{k+1} . \tag{3.3.42}$$

3.3.5 Electron Gas

When the interaction is a Coulomb force, the limit $r \to 0$ gives no difficulty but integrals diverge for $r \to \infty$. We may, nevertheless, calculate them formally.

We consider an assembly of electrons, assuming a uniform positive charge which cancels the charge on the average. The interaction between electrons is

$$\phi(\boldsymbol{r}) = \frac{e^2}{r} \quad (r = |\boldsymbol{r}|), \tag{3.3.43}$$

where the charge of an electron is $-e$. We take periodic boundary conditions and expand $\phi(\boldsymbol{r})$ as

$$\phi(\boldsymbol{r}) = \frac{1}{V} \sum \nu(\boldsymbol{q}) \exp(\mathrm{i}\boldsymbol{q} \cdot \boldsymbol{r}), \tag{3.3.44}$$

where

$$\nu(\boldsymbol{q}) = \begin{cases} \dfrac{4\pi e^2}{q^2} & (\boldsymbol{q} \neq 0) \\ 0 & (\boldsymbol{q} = 0) . \end{cases} \tag{3.3.45}$$

The integral $\int \cdots \int$ in (3.3.42) diverges for large r_{ij}. If we use the Fourier decomposition (3.3.44), the sum over q diverges for small q. But the integral has a factor of the form $\prod(-\beta\phi_{ij})$ and also a factor $1/(k+1)!$. Therefore, if we sum up appropriately, it may converge.

However, it is practically impossible to calculate all the cluster integrals. If we instead take a ring like $\phi_{12}\phi_{23}\phi_{34}\phi_{41}$ connected by the numbering of electrons (ij), we have

$$\int\int \cdots \int \phi_{12}\phi_{23} \cdots \phi_{l1} d\boldsymbol{r}_2 d\boldsymbol{r}_3 \cdots d\boldsymbol{r}_l = \frac{1}{V} \sum_{q} \nu^l(\boldsymbol{q}) . \tag{3.3.46}$$

If we exchange the numbering of electrons, we have a different cluster. But, since both sides of a ring are equivalent, there are $(l-1)!/2$ ring clusters formed by l

electrons. Therefore, if we take only ring clusters, we have the approximation

$$W_{(\mathrm{ring})} = \frac{1}{2V}\sum_{l=2}^{\infty}\frac{(-1)^l\beta^l n^{l-1}}{l}\sum_{q} v^l(\boldsymbol{q})$$

$$= \frac{1}{2nV}\sum_{q}\{-\ln[1+\beta n v(\boldsymbol{q})]+\beta n v(\boldsymbol{q})\} . \tag{3.3.47}$$

Summation over $\boldsymbol{q}$ can be replaced by integration $\int d\boldsymbol{q}/(2\pi)^3$. After performing integration we get

$$W_{(\mathrm{ring})} = \frac{\kappa^3}{12\pi n} , \tag{3.3.48}$$

where

$$\kappa^2 = 4\pi e^2 \beta n . \tag{3.3.49}$$

κ^{-1} has the dimension of length and is called the *Debye length* which characterizes the shielding of a Coulomb force, as we shall see below. The ring approximation yields the equation of state

$$\frac{PV}{Nk_{\mathrm{B}}T} = 1 - \frac{\sqrt{\pi}}{3} e^3 \beta^{3/2} n^{1/2} . \tag{3.3.50}$$

It is to be noted that the deviation from the ideal gas law is proportional to $n^{1/2}$, but not to n. This equation of state was first derived by the theory due to *P. Debye* and *W. Hückel.*

3.3.6 Electrolytes

P. Debye and *W. Hückel* established a theory which takes into account the interaction between ions in a very ingenious way [3.14]. This method is applicable only to Coulomb forces, and is quite intuitive. But it elucidates the physical features of the phenomena. Since ions of the same sign (charge) repel each other and ions of different signs attract, each ion will be surrounded by ions of different signs, and those of the same sign will be kept away, on the average. Such a distribution of ions is called the *ion atmosphere* (Fig. 3.14). Here we shall not smear out the positive charge, and shall consider ions of both signs.

In general, we may consider many kinds of ions. Around an ion of the jth kind, an electric potential ψ_j is set up by the ion and its atmosphere. If there is no external electric field, the electric potential ψ_j (time average) has spherical symmetry if the ion is spherical. If ρ_i denotes the charge density due to the ion atmosphere, the Poisson equation around the ion is

$$\nabla^2\psi_j = -\frac{4\pi\rho_j}{\varepsilon} , \tag{3.3.51}$$

where ε is the dielectric constant of the solution. If there are, on the average, n_i

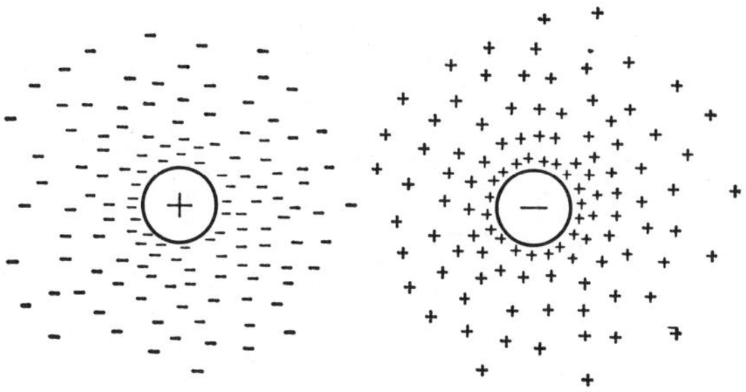

Fig. 3.14. Ion atmosphere

ions of charge e_i, the number density of the ions of the ith kind at a point with electric potential ψ_j is given by the Boltzmann distribution

$$n_i \exp\left(-\frac{e_i \psi_j}{k_B T}\right) \tag{3.3.52}$$

and the total charge density of the ion atmosphere is given by

$$\rho_j = \sum_i e_i n_i \exp\left(-\frac{e_i \psi_j}{k_B T}\right). \tag{3.3.53}$$

If we want to smooth out the positive charges as we did in the preceding section, we have to assume vanishing ψ_j around positive ions, and summation in (3.3.53) is to be taken only over negative ions.

Since electric charges cancel as a whole,

$$\sum_i e_i n_i = 0 . \tag{3.3.54}$$

A little farther from the ion under consideration where

$$\frac{e_i \psi_j}{k_B T} \ll 1 \tag{3.3.55}$$

holds, the Poisson equation is linearized to give

$$\frac{1}{r}\frac{d^2(r\psi_j)}{dr^2} = \kappa^2 \psi_j \tag{3.3.56}$$

with

$$\kappa^2 = \frac{4\pi}{\varepsilon k_B T}\sum_i n_i e_i^2 . \tag{3.3.57}$$

We have the boundary condition that $\psi_j = 0$ for $r \to \infty$. If, for simplicity, we assume that all the ions are hard spheres with diameter σ, then, for $r \to \sigma$, ψ_j tends to the Coulomb force field so that

$$-\frac{d}{dr}\psi_j \longrightarrow \frac{e_j}{\varepsilon r^2} \quad (r \to \sigma) . \tag{3.3.58}$$

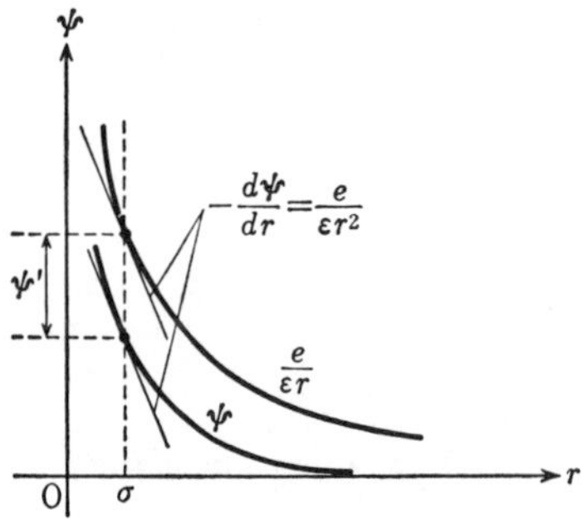

Fig. 3.15. Electric field of an ion atmosphere

Solving (3.3.56) under these conditions, we have (Fig. 3.15)

$$\psi_j = \frac{e_j}{\varepsilon(1+\kappa\sigma)} \frac{\exp[-\kappa(r-\sigma)]}{r} . \tag{3.3.59}$$

If we smooth out the positive charges, in (3.3.57) for κ, summation extends over negative charges only, and if negative charges are electrons, we have the same κ as (3.3.49). As we see from (3.3.59), κ^{-1} (the Debye length) stands for the shielding length of the ion atmosphere.

Let G_e be the thermodynamic potential due to the interaction between the ion under consideration and its ion atmosphere. This is obtained as the work required to charge up ions from the virtual state with vanishing charges. When the ion charge is e_j, the electric potential $\psi'(e_j)$ due to the atmosphere at the position of the ion is obtained when we subtract the electric potential of the ion itself from ψ_j (Fig. 3.15):

$$\begin{aligned}\psi'(e_j) &= \lim_{\gamma\to\sigma}\left[\psi(e_j) - \frac{e_j}{r}\right] \\ &= -\frac{e_j\kappa}{\varepsilon(1+\kappa\sigma)} .\end{aligned} \tag{3.3.60}$$

Suppose that charges of ions are λe_j $(0 \leqslant \lambda \leqslant 1)$, and we add small amounts of charge $d(\lambda e_j)$ to all the ions. Since κ is also multiplied by λ, the work of charging up all the ions is given as

$$\begin{aligned}G_e &= \sum_j N_j \int_0^1 \psi'(\lambda e_j) d(\lambda e_j) \\ &= -\sum_j \frac{N_j e_j^2 \kappa}{\varepsilon} \int_0^1 \frac{\lambda^2 d\lambda}{1+\kappa\sigma\lambda} .\end{aligned}$$

When the ions are sufficiently dilute, $\kappa\sigma \ll 1$, we have

$$G_e = -\sum_j \frac{N_j e_j^2 \kappa}{3\varepsilon} , \tag{3.3.61}$$

where $N_j = n_j V$ is the total number of ions of the jth kind.

NW in the preceding section corresponds to $-G_e/k_B T$, so that in the Debye-Hückel theory we have

$$W_{DH} = \frac{e_j^2 \kappa}{3\varepsilon k_B T}, \tag{3.3.62}$$

where positive charges are smoothed out, so that the summation in (3.3.61) is extended over negative charges only and in addition, only one kind of negative ion is assumed ($N = N_j$). When we put $\varepsilon = 1$, we see that W_{DH} coincides with $W_{(\text{ring})}$.

Thus, it has been shown that the ring approximation is equivalent to the Debye-Hückel approximation, and vice versa. If we want to improve the results of the Debye-Hückel theory, we will have to take into account diagrams other than rings. It is not at all clear what kind of diagrams are necessary for improvement.

The pressure in (3.3.50) corresponds, in the case of electrolyte solutions, to the osmotic pressure due to the solute ions.

4. Phase Transitions

A gas whose equation of state is described by the Boyle-Charles law and a paramagnetic substance which obeys Curie's law are examples of so-called ideal systems. These systems are composed of elements with negligible interactions and the treatment of these systems can be reduced essentially to that of a single element. The harmonically vibrating lattices have strong interactions among particles, but they are ideal systems in view of the existence of normal modes or phonons. In contrast to these ideal systems, systems which are by no means reducible to ideal systems exist and thus have strong interactions among constituent elements which can never be ignored. They are sometimes called cooperative systems and exhibit, among other things, a cooperative phenomenon called a phase transition. For example, a gas condenses to the liquid state by compression or by cooling, and a paramagnetic substance becomes ferromagnetic by cooling below the Curie temperature. The ideal Bose gas undergoes a Bose condensation, because it can be regarded effectively as a cooperative system having attractive interactions among atoms by virtue of the symmetry of the wave function. Through a phase transition, a substance acquires a new structure or a new property which is absent before the phase transition. Sometimes the biological function of a biomaterial can be regarded as a result of a phase transition.

As far as thermal equilibrium is concerned, we have a general prescription for writing down the partition function, according to the Gibbs statistical mechanics, provided that the elements of the system are known. The partition function can be used to give any equilibrium property of this system, when suitably treated. The real systems, however, are usually very complicated. Therefore, in order to obtain a concrete result, we turn to tractable particular models.

In this chapter we do not discuss phase transitions in real substances. Our aim is to elucidate the fundamental mechanisms which give rise to a phase transition. To do this, we first present some of the typical models [4.1–13].

4.1 Models

4.1.1 Models for Ferromagnetism

We assume that each lattice point in a crystal has an atom with a spin. Let the spin in the ith lattice point be s_i, then the Hamiltonian of this spin system in a magnetic field H applied in the z-direction is

$$\mathscr{H} = -2\sum_{\langle ij\rangle} J_{ij}\boldsymbol{s}_i\cdot\boldsymbol{s}_j - g\mu_B H\sum_i s_i^z\,, \tag{4.1.1}$$

where $\sum$ stands for the sum over the pairs of spins. The quantities s^z, g and μ_B are the z-components of the spin operator s, Landé's factor and the Bohr magneton, respectively. J_{ij} is the exchange integral which depends on the distance between the ith and jth spins and can be assumed to be zero for pairs other than the nearest ones; $J_{ij} > 0$ for ferromagnetic interaction and $J_{ij} < 0$ for antiferromagnetic. The scalar product $\boldsymbol{s}_i\cdot\boldsymbol{s}_j$ can be expressed in terms of their components as

$$\boldsymbol{s}_i\cdot\boldsymbol{s}_j = s_i^x s_j^x + s_i^y s_j^y + s_i^z s_j^z\,. \tag{4.1.2}$$

If the magnitude of s equals $\frac{1}{2}$, the components are expressed by Pauli matrices as

$$\begin{aligned} s_j^x &= \frac{1}{2}(\sigma_x)_j = \frac{1}{2}\begin{bmatrix} 0 & 1 \\ 1 & 0 \end{bmatrix}_j \\ s_j^y &= \frac{1}{2}(\sigma_y)_j = \frac{1}{2}\begin{bmatrix} 0 & -i \\ i & 0 \end{bmatrix}_j \\ s_j^z &= \frac{1}{2}(\sigma_z)_j = \frac{1}{2}\begin{bmatrix} 1 & 1 \\ 0 & -1 \end{bmatrix}_j\,. \end{aligned} \tag{4.1.3}$$

This interaction is isotropic with respect to the x, y, z-components of different spins. This is called the Heisenberg model. When the x- and y-components of the interactions are negligible, the Hamiltonian reduces to

$$\mathscr{H} = -2\sum_{\langle ij\rangle} J_{ij}s_i^z s_j^z - g\mu_B H\sum_i s_i^z\,. \tag{4.1.4}$$

In this case, s_i^z takes $\frac{1}{2}$ or $-\frac{1}{2}$ and the quantum mechanical effects of the commutation properties of the spin operators no longer have to be taken into consideration. This model is called the Ising model (Fig. 4.1). On the other hand, when z-components can be ignored and x- and y-components have anisotropic contributions to the Hamiltonian, then we have

$$\mathscr{H} = -2\sum_{\langle ij\rangle} J_{ij}[(1+\eta_{ij})s_i^x s_j^x + (1-\eta_{ij})s_i^y s_j^y] - g\mu_B H\sum_i s_i^z\,, \tag{4.1.5}$$

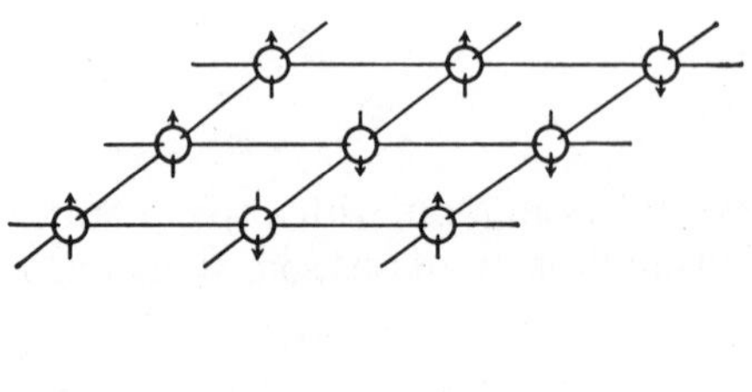

Fig. 4.1. Two-dimensional Ising model

where η_{ij} is a constant which depends on the distance between the ith and jth spins. This is called the XY model.

We have not yet found a real system approximately represented by an XY model in contrast to the situation for the Ising model which can be applied to a real system having a strong anisotropy in one direction. However, the XY model is of theoretical interest because it is an exactly soluble model in a one-dimensional case [4.14].

4.1.2 Lattice Gases

The lattice point or the cell of a crystal is assumed to be either occupied by an atom or not. Let p_i (= 1, or 0) be the state of the ith site, according to whether it is occupied ($p_i = 1$) or not ($p_i = 0$), and let $-2\varepsilon_{ij}$ be the interaction energy between two atoms on the ith and jth sites, respectively. Then the interaction energy can be written as

$$E_{\mathrm{p}} = -\sum_{\langle ij \rangle} 2\varepsilon_{ij} p_i p_j \,. \tag{4.1.6}$$

This system is called a lattice gas (Fig. 4.2). This is also considered as a lattice model for solutions or alloys, when the lattice points are regarded as occupied by either atoms A or B. In fact let, for example, $p_i(A) = 1$ and $p_i(B) = 0$ if the ith site is occupied by an A atom and thus not by a B atom; then

$$\begin{aligned} E_{\mathrm{p}} = \sum_{\langle ij \rangle} \{ & \varepsilon_{ij}(AA) p_i(A) p_j(A) + \varepsilon_{ij}(AB)[p_i(A) p_j(B) + p_i(B) p_j(A)] \\ & + \varepsilon_{ij}(BB) p_i(B) p_j(B) \} \,, \end{aligned} \tag{4.1.7}$$

where $\varepsilon_{ij}(AB) = \varepsilon_{ij}(BA)$ stands for the interaction energy between two atoms of A at the ith lattice point and B at the jth lattice point. Since we have

$$p_i(A) + p_i(B) = 1 \,, \tag{4.1.8}$$

(4.1.7) can be written as

$$\begin{aligned} E_{\mathrm{p}} = & \sum [2\varepsilon_{ij}(AB) - \varepsilon_{ij}(AA) - \varepsilon_{ij}(BB)] p_i(A) p_j(B) \\ & + \text{terms independent of the configurations of } A \text{ and } B \text{ atoms} \\ & \quad \text{on the lattice} \,. \end{aligned} \tag{4.1.9}$$

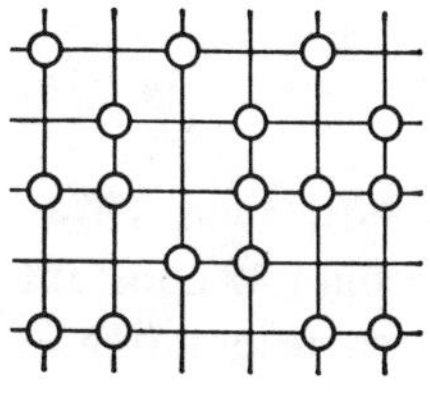

Fig. 4.2. Two-dimensional lattice gas. Atoms may also be sited at the center of each cell

Note that ε_{ij} defined by $\varepsilon_{ij} = \frac{1}{2}[2\varepsilon_{ij}(AB) - \varepsilon_{ij}(AA) - \varepsilon_{ij}(BB)]$ is the increase of energy when an AB pair is created in the solution. Thus, when A and B are regarded as atoms and vacancies respectively, (4.1.9) is equivalent to (4.1.6) and can be applied, *mutatis mutandis*, to the problem of alloys and solutions.

4.1.3 Correspondence Between the Lattice Gas and the Ising Magnet

Let the number of lattice points and atoms in a lattice gas be N and n, respectively; then we have

$$\sum_{i=1}^{N} p_i = n \, . \tag{4.1.10}$$

The canonical partition function Z_{p} and the grand canonical partition function Ξ_{p} are, respectively,

$$Z_{\text{p}} = j_{\text{p}}^n \sum_{\sum p_i = n} \exp(-\beta E_{\text{p}}) \tag{4.1.11}$$

$$\begin{aligned} \Xi_{\text{p}} &= \sum_{n=0}^{N} \exp(+\beta\mu_{\text{p}} n) j_{\text{p}}^n \sum_{\Sigma p_i = n} \exp(-\beta E_{\text{p}}) \\ &= \sum_{\{p_i\}} \exp\left\{-\beta\left[E_{\text{p}} - (\mu_{\text{p}} + kT\ln j_{\text{p}})\sum_i p_i\right]\right\} , \end{aligned} \tag{4.1.12}$$

where j_{p} is the contribution to the partition function from the vibration of an atom on a lattice point, which is assumed to be independent of the configuration surrounding the atom; μ_{p} is the chemical potential and the sum is taken over all the values of $p_i = 1$, and 0 ($i = 1, 2, \ldots, N$).

Next consider an Ising magnet. If we put $2s_i^z = \sigma_i$ then (4.1.4) becomes

$$\begin{aligned} E_{\text{s}} &= -\sum_{\langle ij \rangle} J_{ij}\sigma_i\sigma_j/2 - mH\sum \sigma_i \\ m &= g\mu_{\text{B}}/2 \end{aligned} \tag{4.1.13}$$

and the partition function Ξ_{s} is

$$\Xi_{\text{s}} = \sum_{\{\sigma_i\}} \exp(-\beta E_{\text{s}}) \, , \tag{4.1.14}$$

where we use the symbol Ξ_{s} instead of Z_{s} to clarify the correspondence with the lattice gas. The magnetization is given by

$$M = \sum_i m\sigma_i \, . \tag{4.1.15}$$

Equation (4.1.15) corresponds to the condition (4.1.10) in the lattice gas. However, the sum in (4.1.14) over all $\sigma_i = \pm 1$ is taken without keeping M constant and hence, Ξ_{s} in the spin system is the grand partition function corresponding to

Ξ_p in the lattice gas. To see this relation more closely, we put

$$p_i = (\sigma_i + 1)/2 . \tag{4.1.16}$$

Then $p_i = 1$ or 0 according to whether $\sigma_i = 1$ or -1, respectively, and we have

$$\begin{aligned} E_p - (\mu_p + kT\ln j_p)\sum_i p_i \\ = -\frac{1}{2}\sum_{\langle ij\rangle} \varepsilon_{ij}\sigma_i\sigma_j - \frac{1}{2}(\varepsilon_0 + \mu_p + kT\ln j_p)\sum_i \sigma_i \\ - \frac{N}{2}\left(\frac{\varepsilon_0}{2} + \mu_p + kT\ln j_p\right), \end{aligned} \tag{4.1.17}$$

where we define

$$\varepsilon_0 = \sum_{\substack{j=1 \\ j\neq i}}^{N} \varepsilon_{ij} . \tag{4.1.18}$$

The sum is considered convergent for $N \to \infty$, by assuming that ε_{ij} approaches 0 sufficiently rapidly for pairs a long distance apart and thus, ε_0 is independent of i in a large system ($N \to \infty$). Then (4.1.12) can be rewritten as

$$\left.\begin{aligned} \Xi_p &= \exp\left[\frac{\beta N}{2}\left(\frac{\varepsilon_0}{2} + \mu_p + kT\ln j_p\right)\right]\cdot \Xi_p' \\ \Xi_p' &= \sum_{\{\mu_i\}} \exp\left(-\frac{E_p'}{kT}\right) \end{aligned}\right\} \tag{4.1.19}$$

$$E_p' = -\left[\sum_{\langle ij\rangle} \varepsilon_{ij}\sigma_i\sigma_j/2 + (\varepsilon_0 + \mu_p + kT\ln j_p)\sum_i \sigma_i/2\right], \tag{4.1.20}$$

where Ξ_p is related to the pressure P by

$$\Xi_p = \exp\left(\frac{PN}{kT}\right), \tag{4.1.21}$$

where the volume per lattice point is assumed to be 1. The average number n of atoms in a lattice gas is given by

$$kT\,\partial\ln\Xi_p'/\partial\mu_p = \left\langle \sum_i \sigma_i \right\rangle \Big/ 2 = n - N/2 . \tag{4.1.22}$$

On the other hand, Ξ_s of the Ising system (4.1.14) gives the Gibbs free energy G_s of the system in the presence of the magnetic field H,

$$\left.\begin{aligned} G_s &= E_s - TS - HM = \mu_s N \\ \Xi_s &= \exp\left(-\frac{G_s}{kT}\right) = \exp\left(-\frac{\mu_s N}{kT}\right) \end{aligned}\right\}, \tag{4.1.23}$$

Table 4.1. Correspondence between the Ising system and the lattice gas

	Ising system	Lattice gas
Interaction energy	J_{ij}	ε_{ij}
Magnetization and number of atoms	M/m	$2n - N$
Magnetic field and chemical potential	mH	$(\varepsilon_0 + \mu_p + kT\ln j_p)/2$
Chemical potential and pressure	$-\mu_s$	$P - (\varepsilon_0/2 + \mu_p + kT\ln j_p)/2$
	$mH - \mu_s$	$P + \varepsilon_0/4$

where μ_s is the chemical potential and the magnetization M is given by

$$kT\partial \ln \Xi_s/\partial H = m\left\langle \sum_i \sigma_i \right\rangle = M . \quad (4.1.24)$$

The results above suggest a correspondence between the Ising system and the lattice gas through Ξ_s and Ξ'_p.

In Table 4.1, the first and the third lines come from the comparison between (4.1.13, 20), the second between (4.1.24) and (4.1.22) and the fourth from (4.1.23) and (4.1.19–21). Consequently, the derivation of the equation of state of a lattice requires the Ising system in the presence of a magnetic field, and vice versa.

Below the Curie point we have a spontaneous magnetization whose value per spin divided by m, $I = M/mN$, then has a finite value I_0 when we let $H \to 0+$, or $-I_0$ for $H \to 0-$. The state $H = 0$ is thus singular. The corresponding state in the lattice gas is given by $y_p = 1$, as seen from Table 4.1 and (4.1.28), and gives rise to the coexistence of the liquid and the gas phases with densities

$$\left(\frac{n}{N}\right)_l = \frac{1}{2}(1 + I_0) \quad \text{and} \quad \left(\frac{n}{N}\right)_g = \frac{1}{2}(1 - I_0) ,$$

respectively. In the two-dimensional Ising system on a square lattice with interaction J between nearest-neighbor spins, *Yang* [4.15] showed that $I_0 = (1 - 1/\sinh^4 2L)^{1/8}$ [$L = J/2kT$, see (4.4.58)] and through this relation, the densities of the coexisting liquid and gas phases are determined as a function of L. The pressure P is given by this relation together with *Onsager's* solution

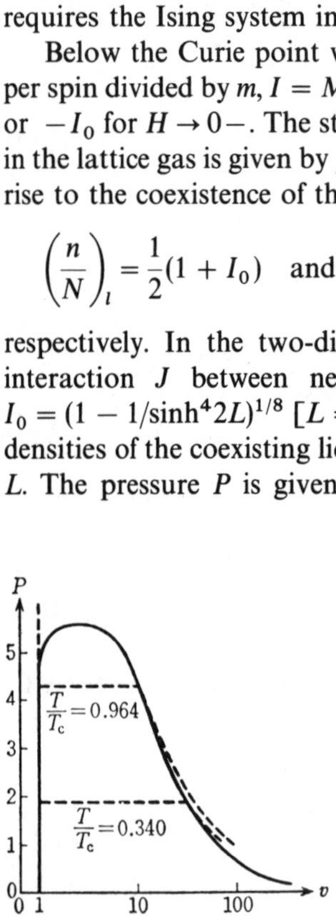

Fig. 4.3. Pressure vs. specific volume in a lattice gas [4.15]

(4.4.49) [4.16]

$$P/kT = -2L + \ln(2\cosh 2L) + \frac{1}{2\pi}\int_0^{\pi} \ln\frac{1}{2}\left(1 + \sqrt{1 - \kappa_1^2 \sin^2\phi}\right) d\phi\,,$$

where

$$\kappa_1 = 2\sinh 2L/\cosh^2 2L\,. \tag{4.1.25}$$

The coexistence curve is shown in Fig. 4.3, where T_c is the critical temperature. The isotherms in single phases have not yet been obtained exactly, but approximate ones are shown in broken lines.

4.1.4 Symmetric Properties in Lattice Gases

Let the fugacity be ξ, defined by

$$\exp[\beta(\mu_p + kT\ln j_p)] = \xi\,. \tag{4.1.26}$$

The grand partition function Ξ'_p defined by (4.1.19) is written

$$\Xi'_p(y_p) = \sum_{\{\sigma_i\}=\pm 1} \exp\left[(\beta/2)\sum_{\langle ij\rangle} \varepsilon_{ij}\sigma_i\sigma_j\right] y_p^{\Sigma\sigma_i} \tag{4.1.27}$$

$$y_p = \exp\left(\frac{\beta}{2}\varepsilon_0\right)\xi^{1/2}\,. \tag{4.1.28}$$

Since (4.1.27) is to be invariant when the signs of all σ_i are changed ($\sigma_i \to -\sigma_i$), we have

$$\Xi'_p(y_p) = \Xi'_p\left(\frac{1}{y_p}\right)\,. \tag{4.1.29}$$

Furthermore, we have from (4.1.22)

$$2\left(n - \frac{N}{2}\right) = \left\langle \sum_i \sigma_i \right\rangle = 2kT\frac{\partial \ln \Xi'_p}{\partial \mu_p} = \frac{\partial \ln \Xi'_p}{\partial \ln y_p} \tag{4.1.30}$$

and by differentiating this again

$$\begin{aligned} 2\partial(n - N/2)/\partial \ln y_p &= (\Xi'_p)^{-1}\partial^2\Xi'_p/\partial \ln y_p^2 - [(\Xi'_p)^{-1}\partial\Xi'_p/\partial \ln y_p]^2 \\ &= \langle(\textstyle\sum\sigma_i)^2\rangle - (\langle\textstyle\sum\sigma_i\rangle)^2 \\ &= \langle(\textstyle\sum\sigma_i - \langle\textstyle\sum\sigma_i\rangle)^2\rangle \end{aligned} \tag{4.1.31}$$

which is always positive, where use is made of the relation

$$\frac{1}{\Xi'_p}\frac{\partial^2\Xi'_p}{\partial \ln y_p^2} = \frac{1}{\Xi'_p}\sum(\textstyle\sum\sigma_i)^2 \exp\left(-\frac{E'_p}{kT}\right) = \langle(\textstyle\sum\sigma_i)^2\rangle\,. \tag{4.1.32}$$

These relations lead to the following statements: (i) Ξ'_p and $n - N/2$ are even and odd functions of $\ln y_p$, respectively; (ii) Eq. (4.1.31) shows that $n - N/2$ is an

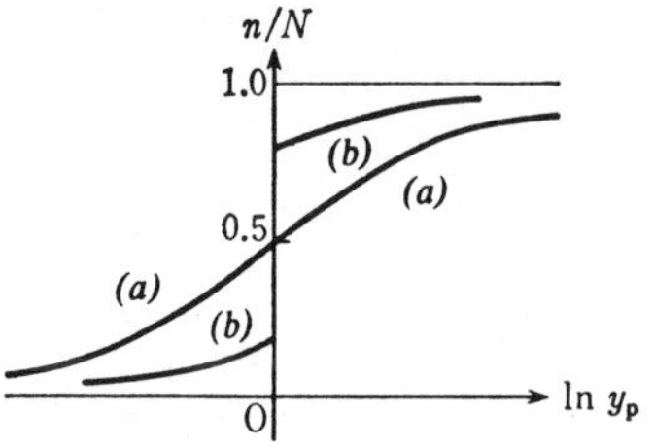

Fig. 4.4. n/N and $\ln y_\mathrm{p}$ in a lattice gas

increasing function of $\ln y_\mathrm{p}$, and thus we have the relation between n/N and $\ln y_\mathrm{p}$ shown in Fig. 4.4. The curve (a) is obtained when we assume that n/N is analytic in the region $(-\infty, \infty)$ of $\ln y_\mathrm{p}$. If a singular point is to be expected to exist, we have curve (b), because the singular point should occur at $\ln y_\mathrm{p} = 0$ and this is in accordance with *Lee-Yang's theorem* to be discussed later [4.17]. At sufficiently high temperatures, from (4.1.27) we have

$$\Xi'_\mathrm{p}(y_\mathrm{p}) = \left(y_\mathrm{p} + \frac{1}{y_\mathrm{p}}\right)^N$$

which is just the expression for free spin systems and n/N is analytic for $y_\mathrm{p} \neq 0, \infty$.

The Ising system for $H = 0$ has a special symmetry. In particular, we assume that the interaction exists between nearest neighbors only and further, that the total lattice is composed of two sublattices α and β whose lattice points are mutually nearest neighbors. Then the partition function (4.1.14) of the Ising system in the case $H = 0$ is invariant under the reversal of all the signs of J_{ij}'s, because the energy (4.1.13) is invariant under the reversal of both $J_{ij} \to -J_{ij}$ and $\sigma_i \to -\sigma_i$ for either the α- or β-lattice. Consequently, the partition function for a ferromagnet with $J_{ij} > 0$ and for an antiferromagnet with $J_{ij} < 0$ turns out to be the same. Therefore, the behavior of specific heat with respect to temperature is the same for both cases, and if some singularity is found in one of the two, then the other substance has a singularity at the same temperature.

Let N_+^α, N_-^α be the numbers of spins of $+1$ and -1, respectively, on the α-lattice and similarly, N_+^β and N_-^β are defined on the β-lattice, then the magnetization in the ferromagnet is proportional to

$$\langle M\rangle/m = \left(\sum_i \sigma_i\right) = (N_+^\alpha + N_+^\beta) - (N_-^\alpha + N_-^\beta)\,, \tag{4.1.33}$$

but the corresponding quantity in the antiferromagnet is

$$(-N_+^\alpha + N_+^\beta) - (-N_-^\alpha + N_-^\beta) = (N_+^\beta + N_-^\alpha) - (N_+^\alpha + N_-^\beta)\,. \tag{4.1.34}$$

Equations (4.1.33, 34) have the same functional form. The ordered state in the antiferromagnet of the lowest energy is such that the spins on the α-lattice are all -1, and the spins on the β-lattices are all $+1$, and the long-range order in

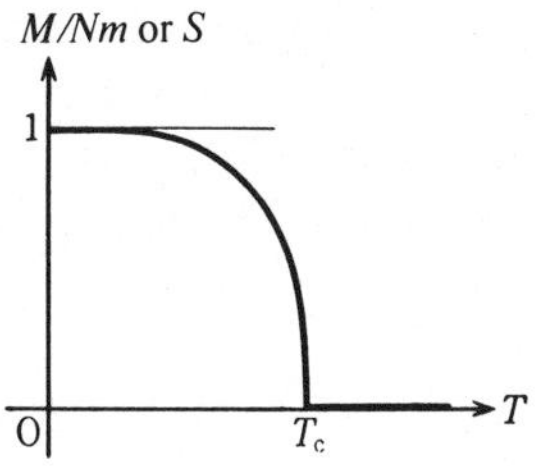

Fig. 4.5. Spontaneous magnetization in the ferromagnet and long-range order in the antiferromagnet

the antiferromagnet is defined by

$$S = \frac{1}{N}\langle (N_+^\beta + N_-^\alpha) - (N_+^\alpha + N_-^\beta)\rangle \; .$$

Thus, $\langle M\rangle/Nm$ and S are expressed by the same curve as shown in Fig. 4.5. T_c is the Curie temperature in the ferromagnet and the Néel temperature in the antiferromagnet.

4.2 Analyticity of the Partition Function and Thermodynamic Limit

4.2.1 Thermodynamic Limit

The system in thermal equilibrium is usually described by a small number of thermodynamic variables. One of the properties shared by the extensive thermodynamic variables such as volume, entropy, free energy, etc., is that they are proportional to the number of elements (molecules, spins, etc.) when the intensive variables such as temperature and pressure are kept constant. Furthermore, thermodynamic variables are stepwise analytic and at the point of a phase transition, some of the thermodynamic quantities change more or less abruptly. We can say that at the phase transition, thermodynamic quantities lose analyticity with respect to state variables such as temperature and density, etc.

On the other hand, statistical mechanics introduces many kinds of ensembles, as described in Chap. 1, and we can derive through any of the ensembles relevant thermodynamic quantities which should be identical irrespective of the ensembles adopted.

These facts suggest that some restrictions exist on the nature of the interactions between elements and that the system should be made sufficiently large, i.e., $N \to \infty$, $V \to \infty$ on keeping the density N/V constant. This limit is called the *thermodynamic limit.*

The necessity of taking the thermodynamic limit is two-fold. One is to guarantee homogeneity of the system, or in other words, to extinguish the effects of the boundary. The second is to obtain the singularity in the partition function or in the thermodynamic quantities which gives rise to the phase transition.

To see the latter more concretely, consider an Ising system which has the energy $-(J/2)\,\sigma_i\sigma_j$ between two nearest-neighbor spins i and j. The partition function Ξ_s of this system with N spins is given by

$$\Xi_s = \sum_{\sigma_1=\pm 1} \cdots \sum_{\sigma_N=\pm 1} \exp\left[\frac{J}{2kT}\left(\sum_{i,j}\sigma_i\sigma_j\right)\right] . \tag{4.2.1}$$

This function, when considered as a function of T, is analytic except at $T=0$, because this is a sum of finite terms of exponential functions which are analytic except at $T=0$. This means that as long as N is finite, no singularity or no phase transition can exist. In real systems, we always deal with finite systems and thus the point of phase transition is not a singular point in a mathematically strict sense. However, the unusual behavior found at the so-called phase transition in real systems should be understood as being due to the singularity of this system in the thermodynamic limit. This interpretation of the phase transition occurring in real systems is reasonable, because our recognition is never exact.

The homogeneity of the system requires restrictions on the interaction energy between elements (molecules). In the bulk of a material far from the surface, the effect of the surface must vanish. To see this, we group the molecules of the system into I and II, and then the energy of the total system can be written as

$$U(\mathrm{I}, \mathrm{II}) = U(\mathrm{I}) + U(\mathrm{II}) + W , \tag{4.2.2}$$

where W is the interaction energy of I and II. Let r be the smallest distance between two molecules, one from I and another from II (Fig. 4.6); then the condition

$$|W| \leqq A n_1 n_2 r^{-q} \tag{4.2.3}$$

for large r is sufficient to suppress surface effects, where n_1 and n_2 are the numbers of molecules in I and II, respectively, $A \geqq 0$ and q is larger than the dimensionality ν of this system ($q > \nu$).

Furthermore, to avoid an accumulation of molecules in a small volume of space, the potential energy $U(n)$ of a system of n molecules must satisfy

$$U(n) \geqq -nB \tag{4.2.4}$$

with $B \geqq 0$. If a lower limit of the interaction energy exists, the condition (4.2.4) is satisfied. This is sufficient to make the grand partition function converge. To

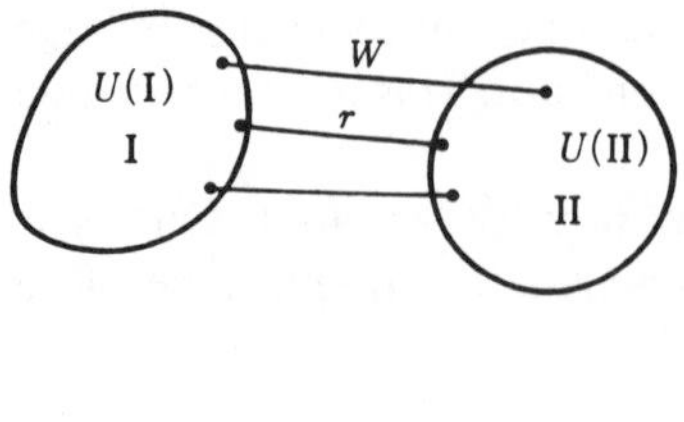

Fig. 4.6. The interaction energy W and the shortest distance r between two groups of molecules

see this, consider a system of n monatomic molecules of one species, then the partition function in a canonical ensemble is given by

$$Z = \frac{1}{n!}\left(\frac{2\pi mkT}{h^2}\right)^{3n/2} \int_V \exp[-\beta U(n)]\, d\tau_n \, , \tag{4.2.5}$$

where

$$d\tau_n = dx_1 dy_1 dz_1 \cdots dx_n dy_n dz_n$$

and the integrations are taken with respect to the coordinates of n molecules over volume V; the grand partition function is

$$\Xi = 1 + \sum_{n=1}^{\infty} \frac{\xi^n}{n!} \int \exp[-\beta U(n)]\, d\tau_n \, , \tag{4.2.6}$$

where ξ is the fugacity

$$\xi = \left(\frac{2\pi mkT}{h^2}\right)^{3/2} \lambda \, , \tag{4.2.7}$$

where λ is the absolute activity. By virtue of the inequality (4.2.4) we have

$$\Xi \leqq 1 + \sum \frac{\xi^n}{n!} V^n \exp(n\beta B) = \exp[\xi V \exp(\beta B)] \tag{4.2.8}$$

and this proves convergence.

The thermodynamic functions such as pressure, internal energy or entropy, etc., are obtained by taking the thermodynamic limit. For example, pressure is given in a canonical ensemble by

$$\frac{P}{kT} = \lim \frac{1}{V} \frac{\partial \ln Z}{\partial V} \tag{4.2.9}$$

and in a grand canonical ensemble by

$$\frac{P}{kT} = \lim \frac{1}{V} \ln \Xi \, . \tag{4.2.10}$$

One can show the existence of these thermodynamic limits, and the thermodynamic quantities thus derived satisfy the thermodynamic condition, mentioned at the beginning of this section, provided that the molecular interaction has the properties discussed above. Furthermore, these thermodynamic quantities can also be shown as being thermodynamically stable. Thus, for example, pressure is a nonincreasing function of volume at a constant temperature. Thermodynamic functions derived by means of various statistical ensembles are identical, and thus the pressures given by (4.2.9, 10) are the same. We do not enter into mathematical details of proving this identity, but this is confirmed in the thermodynamic limit. Thus, in an essentially finite system, different results are obtained from different ensembles. For example, the force K acting between the ends of a flexible rubber-like molecule is given by

$$\frac{x}{nb} = L\left(\frac{Kb}{kT}\right), \tag{4.2.11}$$

where L is the Langevin function $L(y) = \coth(y - y^{-1})$ and x is the end-to-end distance of the polymer molecule which is composed of n-bonds of length b connected successively and randomly. Equation (4.2.11) can be derived by making use of the constant force ensemble [corresponding to the constant pressure ensemble (T–P ensemble) in gases]. If one uses the constant end-to-end distance ensemble [corresponding to a T–V ensemble in gases], the result is different from (4.2.11) and reduces to (4.2.11) only when $n \to \infty$.

4.2.2 Cluster Expansion

In Sect. 3.3.4, the cluster expansion method of imperfect gases was discussed. Equation (3.3.28) is an expansion in powers of the fugacity ξ,

$$\frac{P}{kT} = \sum_{l=1}^{\infty} b_l \xi^l, \tag{4.2.12}$$

where b_l is the cluster integral. b_l is considered as being a function determined by the intermolecular potential and temperature. This is true when we let $V \to \infty$, keeping l constant. The infinite power series (4.2.12) is then obtained on letting $N \to \infty$. *Mayer* [4.18] examined the condensation phenomenon by studying the singularity of the function continued analytically from the power series (4.2.12). However, the limit to make first $V \to \infty$ and then $N \to \infty$ is not necessarily identical to the thermodynamic limit. It is now believed that the singularity of (4.2.12) is not the condensation point but the limit of supersaturation.

4.2.3 Zeros of the Grand Partition Function

Again consider the grand partition function (4.2.6). Let the integration over the coordinate space be Q_n. If the intermolecular potential has a hard core, there is an upper limit $N(V)$ in the number of molecules included in a finite volume V. Thus, (4.2.6) is essentially a polynomial

$$\Xi(V) = 1 + \sum_{n=1}^{N(V)} \frac{\xi^n}{n!} Q_n \tag{4.2.13}$$

which has $N(V)$ roots ξ_i [$i = 1, 2, \ldots, N(V)$] in the complex region of ξ and, therefore, can be written as

$$\Xi(V) = \prod_{i=1}^{N(V)} \left(1 - \frac{\xi}{\xi_i}\right). \tag{4.2.14}$$

Pressure is thus given by

$$\frac{P}{kT} = \lim_{V\to\infty} \frac{1}{V} \sum_{i=1}^{N(V)} \ln\left(1 - \frac{\xi}{\xi_i}\right). \tag{4.2.15}$$

As long as V is finite, ξ_i are not positive real, because the coefficients of ξ^n in (4.2.13) are always positive. Consequently, P has no singularity on the positive real axis of ξ. This is in accordance with what we have said in Sect. 4.2.1. On letting $V \to \infty$, the roots ξ_i will be distributed continuously and its limiting distribution will yield non-zero density at ξ_0 on the positive real axis. ξ_0 thus obtained is a singular point which is identified as a condensation point. This is the idea of *Yang* and *Lee* [4.19]. Whether this limiting distribution exists in real systems or not is unknown. In lattice gases, however, *Lee* and *Yang* [4.17] proved that when ε_{ij} in (4.1.7) is positive, or in magnetism when ferromagnetic interaction is concerned, the zeros of the grand partition function lie on a circle $|y^{\mathrm{P}}| = 1$, irrespective of the magnitude of N. The number of zeros increases with N, and if its limiting distribution becomes continuous on an arc intersecting the positive real axis, this intersection gives the phase transition. This is what is surmised in Sect. 4.1.4. In the anisotropic ferromagnetic Heisenberg model, it is proved that the zeros lie on a circle $|y^{\mathrm{P}}| = 1$, just like in the Ising model [4.20].

The distribution of zeros in antiferromagnetic interaction is unknown as a general theorem, but for finite N, this distribution can be studied on a computer. According to the numerical results, the distribution is not simple and varies from one case to another.

Now let us go back to (4.2.14, 15) and put $\xi_k = r_k \exp(\mathrm{i}\theta_k)$ (Fig. 4.7). Taking ξ as positive real we have

$$\ln\left(1 - \frac{\xi}{\xi_k}\right) = \ln \frac{(\xi^2 - 2\xi r_k \cos\theta_k + r_k^2)^{1/2}}{r_k} + \text{imaginary part}\,. \tag{4.2.16}$$

The distance between P and P_k in Fig. 4.7 is equal to $(\xi^2 - 2\xi r_k \cos\theta_k + r_k^2)^{1/2}$; the real part of (4.2.16) is just the logarithmic potential at P with charge 1/2 at P_k (potential at P from the charge distribution of line density 1/2 on a straight line through P_k perpendicular to the figure). According to this interpretation, the pressure P at ξ is proportional to the logarithmic potential from the charge 1/2 at all the zeros on the complex plane. If, in the thermodynamic limit, these charges lie continuously on a line (Fig. 4.8) which intersects the real axis at $\xi_0 > 0$, then the potential itself changes continuously on crossing ξ_0 from the left

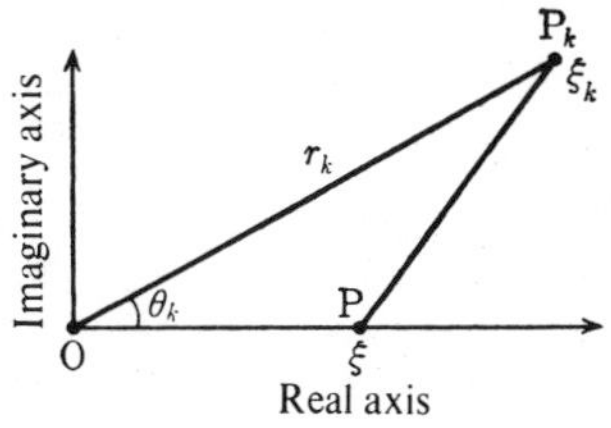

Fig. 4.7. Points P for real ξ and P_k for $\xi_k = r_k \exp(\mathrm{i}\theta_k)$

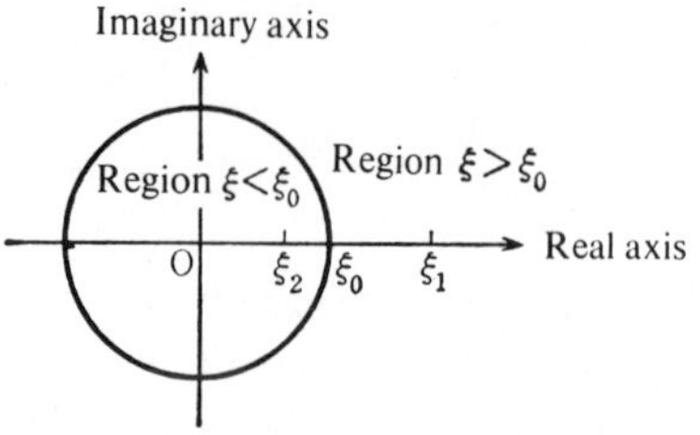

Fig. 4.8. Singular points $\xi_2(<\xi_0)$ and $\xi_1(>\xi_0)$ on the real axis

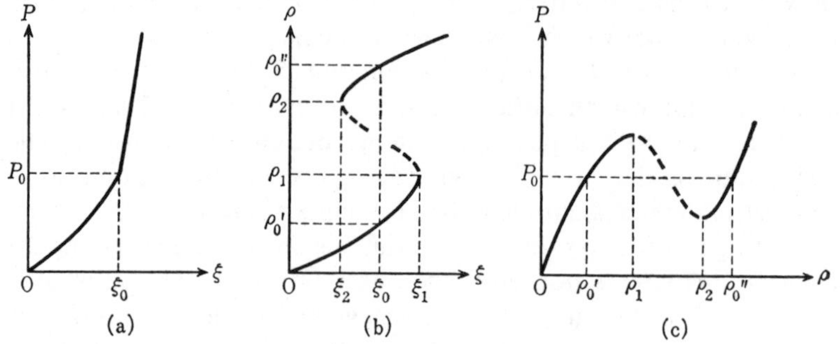

Fig. 4.9. Relations between P and ξ (**a**), ρ and ξ (**b**), and P and ρ (**c**)

(gas) to the right (condensed state), but the electric field changes abruptly. The electric field is (minus) the gradient of the potential and is interpreted as proportional to the density

$$\rho = \xi \frac{\partial}{\partial \xi}\left(\frac{P}{kT}\right).$$

Consequently, the density changes discontinuously at ξ_0, which is the manifestation of a phase transition.

Even if the roots ξ_i are distributed continuously on a line, this line is not necessarily the natural boundary of the function defined by the rhs of (4.2.15), because analytical continuation is sometimes possible over ξ_0 from left or from right. The singular point ξ_1 (ξ_2) nearest ξ_0 on the real axis of the function analytically continued from the region $\xi < \xi_0$ ($\xi > \xi_0$) is then the limit of the metastable region (Figs. 4.8, 9).

When the magnetic field is absent, we cannot apply this argument of the distribution of zeros in the complex y-plane. In this case, the complex plane of inverse temperature is useful, as will be discussed in Sect. 4.4.3.

4.3 One-Dimensional Systems

There are some exactly soluble models in one dimension [4.21].

4.3.1 A System with Nearest-Neighbor Interaction

Molecules numbered 1, 2, . . . , N, are arranged on a straight line consecutively and interaction forces act between two neighboring molecules.[1]

After performing integration in momentum space, the partition function of this system becomes

$$Z(T, L, N) = \left(\frac{2\pi mkT}{h^2}\right)^{N/2} \int \cdots \int \exp\left[-\beta \sum_{1 \leqq i \leqq N-1} u(x_{i+1} - x_i)\right] dx_1 \cdots dx_N \,, \tag{4.3.1}$$

where m is the mass of a molecule, u is the potential of the intermolecular force, L is the length of the interval where N molecules are confined, and the integrations are carried out while keeping the relations $0 \leqq x_1 \leqq x_2 \leqq \cdots \leqq x_N \leqq L$.

In a T–P ensemble, the partition function is

$$\begin{aligned} Y(T, P, N) &= \int_0^\infty Z(T, L, N) \exp(-\beta PL)\, dL \\ &= \left(\frac{2\pi mkT}{h^2}\right)^{N/2} \int \cdots \int \exp\{-\beta P[x_1 + (x_2 - x_1) + \cdots \\ &\quad + (L - x_N)]\} \exp\{-\beta[u(x_2 - x_1) + u(x_3 - x_2) + \cdots \\ &\quad + u(x_N - x_{N-1})]\} dx_1 \cdots dx_N dL \\ &= \left(\frac{2\pi mkT}{h^2}\right)^{N/2} \frac{1}{(\beta P)^2} \left[\int_0^\infty \exp\{-\beta[u(x) + P(x)\, dx]\}\right]^{N-1} . \end{aligned} \tag{4.3.2}$$

The relationship between P and L is obtained from (2.5.4), when L is sufficiently large, as

$$L = -kT \frac{\partial \ln Y(P, T, N)}{\partial P} = -NkT \frac{\partial}{\partial P} \ln F(P) \tag{4.3.3}$$

$$F(P) = \int_0^\infty \exp\{-\beta[u(x) + P(x)]\}\, dx \,. \tag{4.3.4}$$

To proceed further, we have to know the functional form of $u(x)$. If $u(x) > 0$ for small x, and $u(x) \to 0$ sufficiently rapidly for large x, then for $P > 0$ the integral $F(P)$ exists and is a regular and decreasing function of P. Therefore, no singularity in the P–L relation exists and no transition occurs.

---

[1] If the intermolecular force is long range, it is unnatural to assume that the force is acting on neighboring molecules only, even though the integral (4.3.4) converges. This assumption is valid in the case when the molecule has a hard core of diameter δ, and $u(x) = 0$ for $x \geqq 2\delta$.

4.3.2 Lattice Gases

When the molecules are confined on lattice points we are dealing with a lattice gas. As explained in Sect. 4.1.3, a lattice gas is equivalent to an Ising model. The Ising models with nearest-neighbor interaction only will be discussed separately in Sect. 4.4, and it will be shown that in a one-dimensional system no phase transition occurs. Therefore, this is also the case for one-dimensional lattice gases with nearest-neighbor interaction.

Next, we consider a system with interactions among finite numbers of neighbors. For simplicity, consider a system with interactions up to third neighbors (Fig. 4.10), then the energy can be written as

$$E = u(p_1, p_2, p_3, p_4) + u(p_2, p_3, p_4, p_5) + u(p_3, p_4, p_5, p_6) + \cdots, \tag{4.3.5}$$

where $p_i = 1$ or 0, according to whether the ith site is occupied by a molecule or not. And $u(p_i, p_{i+1}, p_{i+2}, p_{i+3})$ is the energy attributed to the molecule (if any) on the ith site by the interactions from molecules on $(i+1)$, $(i+2)$ and $(i+3)$th sites. We introduce a grand partition function and use the matrix method to be described in Sect. 4.4.2 by defining the elements of the matrix as

$$\exp[-\beta u(p_i, p_{i+1}, p_{i+2}, p_{i+3})]\,\xi^{p_i} = w(p_i, p_{i+1}, p_{i+2} \,|\, p_{i+1}, p_{i+2}, p_{i+3}) ; \tag{4.3.6}$$

the number of states of three consecutive sites is $2^3 = 8$ and they are numbered as $k = 1, 2, 3, \ldots, 8$, as follows

$$k = 1{:}000,\ 2{:}001,\ 3{:}010,\ 4{:}011,\ 5{:}100,\ 6{:}101,\ 7{:}110,\ 8{:}111 . \tag{4.3.7}$$

Then we have a matrix W of the form

$$W = \begin{bmatrix} \times & \times & 0 & 0 & 0 & 0 & 0 & 0 \\ 0 & 0 & \times & \times & 0 & 0 & 0 & 0 \\ 0 & 0 & 0 & 0 & \times & \times & 0 & 0 \\ 0 & 0 & 0 & 0 & 0 & 0 & \times & \times \\ \times & \times & 0 & 0 & 0 & 0 & 0 & 0 \\ 0 & 0 & \times & \times & 0 & 0 & 0 & 0 \\ 0 & 0 & 0 & 0 & \times & \times & 0 & 0 \\ 0 & 0 & 0 & 0 & 0 & 0 & \times & \times \end{bmatrix}, \tag{4.3.8}$$

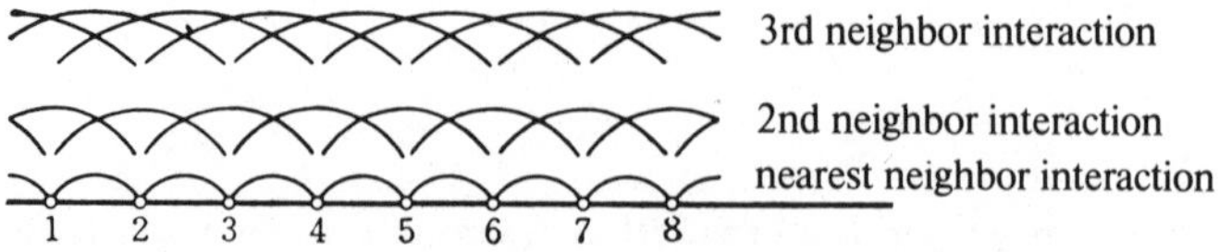

Fig. 4.10. Long-range interactions between atoms

where × stands for a matrix element which is not equal to 0. Some of the matrix elements are zero [for example, the (2, 1) element $w(2\,|\,1) \equiv w(001|000)$], because in the case of $w(2|1)$, $p_{i+2}(=1)$ in the state 2 is different from $p_{i+2}(=0)$ in the state 1. On identifying the $(N+1)$th site with the 1st site, the grand partition function Ξ is given by

$$\Xi = \operatorname{tr} W^N = \sum_{i=1}^{8} \lambda_i^N , \tag{4.3.9}$$

where λ_i $(i = 1, 2, \ldots, 8)$ are the eigenvalues of W. If the eigenvalue of the largest absolute value (in short, largest eigenvalue) (say λ_1) is positive and does not become degenerate even on changing $\beta = 1/kT$, we have

$$\frac{P}{kT} = \lim_{N\to\infty} \frac{1}{N} \ln \Xi = \ln \lambda_1 . \tag{4.3.10}$$

Since λ_1 is an analytic function of β, no singularity occurs as a function of temperature[2] and N(length). Positivity and nondegeneracy of the largest eigenvalue are guaranteed by the theorem of Perron, according to which the largest eigenvalue of a matrix whose elements are all positive is not degenerate and positive. Further, the eigenvector of the largest eigenvalue has positive components. In the case of W given by (4.3.8), it is shown that all the matrix elements of W^4 are positive. Thus we can apply this theorem to W^4 and Perron's theorem holds also for W. When we obtain a matrix of all positive elements through several time multiplications of the matrix, we can say that there is always direct or indirect connectivity of any two states by intercalating an appropriate number of states in between. The matrix is then irreducible. Perron's theorem still holds for an irreducible matrix with nonnegative elements (theorem of Frobenius).

4.3.3 Long-Range Interactions

In a one-dimensional system, no phase transition can be expected if the interaction is short-ranged, as has been mentioned above, but an infinitely long-range interaction can give rise to the phase transition. To see this, consider a system with interaction $J(n)$ or $\varepsilon(n)$ between two spins or particles of distance n in the unit of the lattice constant.

When $\varepsilon(n)$ is constant (say ε) irrespective of n in a lattice system, the total energy E_p is given by (4.1.6):

$$E_p = -2\varepsilon \sum_{\langle ij \rangle} p_i p_j = -\varepsilon N(N-1) \simeq -\varepsilon N^2 \tag{4.3.11}$$

---

[2] If the largest eigenvalues are degenerate at β_c $(\lambda_1 = \lambda_2)$, and on one side of β (say $\beta < \beta_c$) $\lambda_1 > \lambda_2$ and on the other side $(\beta > \beta_c)$ $\lambda_2 > \lambda_1$, then the thermodynamical state changes from λ_1 to λ_2 at β_c on increasing β. β_c is the point of phase transition.

which shows that the total energy is proportional to N^2 and does not satisfy the thermodynamic condition (extensive property). One can remedy this if ε is taken as ε/N. This model is equivalent to the Weiss approximation or the Bragg–Williams approximation to be discussed later (Sect. 4.5.1). The equation of state obtained has a part of positive slope ($\partial P/\partial v > 0$) in contradiction to thermodynamic stability. This is due to the assumption of a uniform distribution of particles over the volume, and thus can yield a phase transition irrespective of the dimensionality, if combined with Maxwell's rule. The interaction in this model implies that it is infinitely long-ranged with negligible magnitude.

Now examine the condition for the phase transition in more detail in connection with the n-dependence of $J(n)$ for large n. Assume $J > 0$ (4.1.5) and put

$$M_0 = \sum_{n=1}^{\infty} J(n) . \qquad (4.3.12)$$

If M_0 becomes infinite, the difference between the lowest (all $\mu_i = 1$) and the second lowest (one μ_i only is inverted) levels are infinitely large. The system is always in the lowest level and does not undergo any phase transition. M_0 must be finite for the occurrence of phase transition. Furthermore, *Ruelle* [4.22] showed that if

$$M_1 = \sum_{n=1}^{\infty} nJ(n) \qquad (4.3.13)$$

is finite, the system is always in a disordered state and undergoes no phase transition. This can be understood as follows. The boundary of two infinitely long ordered states $+ + + \cdots$ and $- - - \cdots$ has the finite energy M_1. Therefore, the boundaries of two ordered states of finite lengths $+ + + \cdots +$ and $- - - \cdots -$ have an energy less than M_1. This means that this system cannot be in an ordered state, since it will have many boundaries of this kind at finite temperatures. *Dyson* [4.23] proved that if $J(n)$ ($\geqq 0$) decreases monotonically with increasing n, then M_0 as well as K defined by

$$K = \sum_{n=1}^{\infty} [\ln\ln(n+4)][n^3 J(n)]^{-1} \qquad (4.3.14)$$

remain finite and a phase transition occurs at a finite temperature. Furthermore, even in the case where M_1 diverges, no phase transition occurs provided that

$$(\ln\ln N)^{-1} \sum_{n=1}^{N} nJ(n) \to 0 \quad \text{for } N \to \infty . \qquad (4.3.15)$$

Baker [4.24] considered a system with an interaction

$$J(n) = \alpha \exp(-\gamma n) , \qquad (4.3.16)$$

where M_0 and M_1 are finite for positive γ ($\gamma > 0$). If we put $\alpha = \alpha_0 \gamma$ and let $\gamma \to 0$, then $M_0 = \alpha$, but M_1 becomes infinite. Consequently, a phase transition can take place. Almost the same model was considered by *Kac* et al. [4.25]. The

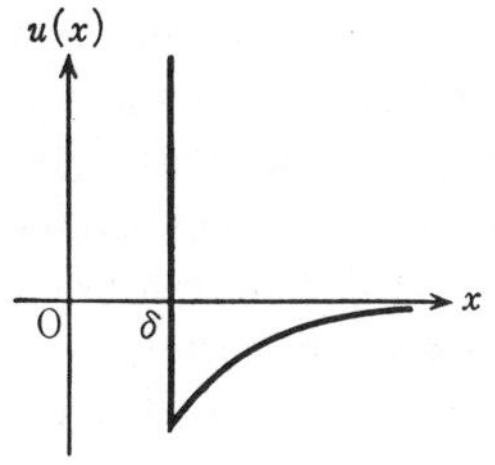

Fig. 4.11. Intermolecular force with a hard core and a long range attraction

interaction has a hard core of diameter δ and an exponentially decreasing potential with increasing distance outside this hard sphere (Fig. 4.11), i.e.,

$$u(x) = \begin{cases} \infty, & x < \delta \\ -\alpha \exp(-\gamma x), & x \geqq \delta \,. \end{cases} \tag{4.3.17}$$

By making use of the fact that a Gaussian–Markovian process has a correlation function of the type (4.3.16), an exact equation of state of this system can be obtained and by putting $\alpha = \alpha_0 \gamma$ and letting $\gamma \to 0$, the van der Waals equation of state is derived

$$P = \frac{kT}{l - \delta} - \frac{\alpha_0}{l^2} \tag{4.3.18}$$

together with Maxwell's rule. In other words, in the limit of an infinitely long-range interaction of exponential type and infinitely small magnitude, Van der Waals type equation of state results in a one-dimensional system.

It would be appropriate here in the section devoted to one-dimensional systems to mention the derivation of van der Waal's equation in a higher dimension. *van Kampen* [4.26] developed an approximate derivation of van der Waal's equation in three-dimensional systems by dividing the volume into a large number of cells which are small compared with the range of long-range attractive forces but are large enough to contain many particles. The distribution of particles over the cells is not assumed to be uniform but is determined by minimizing the free energy. This nonuniform distribution is different in the Bragg-Williams model mentioned in connection with (4.3.11), and can yield the van der Waals loop together with Maxwell's rule. By combining this idea with the *Kac*, *Uhlenbeck* and *Hemmer* one-dimensional treatment, *Lebowitz* and *Penrose* [4.27] were able to rigorously derive a van der Waals equation in a higher dimensional system.

4.3.4 Other Models

The one-dimensional XY model can be treated exactly, but we do not show the details here. Other important one-dimensional systems are found in biological macromolecules [4.28]. The helix-coil transitions encountered in DNA (deoxyribonucleic acid) and polypeptides can be regarded as applications of Ising systems. The models in polypeptides are equivalent to the lattice gas model with short-range interaction discussed in Sect. 4.3.2 and exhibit no singularity. Therefore the helix-coil transitions in polypeptides are diffuse and are not mathematically rigorous transitions. On the other hand, DNA models sometimes introduce a long-range interaction between two base pairs bonded through the random coil conformation. This gives rise to a phase transition in its rigorous meaning. The three-dimensional structures of globular proteins are also considered as the ordered structures acquired by the conformational phase transitions.

4.4 Ising Systems

4.4.1 Nearest-Neighbor Interaction

We assume that the interaction energy of an Ising system with interaction between nearest neighbors only is given by

$$\sum_{\langle ij\rangle} -\frac{J}{2}\mu_i\mu_j\,, \tag{4.4.1}$$

where the pairs $\langle ij\rangle$ are taken between nearest neighbors and J is assumed positive ($J>0$) in the following, unless stated otherwise, so that the lower energy state has parallel spins. The value of each spin is either $+1$ or -1 ($\mu_i = \pm 1$). The partition function in the absence of an external magnetic field is

$$Z = \sum_{\mu_1=\pm 1}\cdots\sum_{\mu_N=\pm 1}\prod_{\langle ij\rangle}\exp(L\mu_i\mu_j) \tag{4.4.2}$$

$$= \sum_{\mu_1=\pm 1}\cdots\sum_{\mu_N=\pm 1}\prod_{\langle ij\rangle}(\cosh L + \mu_i\mu_j \sinh L)\,, \tag{4.4.3}$$

where $L = J/2kT$ and use is made of the identity valid for $\mu\mu' = \pm 1$:

$$\exp(L\mu\mu') = \cosh L + \mu\mu' \sinh L\,. \tag{4.4.4}$$

a) *One-Dimensional Systems*

Number the spins from left to right as 1, 2, . . . , N. The partition function is then written as

$$Z = (\cosh L)^N \sum_{\mu_1=\pm 1}\cdots\sum_{\mu_N=\pm 1}\prod_{i=1}^{N}(1 + \mu_i\mu_{i+1}\tanh L) \tag{4.4.5}$$

which can be reduced to a closed form by expanding the product of the rhs and taking account of the fact that $\sum_{\mu_i=\pm 1}\mu_i = 0$ and $\mu_i^2 = 1$. When both ends are open, we may put $\mu_{N+1} = 0$ and the sum over $\mu_i = \pm 1$ of the products of any number of $\mu_i\mu_{i+1}$ is zero and hence, $Z = (2\cosh L)^N$ is obtained. For the periodic condition $\mu_{N+1} = \mu_1$, only the term $\mu_1\mu_2\cdot\mu_2\mu_3\cdot\cdots\cdot\mu_N\mu_1 (=1)$ is nonzero and hence $Z = (2\cosh L)^N + (2\sinh L)^N$. Since $\cosh L > |\sinh L| > 0$, in either case when N is large, we have

$$Z = (2\cosh L)^N \, . \tag{4.4.6}$$

This is a smooth function of L or of temperature. Thus no phase transition can be expected.

b) *Many-Dimensional Systems*

The partition function is

$$Z = (\cosh L)^s \sum_{\mu_1 = \pm 1} \cdots \sum_{\mu_N = \pm 1} \prod_{\langle ij \rangle} (1 + u\mu_i\mu_j) \, , \tag{4.4.7}$$

where

$$u = \tanh L \tag{4.4.8}$$

and s is the total number of nearest-neighbor pairs $\langle ij \rangle$ and is equal to $cN/2$ when the surface effect is ignored, c being the number of nearest neighbors of a lattice point. Expanding the rhs of (4.4.7) by writing $\mu_i\mu_j$ as $(\mu_i\mu_j)$ to avoid confusion, we have

$$Z = (\cosh L)^s \sum \cdots \sum \left[1 + u \sum_{\langle ij \rangle} (\mu_i\mu_j) + u^2 \sum_{\langle ij \rangle} \sum_{\langle kl \rangle} (\mu_i\mu_j)(\mu_k\mu_l) + \cdots \right] , \tag{4.4.9}$$

where the same pairs do not appear twice in the coefficient $(\mu_i\mu_j)(\mu_k\mu_l) \cdots$ of u^n. It is convenient to represent each term of the coefficient of u^n in (4.4.9) by a figure of bonds connecting the spin pairs $(\mu_i\mu_j)$. For example, any term in the coefficient of u^7 is represented by a figure with 7 bonds and Fig. 4.12 represents one of the terms, $(\mu_1\mu_2)(\mu_3\mu_4)(\mu_1\mu_5)(\mu_2\mu_6)(\mu_3\mu_7)(\mu_5\mu_6)(\mu_{10}\mu_{11})$, whose contribution to Z vanishes since $\sum_{\mu_i = \pm 1} \mu_i = 0$. In general, a term which contains an odd number of μ_i vanishes, but a term which is a product of even numbers of μ_i's does not, just as in Fig. 4.13, where the figure is composed of closed polygons that do not share sides. Since a figure of this sort makes a contribution $2^N u^n$ to

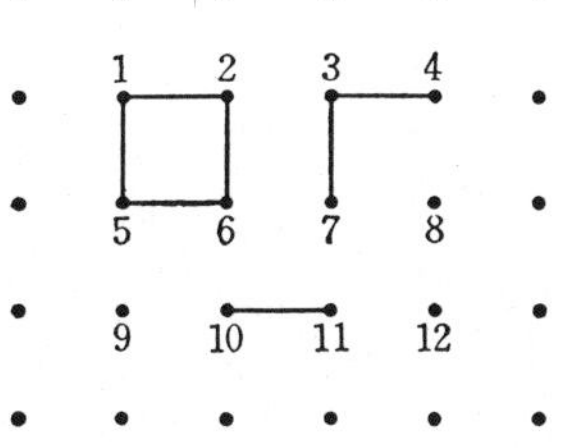

Fig. 4.12. An example of spin pairs whose contribution vanishes

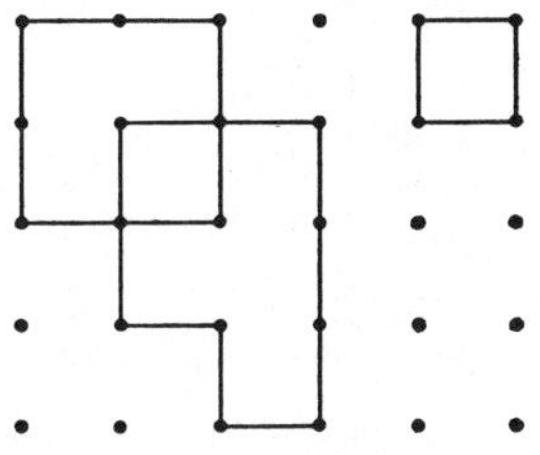

Fig. 4.13. An example of spin pairs whose contribution does not vanish

the partition function, we have

$$Z = 2^N(\cosh L)^s\left(1 + \sum_n \Omega_n u^n\right), \tag{4.4.10}$$

where Ω_n is the number of figures with n bonds composed exclusively of polygons. In the square lattice (2-dimensional) or in the simple cubic lattice (3-dimensional) n should be even. In a one-dimensional lattice, $\Omega_n = 0$ for all $n > 0$, with the exception of Ω_N under periodic conditions.

The partition function can be rewritten in another way. Let the number of antiparallel spin pairs ($\uparrow\downarrow$) be r and the number of parallel spin pairs be $s - r$. Then, when r is specified,

$$\sum_{\langle ij\rangle} \mu_i\mu_j = (s - r) - r = s - 2r\,. \tag{4.4.11}$$

Hence (4.4.2) is written

$$Z = 2\exp(sL)\left[1 + \sum_r \omega_r \exp(-2Lr)\right], \tag{4.4.12}$$

where ω_r is the number of configurations with r antiparallel spin pairs, and the coefficient 2 comes from the contribution by inverting all the spins. When $J > 0$, $\exp(-2Lr)$ becomes small for low temperatures, therefore (4.4.12) can be regarded as a low temperature expansion. On the other hand (4.4.10) is regarded as a high temperature expansion. In either case this expansion can be obtained, at least for lower terms, by examining Ω_n or ω_n in detail.

c) *Two-Dimensional Systems*

In a 2-dimensional lattice, let us put $\circ$ at the center of every cell as is done in Fig. 4.14. These $\circ$'s constitute a dual lattice. As is easily understood, the dual lattice of a square lattice is a square lattice of the same form, and the dual lattice of a honeycomb lattice is a triangular lattice and vice versa. Of course the original lattice and the dual lattice are reciprocal.

Put $\uparrow$ spins at the dual lattice point inside the polygons in the figures of n bonds considered above and put $\downarrow$ spins outside the polygons, so that the

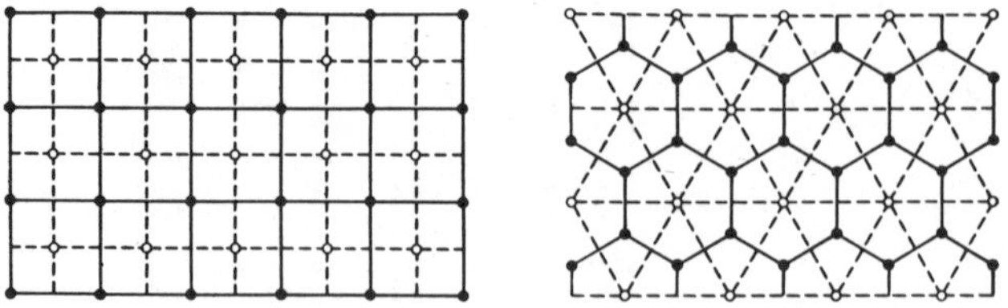

(a) (b)

Fig. 4.14 a, b. Dual lattices. **(a)** A square lattice $\leftrightarrow$ a square lattice; **(b)** a honeycomb lattice $\leftrightarrow$ a triangular lattice

connecting pairs of antiparallel spins on the dual lattice intersect the n bonds of the original lattice. In this way we have a 1:1 correspondence between the configurations on the two lattices. Since a figure of n bonds in the original lattice corresponds to a figure with n antiparallel spin pairs in the dual lattice, we have

$$\Omega_n = \omega_n^* , \tag{4.4.13}$$

where * means the dual lattice, and

$$\omega_n = \Omega_n^* \tag{4.4.14}$$

by virtue of the reciprocity of the two lattices. Obviously, we have

$$s^* = s . \tag{4.4.15}$$

Consequently, while the partition function of the original lattice is given by

$$Z(T) = 2^N (\cosh L)^s \left\{ 1 + \sum_n \Omega_n (\tanh L)^n \right\} , \tag{4.4.16}$$

the partition function of the dual lattice is

$$Z^*(T) = 2 \exp(sL) \left[1 + \sum_n \Omega_n \exp(-2Ln) \right] . \tag{4.4.17}$$

Now we restrict ourselves to the case $J > 0$ $(L > 0)$ and define L^* and a temperature T^* by

$$\exp(-2L^*) = \tanh L \tag{4.4.18}$$

and

$$T^* = J/2kL^* , \tag{4.4.19}$$

which are uniquely determined for $L > 0$. Then we have several symmetric relations such as

$$\left. \begin{array}{l} \exp(-2L) = \tanh L^* \\ \sinh 2L \sinh 2L^* = 1 \\ \tanh 2L^* \cosh 2L = \tanh 2L \cosh 2L^* = 1 \end{array} \right\} . \tag{4.4.20}$$

In terms of T^* we obtain

$$Z(T) = 2^{N-1} (\cosh L)^s \exp(-sL^*) Z^*(T^*) \tag{4.4.21}$$

which is rewritten in a form which becomes symmetric for large N

$$\frac{Z(T)}{2^{(N-1)/2} (\cosh 2L)^{s/2}} = \frac{Z^*(T^*)}{2^{(N^*+1)/2} (\cosh 2L^*)^{s/2}} , \tag{4.4.22}$$

where use is made of the Euler's theorem of polyhedra

$$N + N^* = s \tag{4.4.23}$$

valid for a periodic boundary condition, or on a torus.

d) *Curie Point*

Lower T corresponds to higher T^*. On raising the temperature, if the singularity of $Z(T)$, if any, is found at $T = T_c$, then the singularity of $Z^*(T^*)$ lies at $T^* = T_c^*$ corresponding to T_c.

Since a square lattice is self-dual, Z is equal to Z^*, and if one and only one singular point exists, then $T_c = T_c^*$ ($L_c = L_c^*$) gives the Curie point. In fact, *Onsager* [4.16] showed rigorously that a Curie point exists in a square lattice, as will be mentioned later. Consequently the Curie temperature is given rigorously by [4.29]

$$\sinh^2 2L_c = 1 \tag{4.4.24}$$

or

$$L_c = \frac{J}{2kT_c} = 0.4407 . \tag{4.4.25}$$

Some complications arise in triangular lattices or honeycomb lattices whose dual lattices are different from the original ones. The results are as follows:

$$\exp(4L_c) = 3, \qquad L_c = 0.2747 \text{ for a triangular lattice}$$

$$\cosh 2L_c = 2, \qquad L_c = 0.6585 \text{ for a honeycomb lattice .}$$

To write them in a unified fashion, it is convenient to use the Gudermannian angle g by the relation $\cosh 2L_c = \sec g$, then the Curie point is given by $g = \pi/c$, where c is the number of nearest neighbors. The value of L_c increases with decreasing c. The approximate treatment such as the Bethe approximation also derives a similar relationship between T_c and c.

4.4.2 Matrix Method

a) *One-Dimensional Ising System*

The matrix method is a powerful tool for evaluating the partition function as already described in Sect. 4.3.2. The partition function for a one-dimensional system in the presence of a magnetic field can easily be obtained, and the two-dimensional Ising system in the absence of a magnetic field has been solved rigorously by *Onsager* [4.16]. The Hamiltonian in the one-dimensional case is

$$\mathscr{H} = -\sum_{i=1}^{N} \sigma_i mH - \frac{J}{2}\sum_{i=1}^{N} \sigma_i\sigma_{i+1} , \tag{4.4.26}$$

where m is the magnitude of the magnetic moment of a spin and the cyclic condition $\sigma_1 = \sigma_{N+1}$ is imposed (Fig. 4.15)

$$Z = \sum_{\sigma_1=\pm 1} \cdots \sum_{\sigma_N=\pm 1} K(\sigma_1, \sigma_2)K(\sigma_2, \sigma_3) \cdots K(\sigma_N, \sigma_1) \tag{4.4.27}$$

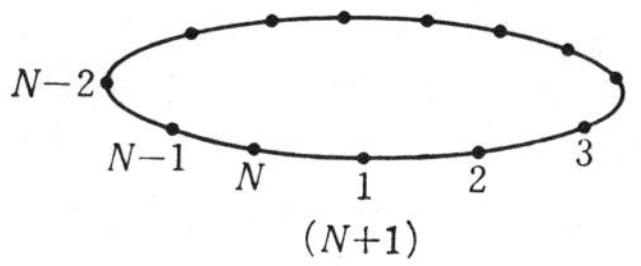

Fig. 4.15. Cyclic condition (ring)

where

$$K(\sigma_i, \sigma_{i+1}) = \exp\left(C\frac{\sigma_i + \sigma_{i+1}}{2} + L\sigma_i\sigma_{i+1}\right) \tag{4.4.28}$$

and

$$C = \frac{mH}{kT}, \quad L = \frac{J}{2kT}. \tag{4.4.29}$$

Equation (4.4.27) can also be written in a matrix formulation as

$$Z = \mathrm{tr}\{K^N\} = \lambda_1^N + \lambda_2^N, \tag{4.4.30}$$

where

$$K = \begin{bmatrix} \exp(C + L) & \exp(-L) \\ \exp(-L) & \exp(-C + L) \end{bmatrix} \tag{4.4.31}$$

and λ_1 and λ_2 are the eigenvalues of the matrix K. They are real, because (4.4.31) is a symmetric real matrix and if $|\lambda_1| > |\lambda_2|$, (4.4.30) is asymptotically approximated as λ_1^N for large N. In the present case

$$\lambda_{1,2} = \exp L \cosh C \pm \sqrt{\exp(2L)\sinh^2 C + \exp(-2L)} \tag{4.4.32}$$

and consequently,

$$Z = [\exp L \cosh C + \sqrt{\exp(2L)\sinh^2 C + \exp(-2L)}]^N. \tag{4.4.33}$$

The magnetization $M = Nm\langle\mu\rangle$ is given by

$$M = -\frac{\partial}{\partial H}(-kT\ln Z) = \frac{Nm\sinh C}{[\exp(-4L) + \sinh^2 C]^{1/2}}. \tag{4.4.34}$$

This quantity is zero for $H \to 0$ $(C \to 0)$. Therefore, no spontaneous magnetization is found in the present case. The susceptibility χ is for $H \to 0$,

$$\chi = \frac{Nm^2}{kT}\exp\left(\frac{4J}{kT}\right) \tag{4.4.35}$$

which gives Curie's law for $kT \gg |J|$.

These results show that any ordered state (ferromagnetism, antiferromagnetism) cannot be realized in the one-dimensional Ising system irrespective of $J > 0$ or $J < 0$. This is what has been mentioned in Sect. 4.3.2 but we can understand it as follows. Consider the case $J > 0$. When an ordered up-spin state is established in a region of a linear array of spins, each end of this region has a down-spin neighbor. By virtue of this single spin, the ordered state of this region cannot affect the arrangements of the spins outside this ordered region. Consequently, the ordered state cannot propagate over the whole region due to the presence of a single down spin. On the other hand, in 2- or 3-dimensional systems, the effect of the ordered up-spin region upon the neighboring spins cannot be extinguished by a single down spin lying on the boundary because a spin state will be affected by the surrounding spins and a single down spin is not enough to stop the propagation of the order beyond the already established ordered region. These considerations will give an intuitive understanding that a one-dimensional system does not have spontaneous magnetization (phase transition), while a 2- or 3-dimensional system does.

b) *Two-Dimensional Ising Systems*

The matrix method can be applied to two-dimensional systems as well. Consider a two-dimensional lattice on a torus which is constructed by tiers of rings (Fig. 4.16) where n spins lie on a ring and the $(m+1)$th ring is identified as the 1st ring. Then the total energy is given by

$$\phi(\sigma_1, \sigma_2) + \phi(\sigma_2, \sigma_3) + \cdots + \phi(\sigma_m, \sigma_1), \tag{4.4.36}$$

where σ_i (with n components) denotes the spin state in the ith ring and $\phi(\sigma_i, \sigma_j)$ is the energy between the ith ring and the jth ring in σ_i and σ_j states respectively. The partition function is then given, in terms of a matrix U,

$$U(\sigma, \sigma') = \exp[-\Phi(\sigma, \sigma')/kT], \tag{4.4.37}$$

as

$$Z = \mathrm{tr}\{U^N\}.$$

Let the largest and the second-largest eigenvalues of U be λ_1 and λ_2, and other eigenvalues be smaller than λ_1 in absolute values, then Z is approximated as

$$\begin{aligned} Z &= \lambda_1^N + \lambda_2^N + \cdots \\ &= \lambda_1^N\left[1 + \left(\frac{\lambda_2}{\lambda_1}\right)^N + \cdots\right] \end{aligned} \tag{4.4.38}$$

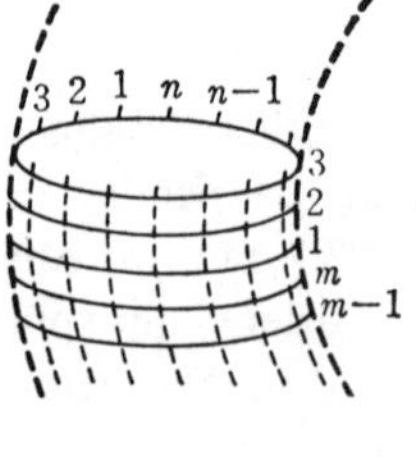

Fig. 4.16. Two-dimensional torus

and the contributions from other eigenvalues are asymptotically ignored. Therefore, we have

$$\log Z = N \log \lambda_1 \tag{4.4.39}$$

irrespective of whether or not the largest eigenvalues are degenerate ($\lambda_1 = \lambda_2$). On the other hand, let us consider the conditional probability $P(\sigma_\nu | \sigma_1)$ of the spin state σ_ν of the νth ring under the spin state σ_1 of the 1st ring, which can be written as

$$P(\sigma_\nu | \sigma_1) = \frac{\sum\limits_{2,3,\ldots,\nu-1} U(1,2)U(2,3)\cdots U(\nu-1,\nu)}{\sum\limits_{2,3,\ldots,\nu} U(1,2)U(2,3)\cdots U(\nu-1,\nu)}, \tag{4.4.40}$$

where we write $U(i, i+1)$ instead of $U(\sigma_i, \sigma_{i+1})$ for simplicity. By virtue of the transformation T which diagonalizes U

$$U = T\Lambda T^{-1}, \tag{4.4.41}$$

where

$$\Lambda = \begin{bmatrix} \lambda_1 & & \\ & \lambda_2 & \\ & & \ddots \end{bmatrix} \tag{4.4.42}$$

is the diagonal matrix ($\lambda_1 \geq \lambda_2 > \cdots$), we can write

$$P(f|0) = \frac{\sum\limits_k T_{0k}\lambda_k^\nu T_{kf}^{-1}}{\sum\limits_{k,f} T_{0k}\lambda_k^\nu T_{kf}^{-1}}, \tag{4.4.43}$$

where the indices 0 and f indicate the spin states of the 1st and the νth layers, respectively. If $\lambda_1 > \lambda_2$, then for large ν the conditional probability

$$P(f|0) = \frac{T_{01}\lambda_1^\nu T_{1f}^{-1}}{\sum\limits_f T_{01}\lambda_1^\nu T_{1f}^{-1}} = \frac{T_{1f}^{-1}}{\sum\limits_f T_{1f}^{-1}} \tag{4.4.44}$$

is independent of the initial state. It implies that no long-range order can exist. On the other hand, if the largest eigenvalues are degenerate, $\lambda_1 = \lambda_2$, we have

$$\begin{aligned} P(f|0) &= \frac{T_{01}\lambda_1^\nu T_{1f}^{-1} + T_{02}\lambda_2^\nu T_{2f}^{-1}}{\sum\limits_f (T_{01}\lambda_1^\nu T_{1f}^{-1} + T_{02}\lambda_2^\nu T_{2f}^{-1})} \\ &= \frac{T_{01}T_{1f}^{-1} + T_{02}T_{2f}^{-1}}{\sum\limits_f (T_{01}T_{1f}^{-1} + T_{02}T_{2f}^{-1})} \end{aligned} \tag{4.4.45}$$

which depends on the initial spin state, and thus exhibits a long-range order. Consequently, the degeneracy of the largest eigenvalue is responsible for the appearance of the long range order, i.e., the phase transition.

The matrix U, however, is a positive matrix and the largest eigenvalue is simple, according to the Perron-Frobenius theorem mentioned in Sect. 4.3, provided that the number of spins in a layer is finite. Therefore, our problem is to search for the largest eigenvalue and its degeneracy in the limit of $n \to \infty$. This was actually done by *Onsager* and others through complicated algebraic treatments, and below a certain temperature T_c the largest eigenvalue is shown to be degenerate. Although several attempts have been made to simplify the original theory, the analyses are not necessarily easy to follow and are still long. We content ourselves by showing only the results.

Exact results are now available for the partition function in the absence of the magnetic field [4.16] and the spontaneous magnetization obtained by introducing an infinitesimally small magnetic field [4.15]. These results have significant features in showing not only the presence of the phase transition in a model which approximates the real substances well, but also in clarifying the shortcomings of the conventional approximate theories to be described in Sect. 4.5, through the logarithmic singularity of the specific heat and the critical indices near the Curie temperature.

Now let us consider a rectangular square lattice with J for lateral interaction and J' for longitudinal interaction, and put

$$\frac{J}{2kT} = L, \quad \frac{J'}{2kT} = L' . \tag{4.4.46}$$

The partition function is then given by

$$\frac{1}{N}\ln Z = \frac{1}{2}\ln(4\cosh 2L\cosh 2L') + \frac{1}{2\pi^2}\int_0^\pi\int_0^\pi \ln(1 - 2\kappa\cos\omega - 2\kappa'\cos\omega')\,d\omega\,d\omega' , \tag{4.4.47}$$

where

$$2\kappa = \frac{\tanh 2L}{\cosh 2L'}, \quad 2\kappa' = \frac{\tanh 2L'}{\cosh 2L} .$$

In a regular square lattice ($J = J'$, $L = L'$, $\kappa = k'$),

$$\frac{1}{N}\ln Z = \ln(2\cosh 2L) + \frac{1}{2\pi^2}\int_0^\pi\int_0^\pi \ln(1 - 4\kappa\cos\omega_1\cos\omega_2)\,d\omega_1\,d\omega_2 \tag{4.4.48}$$

which is transformed to

$$\frac{1}{N}\ln Z = \ln(2\cosh 2L) + \frac{1}{2\pi}\int_0^\pi \ln\frac{1}{2}\left(1 + \sqrt{1 - (4\kappa)^2\sin^2\varphi}\right)d\varphi , \tag{4.4.49}$$

where use is made of the identity

$$\int_0^{2\pi} \ln(2\cosh\mu - 2\cos\omega)\,d\omega = 2\pi\mu$$

and μ is defined by the relation

$$\cosh\mu = \frac{1}{4\kappa|\cos\omega_1|} = y, \quad \mu = \cosh^{-1} y = \ln(y + \sqrt{y^2 - 1})\,. \tag{4.4.50}$$

The energy E is given by

$$E = -\frac{\partial \ln Z}{\partial(1/kT)} = -J\frac{\partial \ln Z}{2\partial L} = -\frac{NJ}{2}\coth 2L\left(1 + \frac{2}{\pi}\kappa_1'' K_1\right), \tag{4.4.51}$$

where K_1 is the complete elliptic integral

$$K_1 = K(\kappa_1) = \int_0^{\pi/2} \frac{d\varphi}{\sqrt{1 - \kappa_1^2 \sin^2\varphi}} \tag{4.4.52}$$

with modulus $\kappa_1 = 4\kappa$ and

$$\kappa_1'' = 2\tanh^2 2L - 1, \quad |\kappa_1''| = \sqrt{1 - \kappa_1^2}\,. \tag{4.4.53}$$

The specific heat is thus

$$C = kL^2 \frac{\partial^2 \ln Z}{\partial L^2}$$

$$= Nk(L\coth 2L)^2 \frac{2}{\pi}\left[2K_1 - 2E_1 - (1 - \kappa_1'')\left(\frac{\pi}{2} + \kappa_1'' K_1\right)\right], \tag{4.4.54}$$

where E_1 is the complete elliptic integral of the second kind

$$E_1 = E(\kappa_1) = \int_0^{\pi/2} \sqrt{1 - \kappa_1^2 \sin^2\varphi}\,d\varphi\,. \tag{4.4.55}$$

The specific heat becomes singular at $\kappa_1 = 1$, because $K_1 = \infty$ at $\kappa_1 = 1$ as seen from the elliptic integral K_1. On the other hand, the internal energy E remains finite, by virtue of the factor κ_1''. The Curie point is given by $\kappa_1 = 2\tanh 2L_c/\cosh 2L_c = 1$ or $\sinh 2L_c = 1$, and near this point we have

$$K_1 \approx \ln\left(\frac{4}{\kappa_1'}\right) \approx \ln\left(\frac{2^{1/2}}{|L - L_c|}\right) \tag{4.4.56}$$

which indicates that the specific heat diverges logarithmically on both the lower and higher sides of the Curie temperature:

$$C \approx A \ln|T - T_c|\,. \tag{4.4.57}$$

The coefficient A is the same on both sides. This exact result reveals that the behavior of the specific heat at the Curie point in a 2-dimensional Ising lattice is quite different from the approximate theories.

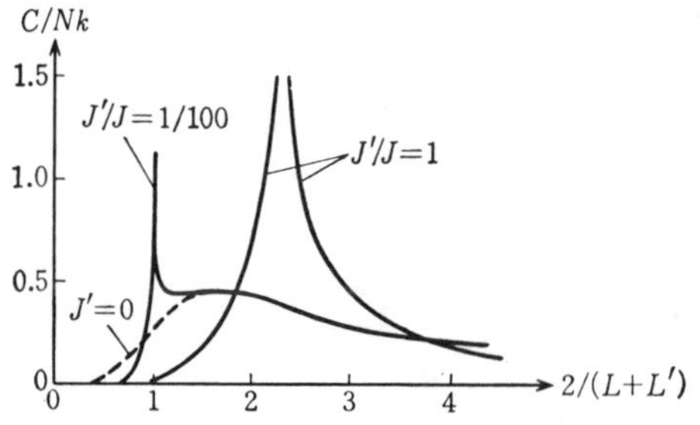

Fig. 4.17. Specific heat of a square Ising lattice [4.16]

Figure 4.17 shows the specific heats of the regular square lattice with $J'/J = 1$, the rectangular square lattice with $J'/J = 1/100$ and the one-dimensional model with $J' = 0$. The abscissa is $4kT/(J + J') = 2/(L + L')$. The integral of the specific heat with respect to temperature is the internal energy which is equal to $-(J + J')$ per spin at 0 K and 0 at an infinitely high temperature. Therefore, the areas under the specific heat for these three cases are the same if the abscissa is taken as the temperature divided by $(J' + J)$.

The spontaneous magnetization is given by the first term of the expansion of the free energy with respect to the magnetic field. *Yang* obtained it exactly by a perturbation method with the result [4.15]

$$M = 0 \quad \text{for } T > T_c \tag{4.4.58}$$

and

$$\frac{M}{Nm} = \left(1 - \frac{1}{\sinh^2 2L \sinh^2 2L'}\right)^{1/8} \quad \text{for } T < T_c\,.$$

Near the Curie temperature

$$M \approx (T_c - T)^{1/8}\,, \tag{4.4.59}$$

as is shown in Fig. 4.18. The critical exponent is $\frac{1}{8}$, not $\frac{1}{2}$ as in molecular field theory.

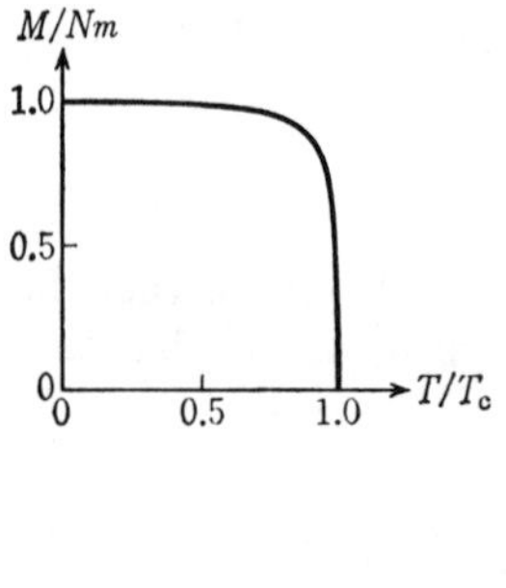

Fig. 4.18. Spontaneous magnetization of a two-dimensional Ising lattice $(J'/J = 1)$ [4.15]

4.4.3 Zeros on the Temperature Plane

The free energy F is given by the partition function Z in a canonical ensemble:

$$-\frac{F}{kT}=\frac{1}{N}\ln Z(\beta, N, V)\,. \tag{4.4.60}$$

Let the zeros of Z regarded as a function of β be β_k ($k = 1, 2, \ldots$). Then we have

$$-\frac{F}{kT}=\frac{1}{N}\sum_{k=1}^{\infty}\ln\left(1-\frac{\beta}{\beta_k}\right)+\text{const.}\,, \tag{4.4.61}$$

where we assumed that the number of zeros is infinite. As long as the system is finite, no zero exists on the positive real axis, but in the limit of an infinitely large system, a zero of β will lie on the positive real axis. This point is considered as the temperature of the phase transition. According tothe expansion (4.4.10), Z, when considered as a function of u, is a polynomial of u for finite N. Let the roots be u_k, then

$$Z = 2^N(\cosh L)^s\prod_k\left(1-\frac{u}{u_k}\right),\quad u=\tanh L\,. \tag{4.4.62}$$

The distribution of the roots will cut the positive real axis on the complex u plane for $N \to \infty$. In the case of a 2-dimensional square lattice, we have an exact Z which gives the distribution of zeros as shown in Fig. 4.19. It is composed of two circles of radius $\sqrt{2}$ centered at 1 and -1. Two circles cut the real axis at $u = \pm(1+\sqrt{2})$, $\pm(\sqrt{2}-1)$. As long as L is real, we have $|u| = |\tanh L| \leqq 1$. Therefore the former two are excluded and $\sqrt{2}-1$ and $1-\sqrt{2}$ are the transition point of ferro- and antiferromagnetism, respectively [4.3]. The value $u=\sqrt{2}-1$ is in accord with the result given by (4.4.25) in the ferromagnetic

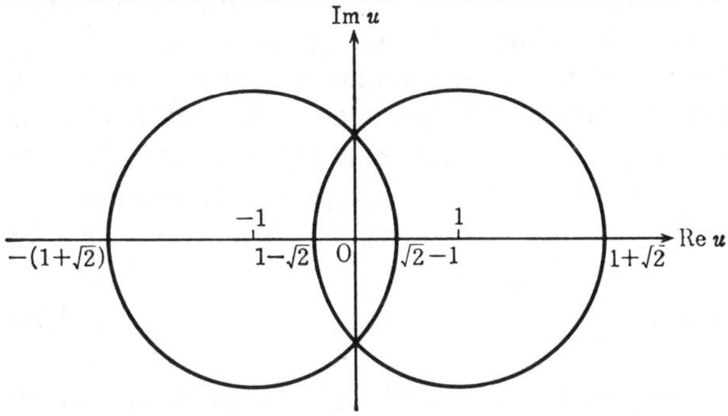

Fig. 4.19. Distribution of zeros of the partition function on the complex u plane in the case of a two-dimensional Ising system

case and $u = 1 - \sqrt{2}$ is for the antiferromagnetic Néel temperature which is obtained from the symmetry property discussed in Sect 4.1.4.

4.4.4 Spherical Model

In an Ising model we have discrete values of spins $|\sigma_i|^2 = 1$ and sum over $\sigma_i = \pm 1$ for each spin. If we relax these restrictions and use the spherical condition instead, we have

$$\sum_{i=1}^{N} \sigma_i^2 = N, \quad -\infty < \sigma_i < \infty .$$

The partition function (4.4.2) becomes

$$Z = \int_{-\infty}^{+\infty} \cdots \int \exp\left(\sum_{\langle ij \rangle} L\sigma_i\sigma_j \right) \delta\left(N - \sum_i \sigma_i^2 \right) \prod_i d\sigma_i$$

which can be written, by making use of the identity

$$\delta(x) = \frac{1}{2\pi} \int_{-\infty}^{+\infty} \exp(\mathrm{i}yx)\, dy ,$$

in a form

$$Z = \frac{1}{2\pi} \int_{-\infty}^{+\infty} \exp(\mathrm{i}yN)\, dy \int_{-\infty}^{+\infty} \cdots \int \left(\prod_i d\sigma_i \right) \exp\left(-\sum_{i,j} a_{ij}\sigma_i\sigma_j \right)$$

which can be calculated rigorously, with a_{ij} appropriately determined. This model shows a phase transition in a 3-dimensional model, where the specific heat is continuous at the Curie temperature but has different slopes on either side of it. The spherical model is considered as a model for the limit of infinitely many $(n \to \infty)$ components of spins.

4.4.5 Eight-Vertex Model

In this section, a short account will be given of the eight-vertex model which was introduced for the ferroelectric transition by *Baxter* [4.30]. In a two-dimensional system, we consider the eight configurations of dipole moments in Fig. 4.20 which are placed on the bonds of a square lattice as indicated by the arrow. These configurations have an even number of arrows pointing into each vertex. The energies $\varepsilon_1, \varepsilon_2, \ldots, \varepsilon_8$ are associated with the vertex configurations. The partition function is thus

$$Z = \sum \exp\left(-\beta \sum_{j=1}^{8} N_j \varepsilon_j \right) , \tag{4.4.63}$$

where N_j is the number of vertices of type j and the summation is over all allowed configurations of arrows. Now we put

$$\omega_j = \exp(-\beta\varepsilon_j) \tag{4.4.64}$$

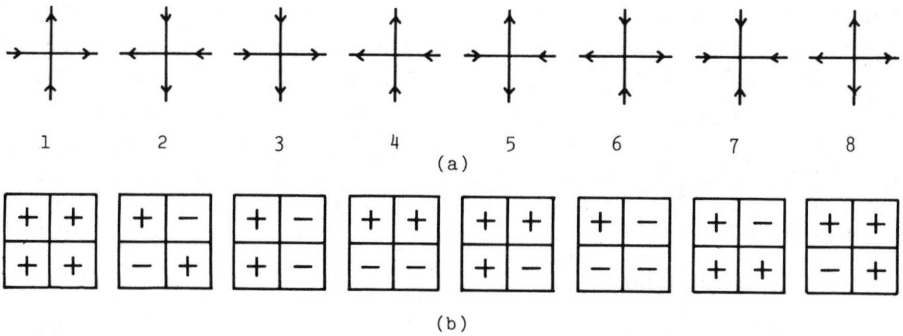

Fig. 4.20. Eight-vertex model (**a**) and corresponding Ising model (b) where + and − stand for up and down spins, respectively

and

$$\begin{aligned}\omega_1 &= \omega_2 = a, \quad \omega_3 = \omega_4 = b \\ \omega_5 &= \omega_6 = c, \quad \omega_7 = \omega_8 = d\end{aligned} \tag{4.4.65}$$

because the model is unchanged by reversing all the arrows.

Consider a horizontal row composed of vertical bonds which is represented by a configuration of arrows $\alpha(\alpha_1, \ldots, \alpha_N)$, where $\alpha_j = +$ or $-$ according to the up or down arrows. There are 2^N possible configurations of arrows on a row. Let the configurations of two consecutive rows be α and α'. Then we can introduce a 2^N by 2^N transfer matrix

$$T(\alpha|\alpha') = \sum \exp(-\beta \sum n_j \varepsilon_j) , \tag{4.4.66}$$

where n_j is the number of vertices of type j in an arrangement of arrows on the horizontal intervening bonds between two rows and the sum is over all the possible arrangements of arrows. Then the partition function can be written in a matrix formulation as

$$Z = \sum_{\alpha_1} \cdots \sum_{\alpha_M} T(\alpha_1|\alpha_2) T(\alpha_2|\alpha_3) \cdots T(\alpha_M|\alpha_1) = \mathrm{tr}\{T^M\} , \tag{4.4.67}$$

where the cyclic boundary condition is imposed. The largest eigenvalue of the matrix T is sufficient for determining the free energy. *Baxter* [4.30] put, by introducing three variables k, η and v,

$$a:b:c:d = \mathrm{sn}(\eta - v):\mathrm{sn}(\eta + v):\mathrm{sn}(2\eta): -k\,\mathrm{sn}(2\eta)\mathrm{sn}(\eta - v)\mathrm{sn}(\eta + v) , \tag{4.4.68}$$

where sn is the elliptic function of modulus k. Then the free energy per vertex f is obtained as follows:

$$-\beta f = -\beta\varepsilon_5 + 2 \sum_{n=1}^{\infty} \frac{\sinh^2[(\tau - \lambda)n](\cosh n\lambda - \cosh n\alpha)}{n \sinh(2n\tau) \cosh(n\lambda)} , \tag{4.4.69}$$

where $\tau = \pi K'/2K$, $\lambda = -i\pi\eta/K$, $\alpha = -i\pi v/K$ and K and K' are the complete elliptic integrals of the first kind of moduli k and $\sqrt{1-k^2}$ [see (4.4.52)]. Equation (4.4.69) holds for real k, λ and α, with the conditions $0 < k < 1$, $|\alpha| < \lambda < \tau$. This restriction is equivalent to $a > 0$, $b > 0$, $d > 0$ and $c > a + b + d$. However, when $c = a + b + d$ occurs at $T = T_c$, f is a sum of an analytic function and a singular function of T. The singular part is proportional to

$$\cot(\pi^2/2\mu)|T - T_c|^{\pi/\mu} \tag{4.4.70}$$

or, if $\pi/2\mu = m$ (an integer), to

$$2\pi^{-1}(T - T_c)^{2m} \ln|T - T_c| , \tag{4.4.71}$$

where $0 < \mu < \pi$ and

$$\cos\mu = (ab - cd)/(ab + cd) . \tag{4.4.72}$$

It should be noted that the critical exponent varies continuously from one to infinity depending on the ratio of coupling constants, in contrast to the universality assumed to hold for short-range interactions (Sect. 4.6).

The eight-vertex model is equivalent to the square Ising model as shown below. Consider an arrangement of spins on a square lattice, as indicated by + or − in Fig. 4.20b, where the dual lattice is represented by solid lines. The arrows are put on the bonds of the dual lattice according to the following rule. If two neighboring spins are parallel, we put an up or right-pointing arrow on the bond between the spins and, if they are antiparallel, we put the down or left-pointing arrows on the bond. The vertex configurations thus obtained from the spin configurations in Fig. 4.20b are those in Fig. 4.20a just above them. The energies of the spin configurations are assumed to exist between diagonally neighboring spins on the crossed bonds,

$$\begin{aligned} \varepsilon_1 &= \varepsilon_2 = -J - J', \quad \varepsilon_3 = \varepsilon_4 = J + J' \\ \varepsilon_5 &= \varepsilon_6 = J' - J, \quad\;\; \varepsilon_7 = \varepsilon_8 = J - J' . \end{aligned} \tag{4.4.73}$$

The lattice composed of crossed bonds is divided into two mutually independent square lattices, each of which is regarded as an Ising spin on a square lattice. Consequently, the value of μ defined by (4.4.72) is $\pi/2$, and one recovers the logarithmic singularity of the specific heat.

An eight-vertex model is also equivalent to a dimer model which concerns the close packing of dimers on the square lattice.

4.5 Approximate Theories

Exact treatments of thermodynamic behavior and especially of phase transitions can be discussed only for quite special systems in 1 and 2 dimensions as described in Sects. 4.3, 4.4. However, the substances of physical interest are usually 3-dimensional. This fact requires the development of approximate theories.

4.5.1 Molecular Field Approximation, Weiss Approximation

The Hamiltonian given by (4.1.4) can be approximated as follows. Nearest-neighbor interaction with a number of nearest neighbors c is assumed and the sum $\sum_j s_j^z$ multiplying s_i^z is replaced by its average, i.e., $c\langle s\rangle$. Then we have

$$\mathscr{H} = -2Jc\sum_i s_i^z\langle s\rangle - g\mu_B H\sum_i s_i^z\,, \tag{4.5.1}$$

where c is the number of nearest neighbors. In the Heisenberg model with a magnetic field H in the z-direction, we can put $\langle s_i^x\rangle = \langle s_i^y\rangle = 0$, thus we reach the same Hamiltonian. The first term of (4.5.1) implies that the effect on the ith spin from its surrounding spins is to apply a magnetic field proportional to their average value $\langle s\rangle$. This kind of approximation thus bears the name of the molecular field or mean field approximation and (4.5.1) in particular is called the Weiss approximation or the Bragg-Williams approximation, which were developed for magnetism or for order-disorder problems in alloys, respectively. Equation (4.5.1) is written as a sum of the Hamiltonian of each spin:

$$\begin{aligned}\mathscr{H} &= \sum_i \mathscr{H}_i\\ \mathscr{H}_i &= (-2Jc\langle s\rangle - g\mu_B H)s_i\end{aligned} \tag{4.5.2}$$

and the partition function is, therefore,

$$\begin{aligned}Z &= Z_i^N,\\ Z_i &= \sum_{s_i}\exp(-\beta\mathscr{H}_i) = 2\cosh\beta\left(Jc\langle s\rangle + \frac{g\mu_B}{2}H\right),\end{aligned} \tag{4.5.3}$$

where $s_i = \pm 1/2$ is assumed. The average value of s_i is

$$\langle s\rangle = \langle s_i\rangle = \sum_{s_i} s_i\frac{\exp(-\beta\mathscr{H}_i)}{Z_i} = \frac{1}{2}\tanh\beta\left(Jc\langle s\rangle + \frac{g\mu_B}{2}H\right). \tag{4.5.4}$$

Figure 4.21 shows that in the case $H = 0$, a nonzero solution exists for

$$T_c \equiv Jc/2k > T\,. \tag{4.5.5}$$

T_c is assigned as the Curie temperature. Equation (4.5.5) implies the existence of the Curie temperature irrespective of the dimensionality of the lattice, which shows the inadequacy of the present approximation in a one-dimensional system. Equation (4.5.4) also gives susceptibility. For $T > T_c$, we have the Curie-Weiss law:

$$\chi = [g\mu_B\partial\langle s\rangle/\partial H]_{H=0} = (g\mu_B)^2/[4k(T - T_c)] \tag{4.5.6}$$

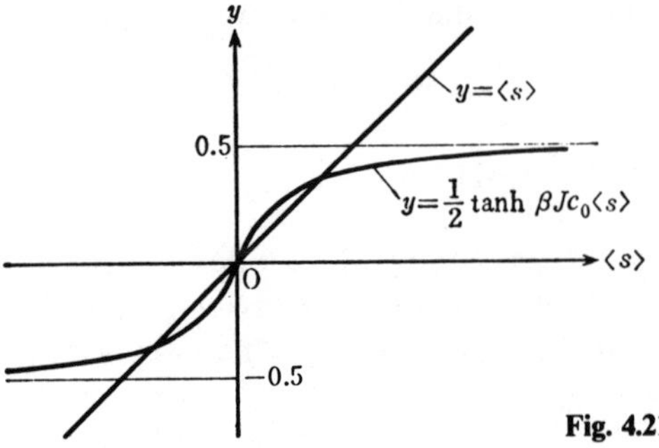

Fig. 4.21. Condition for spontaneous magnetization

for the susceptibility per spin. For $T < T_c$ and $T_c - T \ll T_c$, (4.5.4) yields in the absence of a magnetic field

$$\langle s\rangle = \frac{\sqrt{3}}{2}\left(\frac{T_c - T}{T_c}\right)^{1/2} . \tag{4.5.7}$$

Returning to (4.5.2) and putting

$$\langle s\rangle = \frac{\sum_i s_i}{N} , \tag{4.5.8}$$

we have

$$\mathscr{H} = -\left(\frac{2Jc}{N}\right)\left(\sum_i s_i\right)^2 - g\mu_B H \sum_i s_i . \tag{4.5.9}$$

If all J_{ij}'s in (4.1.4) are put as $2Jc/N$ irrespective of ij pairs, just as we did in the discussion in Sect. 4.3.3, then by virtue of the relation

$$\sum_{\langle ij\rangle} s_i s_j = \frac{1}{2}\sum_i s_i\left(\sum_j s_j - s_i\right) = \frac{1}{2}\left(\sum_i s_i\right)^2 - \frac{N}{2}\frac{1}{4} , \tag{4.5.10}$$

the Hamiltonian becomes equivalent to (4.5.9) for the molecular field approximation when we ignore the constant term of no physical interest.

4.5.2 Bethe Approximation

In the Bethe approximation, the mean fields applied to the spins are considered in a somewhat different way to get a better result than Weiss theory. In rewriting the Hamiltonian (4.1.4), we consider a cluster around a central spin (say i) surrounded by c spins (say k's). They are affected by the mean field H' from their neighboring spins. The Hamiltonian is assumed to be of the following form in the absence of a magnetic field:

$$\mathscr{H} = \sum \mathscr{H}_i$$
$$\mathscr{H}_i = -2J \sum_k^c (s_i s_k + h' s_k), \quad h' = g\mu_B H'/2J \,. \tag{4.5.11}$$

The partition function is thus

$$Z = Z_i^N, \quad Z_i = \sum \mathrm{e}^{-\beta \mathscr{H}_i} = \sum_{s_i = \pm \frac{1}{2}} 2^c \cosh^c \beta J (s_i + h') \,. \tag{4.5.12}$$

The averages of $\langle s_i \rangle$ and $\langle s_k \rangle$ can be calculated as

$$\langle s_i \rangle = \frac{2^c}{2Z_i} \left(\cosh^c \frac{\beta J}{2} (1 + 2h') - \cosh^c \frac{\beta J}{2} (1 - 2h') \right) \tag{4.5.13}$$

and

$$\langle s_k \rangle = \frac{2^c}{2Z_i} \left(\sinh \frac{\beta J}{2} (1 + 2h') \cosh^{c-1} \frac{\beta J}{2} (1 + 2h') - \sinh \frac{\beta J}{2} (1 - 2h') \cosh^{c-1} \frac{\beta J}{2} (1 - 2h') \right) . \tag{4.5.14}$$

We require the equality of these two quantities. The mean field h' is determined by the condition

$$h' = \frac{c-1}{2\beta J} \log \frac{\cosh \beta J (\frac{1}{2} + h')}{\cosh \beta J (\frac{1}{2} - h')} \,. \tag{4.5.15}$$

The ratio $\langle s_i \rangle / \langle s_k \rangle$ becomes

$$\frac{\langle s_i \rangle}{\langle s_k \rangle} \to \tanh \frac{c\beta J}{2} < 1, \quad \text{for large}\, h' \,, \tag{4.5.16}$$

and

$$\frac{\langle s_i \rangle}{\langle s_k \rangle} \to \frac{c}{\coth \frac{\beta J}{2} + (c-1) \tanh \frac{\beta J}{2}} \quad \text{for } h' = 0 \,. \tag{4.5.17}$$

The ratio (4.5.17) is equal to 1 at the temperature $T_c = 1/\beta_c k$ where

$$\tanh \frac{\beta_c J}{2} = \frac{1}{c-1} \tag{4.5.18}$$

is satisfied, or

$$\beta_c = \frac{1}{J} \log \frac{c}{c-2} \,. \tag{4.5.19}$$

When $\beta > \beta_c$, the ratio (4.5.17) is larger than 1, and on increasing h' we can find the value which gives $\langle s_i \rangle / \langle s_k \rangle = 1$. This value h', however, approaches zero when the value of β is reduced to β_c. Furthermore, on reducing β below β_c we cannot find the solution $\langle s_i \rangle = \langle s_k \rangle$ ($\neq 0$). T_c is the critical temperature.

When the temperature is less than T_c but close to it, the average $\langle s_i \rangle$ is given by

$$\langle s_i \rangle = (\tanh \beta_c J)\beta J c h' \tag{4.5.20}$$

and h' is also small and is found to be proportional to $[(T_c - T)/T_c]^{1/2}$ from (4.5.15). Thus we have

$$\langle s \rangle = A(c)\left(\frac{T_c - T}{T_c}\right)^{1/2}, \tag{4.5.21}$$

where the constant $A(c)$ is a function of c, the number of nearest neighbors. The relation (4.5.21) is similar to (4.5.7). If we take the magnetic field into consideration we can calculate the susceptibility as in (4.5.6) for $T > T_c$; it is still proportional to $(T - T_c)^{-1}$ for $T > T_c$.

Although the Bethe approximation and the Weiss approximation have the drawback of predicting phase transitions irrespective of dimensionality, and in particular even in one dimension, they are useful approximations for understanding statistical properties and phase transitions in real systems.

4.5.3 Low and High Temperature Expansions

The high and low temperature expansions in the partition function of Ising systems are given by (4.4.10) and (4.4.12), respectively.

The high temperature expansions are

$$Z^{1/N} = 2(\cosh L)^3(1 + 3u^4 + 22u^6 + 192u^8 + \cdots) \tag{4.5.22}$$

for a simple cubic lattice and

$$Z^{1/N} = 2(\cosh L)^4(1 + 12u^4 + 148u^6 + 1860u^8 + \cdots) \tag{4.5.23}$$

for a body-centered cubic lattice. On the other hand, the low temperature expansions are

$$\frac{1}{N}\ln Z = q^{-3/2}\left(q^6 + 3q^{10} - \frac{7}{2}q^{12} + \cdots\right) \tag{4.5.24}$$

for a simple cubic lattice and

$$\frac{1}{N}\ln Z = q^{-2}\left(q^8 + 4q^{14} - \frac{9}{2}q^{16} + 28q^{20} + \cdots\right) \tag{4.5.25}$$

for a body-centered cubic lattice where $q = \exp(-J/kT)$. Next, consider the high temperature expansion for a Heisenberg model. Assuming nearest-neighbor interactions, the Hamiltonian can be written as

$$\mathscr{H} = -2J \sum_{\langle ij \rangle} s_i \cdot s_j - g\mu_B H \sum_i s_i^z = \mathscr{H}_0 + \mathscr{H}_1 \tag{4.5.26}$$

and the partition function, magnetization M, and susceptibility χ are given by

$$Z = \mathrm{tr}\{\exp(-\beta\mathscr{H})\} \tag{4.5.27}$$

$$M = kT\frac{1}{Z}\frac{\partial Z}{\partial H} \tag{4.5.28}$$

$$\chi = \frac{\partial M}{\partial H} = kT\left[\frac{1}{Z}\frac{\partial^2 Z}{\partial H^2} - \left(\frac{1}{Z}\frac{\partial Z}{\partial H}\right)^2\right]. \tag{4.5.29}$$

Since $\mathscr{H}_0$ and $\mathscr{H}_1$ are commutative, we have

$$\begin{aligned} Z &= \mathrm{tr}\{\exp(-\beta\mathscr{H}_0 - \beta\mathscr{H}_1)\} \\ &= \mathrm{tr}\{\exp(-\beta\mathscr{H}_0)\exp(-\beta\mathscr{H}_1)\} \end{aligned} \tag{4.5.30}$$

which can be expanded in a power series of β. The final result for χ is

$$\chi = \frac{N(g\mu_\mathrm{B})^2}{4kT}\left[1 + \frac{c}{2}\left(\frac{J}{kT}\right) + \frac{c(c-2)}{4}\left(\frac{J}{kT}\right)^4 + \cdots\right], \tag{4.5.31}$$

the higher terms of which can be obtained by tedious calculations requiring special techniques. The method of series expansion is to express χ in a power series of a certain variable y:

$$\chi = \sum_{n=0}^{\infty} a_n y^n , \tag{4.5.32}$$

where $y = \tanh L$ or $\exp(-2L)$ in an Ising model and $y = \beta J$ in a Heisenberg model. If this series diverges at $y = y_\mathrm{c}$, y_c can be identified as corresponding to the Curie temperature. However, it is difficult to find the general term of this series. The radius of convergence r_0 of this series (4.5.32) is given by

$$r_0^{-1} = \lim_{n\to\infty} |a_n|^{1/n} . \tag{4.5.33}$$

In particular, in a positive term series, if $\lim_{n\to\infty} \mu_n \equiv a_n/a_{n-1}$ exists, then $\lim_{n\to\infty}|a_n|^{1/n}$ exists, and $\lim \mu_n$ is equal to $\lim_{n\to\infty}|a_n|^{1/n} = r_0^{-1}$. This provides the ratio method which can give the radius of convergence by numerically calculating the limit of μ_n. However, the ratio method is useless when μ_n changes irregularly and is only valid for positive term series. In these cases, the Padé approximation can be used effectively instead of the ratio method. Let χ be approximated as

$$\chi = \frac{P(y)}{Q(y)} = \frac{p_0 + p_1 y + p_2 y^2 + \cdots + p_L y^L}{1 + q_1 y + q_2 y^2 + \cdots + q_M y^M} . \tag{4.5.34}$$

The coefficients p and q are determined so as to give the correct values a_n up to $n = N$ when this equation is expanded in a power series of y. When the $a_0, a_1, \ldots, a_\mathrm{N}$ in (4.5.32) are given, then by taking $L + M = N$, the p's and q's of the total number $L + M + 1$ can be determined. The singularity of χ comes from

Table 4.2. u_c and γ obtained by the Padé approximation for a simple cubic lattice

$L = M$	2	3	4	5
u_c	0.2151	0.2189	0.21815	0.21818
γ	1.205	1.281	1.2505	1.2518

the zeros of the polynomial $Q(y)$. Taking the critical exponent into consideration (Sect. 4.6.1), we can express

$$\chi = (y - y_c)^{-\gamma} y_c^{\gamma} G(y) \tag{4.5.35}$$

and

$$D(y) = \frac{d \ln \chi(y)}{dy} = -\frac{\gamma}{y - y_c} + \frac{d}{dy} \ln G(y) \tag{4.5.36}$$

which has a simple pole. Therefore, it is better to apply the Padé approximation to $D(y)$ and to find its simple pole and its residue. The critical value u_c of $u = \tanh L$ and the critical exponent γ thus obtained are listed in Table 4.2.

Fisher surmised that in a 3-dimensional Ising model, the value of γ is $5/4 = 1.25$, irrespective of the crystal structures, e.g., a simple cubic lattice, body-centered or face-centered cubic lattices [4.2].

4.6 Critical Phenomena

4.6.1 Critical Exponents

According to experiments, the physical quantities associated with the phase transition vary as $|T - T_c|^{\mu}$ when the critical temperature T_c is approached. The quantities such as μ in the above equations which characterize the phase transition are called critical exponents. They are inherent to the physical quantities under consideration and are supposed to take universal values irrespective of the materials under consideration [an exception is the eight-vertex model (Sect. 4.5.5)]. Moreover, there are simple relations among them. In ferromagnetism, one finds the following relations at the temperature close to the Curie temperature T_c:
Susceptibility

$$\chi = \frac{M}{H} \propto (T - T_c)^{-\gamma} \quad (T > T_c)\,; \tag{4.6.1}$$

spontaneous magnetization

$$M \propto (T_c - T)^{\beta} \quad (T < T_c)\,; \tag{4.6.2}$$

magnetization

$$M \propto H^{1/\delta} \quad (T = T_c) \,. \tag{4.6.3}$$

The specific heat of 3-dimensional materials usually diverges as

$$C(T) \propto \begin{cases} (T - T_c)^{-\alpha} & (T > T_c) \\ (T_c - T)^{-\alpha} & (T < T_c) \,. \end{cases} \tag{4.6.4}$$

If the specific heat is finite at the critical temperature, then $\alpha = 0$. If it diverges like $\ln(1/|T_c - T|)$ as in the 2-dimensional Ising model, we can put $\alpha = 0$, since by virtue of the relation

$$-\ln(T - T_c) = \lim_{\alpha \to 0} \frac{1}{\alpha}(|T - T_c|^{\alpha} - 1) \,,$$

we can regard α as being very small and thus the coefficient $1/\alpha$ as very large. The critical exponents at the liquid-gas critical point are known, too. The densities ρ_l and ρ_g of the liquid state and the gaseous state under the same pressure satisfy the relation

$$\rho_l - \rho_g \propto (T_c - T)^{\beta} \quad (T < T_c) \,. \tag{4.6.5}$$

In accordance with the correspondence between magnetism and the properties of a particle system (Table 4.1), the change in density corresponds to the change in magnetization. Therefore, this β is considered to be the same as the β in spontaneous magnetization. The change in pressure corresponds to the change in magnetic field. Then we have

$$|\rho - \rho_c| \propto |P - P_c|^{1/\delta} \quad (T = T_c) \,. \tag{4.6.6}$$

This δ is considered identical to the δ in the relationship between M and H. Note that compressibility corresponds to susceptibility:

$$\begin{aligned} \kappa_T = -\frac{1}{V}\left(\frac{\partial V}{\partial P}\right)_T &\propto (T - T_c)^{-\gamma} \quad (T > T_c) \\ &\propto (T_c - T)^{-\gamma'} \quad (T < T_c) \,. \end{aligned} \tag{4.6.7}$$

Experiments suggest $\gamma = \gamma'$.

Another important quantity in the critical region is the correlation length. In magnetism, the average size of the region with parallel spins (ordered region) becomes large at temperatures close to T_c. In particle systems, the average size of the clusters of particles becomes large. The average size of this ordered region is the correlation length ξ, which has the relation

$$\xi \propto \begin{cases} |T - T_c|^{-\nu} & (T > T_c) \\ |T - T_c|^{-\nu'} & (T < T_c) \,. \end{cases} \tag{4.6.8}$$

Now we define the correlation function by

$$\begin{aligned} G(\boldsymbol{r}, \boldsymbol{r}') &= \langle (n(\boldsymbol{r}) - \langle n \rangle)(n(\boldsymbol{r}') - \langle n \rangle) \rangle \\ &= n^2(\boldsymbol{r}, \boldsymbol{r}') + \langle n \rangle \delta(\boldsymbol{r} - \boldsymbol{r}') - \langle n \rangle^2 \,, \end{aligned} \tag{4.6.9}$$

where $n(\boldsymbol{r})$ is the density defined by (1.2.3) and $n^2(\boldsymbol{r}, \boldsymbol{r}')$ is the pair distribution function introduced and discussed in (1.2.4). The intensity of scattered light is given by

$$I(k) \propto \iint n^{(2)}(\boldsymbol{r}, \boldsymbol{r}')\exp[-i\boldsymbol{k}(\boldsymbol{r} - \boldsymbol{r}')]d\boldsymbol{r}d\boldsymbol{r}' + V\langle n\rangle ,$$

where the first term represents the interference of the light scattered from different particles and the second term is the intensity of the light scattered from individual particles. The vector $\boldsymbol{k}$ is the difference of the wave vectors of the incident and scattered light and its magnitude is given by

$$k = |\boldsymbol{k}| = \frac{4\pi}{\lambda}\sin\frac{\theta}{2} , \tag{4.6.10}$$

where θ is the angle between the wave vectors and the wavelength λ is assumed to be unchanged by scattering. $I(\boldsymbol{k})$ can also be written as

$$I(\boldsymbol{k}) \propto V[S(\boldsymbol{k}) + (2\pi)^3\langle n\rangle^2\delta(\boldsymbol{k})] , \tag{4.6.11}$$

where $S(\boldsymbol{k})$ is the structure factor familiar in X-ray or light scattering and is the Fourier transform of the correlation function $G(\boldsymbol{r}, \boldsymbol{r}')$ which is regarded as a function of $\boldsymbol{r} - \boldsymbol{r}'$:

$$S(\boldsymbol{k}) = \int \exp(-i\boldsymbol{k}\cdot\boldsymbol{r})G(\boldsymbol{r})d\boldsymbol{r} . \tag{4.6.12}$$

Now let us put

$$G(\boldsymbol{r}) = G_0\frac{\exp(-r/\xi)}{r} , \tag{4.6.13}$$

then

$$S(\boldsymbol{k}) = 4\pi G_0\frac{\xi^2}{1 + k^2\xi^2} . \tag{4.6.14}$$

At the critical point, the correlation length ξ becomes large, letting $\xi \to \infty$, we have

$$S(\boldsymbol{k}) \sim k^{-2} . \tag{4.6.15}$$

This is the result of the form (4.6.13) of the correlation function, and in general, one should put

$$S(\boldsymbol{k}) \sim k^{-2+\eta} \tag{4.6.16}$$

by introducing another critical exponent η. In magnetic substances we can define the correlation function $G(\boldsymbol{r})$ and the structure factor $S(\boldsymbol{k})$ by means of the magnetization density $M(\boldsymbol{r})$ or spin density $\sigma(\boldsymbol{r})$ instead of the particle density $n(\boldsymbol{r})$. Thus (4.6.13) or (4.6.16) can also be applied in magnetism. The critical exponents in various systems are listed in Table 4.3.

Equation (2.7.11) shows that $\chi \equiv M/H$ is equal to the integral of $G(\boldsymbol{r} - \boldsymbol{r}')/kT$. In this expression, one assumes that a uniform magnetic field is applied to the system. We can interpret $G(\boldsymbol{r} - \boldsymbol{r}')/kT$ as the magnetization at a position $\boldsymbol{r}$

Table 4.3. Critical exponents

Physical quantities	χ, k_T	$M, \rho_1 - \rho_g$	$M, \rho - \rho_c$	Specific heat C		$S(k)$	ξ
Exponent	γ, γ'	$\beta(T \leq T_c)$	$\delta(T = T_c)$	$\alpha(T > T_c)$	$\alpha'(T < T_c)$	η	ν
Experiment							
magnetic system	1.30 ~ 1.37	0.33	4.2	$\geqq 0.1$?	O(ln)		
gas–liquid system	> 1.1?	0.33 ~ 0.36	4.2	$\geqq 0.1$?	O(ln)		
Theory							
mean field approximation	1	1/2	3	discon-tinuous	discon-tinuous		
Ising system (2-dimensional)	7/4	1/8	15	O(ln)	O(ln)	0.25	1
Ising system (3-dimensional)	$\simeq 5/4$	$\simeq 5/16$	$\simeq 5.05$	$\simeq 0.1$	$\simeq 0.1$	$\simeq 0.056$	$\simeq 0.638$

caused by the magnetic field of magnitude 1 at $\boldsymbol{r}'$ and is represented by the fluctuation of magnetization as shown in (4.6.9).

Several inequalities exist between critical exponents:
Inequality of Rushbrooke

$$\alpha' + 2\beta + \gamma' \geqq 2 ; \tag{4.6.17}$$

first inequality of Griffith

$$\alpha' + \beta(\delta + 1) \geqq 2 . \tag{4.6.18}$$

These two inequalities are derived from the inequalities required by the thermodynamic stabilities and thus are considered rigorous. By introducing some plausible assumptions, one obtains:
Second inequality of Griffith

$$\gamma(\delta + 1) \geqq (2 - \alpha)(\delta - 1) ; \tag{4.6.19}$$

inequality of Fisher

$$(2 - \eta)\nu \geqq \gamma ; \tag{4.6.20}$$

inequality of Josephson

$$d\nu \geqq 2 - \alpha ; \tag{4.6.21}$$

inequality of Buckingham-Gunton

$$d\frac{\delta - 1}{\delta + 1} \geqq 2 - \eta , \tag{4.6.22}$$

where d is the dimension of the system.

4.6.2 Phenomenological Theory

Sections 4.1–5 were devoted to the study of the mathematical structures of the partition functions and the Hamiltonians required for the appearance of phase transitions. If phase transitions can be comprehensively studied in this way, the critical phenomena are also understood completely. These studies, however, are quite difficult except for some special models. The phenomena of phase transitions are rather common and furthermore, critical phenomena have a universal character independent of the system considered. This suggests that phenomenological descriptions of critical phenomena are possible and are useful for their understanding without recourse to complicated mathematics. One of the approaches to this end is the phenomenological theory of *Landau* on the second-order phase transition where the existence of a phase transition is assumed from the outset. For example, Curie points in magnetism and the critical points of condensation in gas-liquid systems are second-order transition points, at which magnetizations or densities do not change discontinuously. *Landau* developed the phenomenological theory of the second-order phase transition from the change of symmetry at the transition point. He assumed that the free energy could be expanded in powers of an order parameter. In particular, in magnetism, the magnetization M is the order parameter and the free energy can be written in a form

$$F(T, M) = F(T, 0) + AM^2 + BM^4 + \cdots ,$$

where A and B are temperature-dependent, and no odd terms of M are included by virtue of the fact that F must be invariant under the change of the sign of M. In a 3-dimensional Heisenberg model with three components of M, $M^2 = \sum_{i=1}^{3} M_i^2$ is invariant under the transformation of the coordinate system. The coefficient B, on the other hand, will be dependent on the orientation of the vector $\boldsymbol{M}$, but if one takes the orientation of $\boldsymbol{M}$ which minimizes the free energy and keeps the orientation unchanged while varying the magnitude of $\boldsymbol{M}$, we may regard B as simply temperature dependent. The magnitude of the magnetic field H is given by $H = (\partial F/\partial M)_T$; then in the absence of an external field

$$\left(\frac{\partial F}{\partial M}\right)_T = (2A + 4BM^2 + \cdots)M = 0 \tag{4.6.23}$$

which determines M. The isothermal susceptibility χ_T is given by $(\partial M/\partial H)_T$,

$$\chi_T^{-1} = \left(\frac{\partial^2 F}{\partial M^2}\right)_T = 2A + 12BM^2 + \cdots . \tag{4.6.24}$$

In order to get $M \neq 0$ at $T < T_c$, one has to consider the coefficient B which has the opposite sign of A. At temperatures $T > T_c$, $M = 0$ is the unique solution of (4.6.23) which is thermodynamically stable; in other words, F is at a minimum at $M = 0$, and $A > 0$, $B > 0$. The fact that χ_T diverges for $T \to T_c+$ requires that

$A = 0$ at $T = T_c$. These considerations lead to

$$\begin{aligned} A &= A'(T - T_c) + \cdots, & A' &> 0, \\ B(T) &= B(T_c) + B'(T - T_c) + \cdots, & B(T_c) &> 0, \end{aligned} \tag{4.6.25}$$

in the forms expanded in powers of $T - T_c$. Magnetization at $T < T_c$ is thus given by [see (4.6.23)]

$$M = \left[\frac{A'}{2B(T_c)}\right]^{1/2} (T_c - T)^{1/2} + \cdots. \tag{4.6.26}$$

The critical exponents in Landau's model are obtained as follows. Equation (4.6.26) indicates $\beta = 1/2$. At $T > T_c$, one has $M = 0$ which leads to $\gamma = 1$ from (4.6.24). At $T < T_c$, M is substituted into (4.6.24) from (4.6.26) and gives

$$\chi_T^{-1} = 4A'(T_c - T) \tag{4.6.27}$$

which indicates $\gamma' = 1$. Further,

$$H = \left(\frac{\partial F}{\partial M}\right)_T = [2A'(T - T_c) + 4BM^2]M + \cdots \tag{4.6.28}$$

gives rise to $\delta = 3$ at $T = T_c$. The specific heats C_M and C_H are given by

$$\begin{aligned} C_M &= -T\left(\frac{\partial^2 F}{\partial T^2}\right)_M, \\ C_H - C_M &= T\left(\frac{\partial H}{\partial T}\right)_M^2 \chi_T. \end{aligned} \tag{4.6.29}$$

These relations show that C_M and C_H are finite at $T = T_c$ ($\alpha' = \alpha = 0$), and at $T > T_c$, $C_H = C_M$. At $T < T_c$, (4.6.27, 28) give $C_H - C_M = TA'^2/2B(T_c) + O(T_c - T)$ which shows that C_M is continuous but C_H has a finite jump at $T = T_c$.

The critical exponents thus obtained are the same as those in molecular field theory in Table 4.3, but are inconsistent with experiments. This is partly because we take (4.6.25) in the assumed expansion of the free energy. But essentially, this phenomenological theory only takes account of the average quantities, and the inhomogeneity of the system, or in other words, the fluctuations of physical quantities, are not taken into consideration. There is no room for introducing the correlation length ξ. At the critical temperature, ξ becomes large and large fluctuations are observed ($\chi_T = \infty$). This also implies that a spin is affected by remote spins and the approximations, such as those of Weiss and Bethe which only take nearest neighbors into account, are not appropriate. To improve this, we have to consider the density $n(\boldsymbol{r})$ and the pair distribution function $n^{(2)}(\boldsymbol{r}, \boldsymbol{r}')$ in particle systems, or the spin density $\sigma(\boldsymbol{r})$, etc., in magnetism as functions of $\boldsymbol{r}$. To define $\sigma(\boldsymbol{r})$ as continuous function of $\boldsymbol{r}$, one takes a small volume (but large in comparison with the lattice cell) and averages the spin values in this volume. The interaction of the form (4.1.5) is assumed between the nearest neighbors; then from

$$-s_i^z s_j^z = \frac{1}{2}(s_i^z - s_j^z)^2 - \frac{1}{2}(s_i^z)^2 - \frac{1}{2}(s_j^z)^2 \tag{4.6.30}$$

we can represent the first term of the rhs of (4.6.30) as $(\nabla\sigma)^2$ with continuous $\sigma(\boldsymbol{r})$. Therefore, the free energy of this system can be written

$$(k_B T)^{-1}\mathscr{H}(\sigma) = \int d^d\boldsymbol{r}[a_0 + a_2\sigma^2 + a_4\sigma^4 + c(\nabla\sigma)^2 - h\cdot\sigma], \tag{4.6.31}$$

where k_B is the Boltzmann constant and $\mathscr{H}(\sigma)$ is the free energy when $\sigma(\boldsymbol{r})$ is given, but $(k_B T)^{-1}\mathscr{H}(\sigma)$ has usually been called a Hamiltonian in recent literature. The coefficients a_2 and a_4 are those corresponding to A and B in (4.6.25) and include the second and third terms of (4.6.30), while h is proportional to the magnetic fields. This type of Hamiltonian is employed by *Ginzburg* and *Landau* for discussing superconductivity, superfluidity and critical phenomena and is called the Ginzburg-Landau Hamiltonian. The relative probability for the spin density $\sigma(\boldsymbol{r})$ is given by $\exp[-\beta\mathscr{H}(\sigma)]$. If a spin has n components, we have

$$\sigma^2(\boldsymbol{r}) = \sum_{i=1}^{n} \sigma_i(\boldsymbol{r})^2$$

$$(\nabla\sigma)^2 = \sum_{s=1}^{d}\sum_{i=1}^{n}\left(\frac{\partial\sigma_i}{\partial x_s}\right)^2,$$

where x_s $(s = 1, 2, \ldots, d)$ are the components of r in d-dimensional space. Let L be the length of the system of a d-dimensional cube and $\sigma(\boldsymbol{r})$ be expanded in a Fourier series:

$$\sigma(\boldsymbol{r}) = L^{-d/2}\sum_{k<\Lambda} \phi_k \exp(ik\cdot r)\,, \tag{4.6.32}$$

where Λ is the upper limit of k which is determined by the lattice constant. ϕ_k with $k > \Lambda$ are meaningless since they give us information about the arrangement of the spins in a region smaller than the unit cell. Let $\sigma(r)$, which minimizes the Hamiltonian (4.6.31), be $\tilde{\sigma}(r) = \bar{\sigma}$. Then the Fourier components are $\tilde{\phi}_0 = L^{d/2}\bar{\sigma}$ and the Hamiltonian can be expressed as

$$(k_B T)^{-1}\mathscr{H}(\tilde{\sigma}) = L^d(a_0 + a_2\bar{\sigma}^2 + a_4\bar{\sigma}^4 - h\bar{\sigma})\,. \tag{4.6.33}$$

$\bar{\sigma}$ is obtained by putting the derivative of $\mathscr{H}(\tilde{\sigma})$ equal to zero, i.e.,

$$2\bar{\sigma}(a_2 + 2a_4\bar{\sigma}^2) - h = 0\,. \tag{4.6.34}$$

Equation (4.6.33) is equivalent to (4.6.23). Therefore, the critical exponents derived are the same as those in molecular field theory. The Hamiltonian (4.6.31) has its minimum value given by (4.6.33) and fluctuations around it. The

Hamiltonian (4.6.31) can be written in terms of Fourier coefficients as

$$(k_B T)^{-1}\mathscr{H}(\sigma) = a_0 L^d + \sum_{k<\Lambda} \phi_k \phi_{-k}(a_2 + ck^2)$$
$$+ L^{-d} \sum_{kk'k''<\Lambda} a_4(\phi_k \phi_{k'})(\phi_{k''}\phi_{-k-k'-k''})$$
$$- L^{d/2}\phi_0 \cdot h \,, \tag{4.6.35}$$

where the terms $k = 0$ give $(k_B T)^{-1}\mathscr{H}(\bar{\sigma})$. The terms $k \neq 0$ represent fluctuations. This is sometimes called a ϕ^4 model.

In this way one sees that Landau's theory ignores fluctuations, which are assumed to be large near the critical point and can never be ignored, as noticed already. The Hamiltonian which neglects the term fourth order in ϕ is called the Gaussian approximation. Almost certainly this is not a good approximation near the critical point, but we shall take this model for a while.

Let the n components of ϕ_k be ϕ_{ik} ($i = 1, 2, \ldots, n$), then we have

$$\phi_k \phi_{-k} = \sum_i |\phi_{ik}|^2 \,. \tag{4.6.36}$$

The structure factor $S(k)$ is given by

$$S(k) = \frac{1}{L^d}\int \langle \sigma(r)\sigma(r')\rangle \exp[-ik(r - r')]\, dr\, dr'$$
$$= \langle \phi_k \phi_{-k}\rangle = \sum_i \langle |\phi_{ik}|^2\rangle \,. \tag{4.6.37}$$

In the Gaussian approximation, the distribution of $|\phi_{ik}|$ is a Gaussian function of variance $(a_2 + ck^2)^{-1}/2$ and we have

$$S(k) = \frac{n}{2}(a_2 + ck^2)^{-1} \,. \tag{4.6.38}$$

Assuming the form of (4.6.25) of A for a_2, we have

$$S(0) \propto (T - T_c)^{-1}, \quad \lim_{T\to T_c} S(k) = k^{-2}$$

which also yields $\gamma = 1$ [χ is proportional to $S(0)$, see (1.2.14) or (2.7.11)] and $\eta = 0$. The other exponents are $\delta = 3$, and $\beta = 1/2$. By integrating $\exp[-\beta\mathscr{H}(\sigma)]$ over all the Fourier coefficients, we obtain the partition function from which the specific heat can be calculated with the result

$$\alpha = 2 - d/2 \,. \tag{4.6.39}$$

This result only is different from that of the molecular field theory and implies that the specific heat diverges for $T \to T_c$ if $d < 4$, but no anomaly of specific heat is observed for a hypothetical substance of $d > 4$.

In the Gaussian approximation the fluctuation $\langle\overline{\sigma^2}\rangle$ can be obtained by integrating $S(k)$:

$$\langle \sigma^2 \rangle \propto a_2^{(d-2)/2} \propto |T - T_c|^{(d-2)/2} \,.$$

The ratio of the terms of the fourth and second order in the expansion of (4.6.35) is thus

$$a_4\langle\sigma^4\rangle/a_2\langle\sigma^2\rangle \sim a_4\langle\sigma^2\rangle/a_2 \sim a_4|T - T_c|^{(d-4)/2}$$

which becomes large for $d < 4$ ($d = 2, 3$) and implies that the Gaussian approximation fails to hold near T_c. In a hypothetical system of $d > 4$, the simple molecular field approximation is valid.

4.6.3 Scaling

We have mentioned many times that the correlation length is important near the critical point. The definition of (4.6.13), however, is not general because the functional form is of no universal validity. Thus we define

$$\xi^2 = -\frac{1}{2}\frac{1}{S(0)}\left(\frac{d^2S(k)}{dk^2}\right)_{k=0} \tag{4.6.40}$$

by means of the structure factor $S(k)$ which is experimentally measurable. In the above definition, $S(k)$ is assumed to be a function of $k = |\boldsymbol{k}|$, but otherwise we can define direction-dependent ξ. $S(k)$ has a maximum at $k = 0$. The longer the correlation length, the more rapidly $S(k)$ decreases with increasing k (Fig. 4.22). The rhs of (4.6.40) is equal to the inverse of the radius of curvature of a quantity $S(k)/S(0)$ at $k = 0$ and has a dimension of $[k]^{-2}$. Consequently, ξ has a dimension of length and it agrees with ξ in (4.6.14).

The anomalies near the critical point can be understood as the results of infinitely large correlation length. Therefore, we should measure the length in units of ξ and take into account those quantities which are essential in the limit of $\xi \to \infty$.

Let us consider $S(k)$ defined by (4.6.12). It is a function of k as well as of various lengths, $b_1, b_2, \ldots$. Then by virtue of the scaling of length with ξ, $b_i \to b_i/\xi$ and $k \to k\xi$ in accordance with (4.6.10). Consequently,

$$S(k) \to f(k\xi, b_1/\xi, b_2/\xi, \ldots)\,.$$

In the limit of $\xi \to \infty$,

$$S(k) \sim \xi^y[g(k\xi) + \text{higher order of } \xi^{-1}]\,, \tag{4.6.41}$$

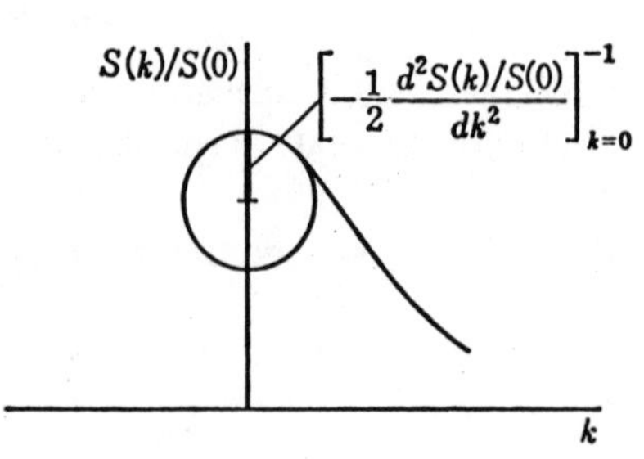

Fig. 4.22. $S(k)/S(0)$ decreases rapidly and the correlation length becomes longer as the radius of curvature of $S(k)/S(0)$ becomes smaller

where y is the lowest order of the powers of ξ^{-1} from the terms of b_i/ξ, in the expression of f. Expression (4.6.16) is obtained in the limit $\xi \to \infty$, provided that

$$g(k\xi) \sim (k\xi)^{-y}$$

and

$$y = 2 - \eta \; . \tag{4.6.42}$$

For spin systems, we have

$$kT\chi = S(0) \sim \xi^{y} g(0) \tag{4.6.43}$$

and

$$\gamma = \nu y \tag{4.6.44}$$

for the exponent of susceptibility (4.6.1), where use is made of (4.6.8). Eliminating y from (4.6.42, 44), one has

$$\gamma = \nu(2 - \eta) \tag{4.6.45}$$

which refers to the equal sign in Fisher's relation (4.6.20).

This method of scaling does not mean a mere change of the unit of length, but is based on the idea that ξ becomes large while the temperature T and the Boltzmann constant k_B are kept unchanged. Consequently, one should use a method which is different from conventional dimensional analysis. Since the magnitude of $k_B T$ is unchanged, the free energy is also unchanged. Let the free energy densities and the volume elements before and after the scaling be F, dV and F', dV', respectively; then we have

$$FdV = F'dV'$$

$$dV' = \xi^{d} dV$$

and thus

$$F = \xi^{-d} F' \cong (T - T_c)^{\nu d} F' \; , \tag{4.6.46}$$

where F' is considered to be independent of ξ for large ξ. Consequently, the specific heat is given by

$$C = -T \frac{\partial^2 F}{\partial T^2} \cong (T - T_c)^{\nu d - 2}$$

and

$$\alpha = 2 - \nu d \tag{4.6.47}$$

for the exponent of specific heat. If we put $M \simeq \xi^{m}$ and $H \simeq \xi^{-h}$, we have from (4.6.43)

$$y = m + h$$

and from (4.6.9) for magnetism and (4.6.41)

$$y = d + 2m \; .$$

These two relations and (4.6.42) yield

$$m = (2 - \eta - d)/2, \quad h = (2 - \eta + d)/2$$

and

$$\delta = -\frac{h}{m} = \frac{d + 2 - \eta}{d - 2 + \eta}, \quad \beta = -\nu m = -\frac{\nu}{2}(2 - \eta - d) . \tag{4.6.48}$$

Instead of using ξ which becomes infinite at the critical temperature, we have another method of scaling. Let the length be measured by a unit of length s times larger. Then ξ becomes $1/s$ times. On the other hand, $\xi \to \infty$ when $T \to T_c$. Thus, one can put

$$t = |T_c - T|/T_c$$

and scale t by

$$t' = s^{1/\nu} t \tag{4.6.49}$$

in accordance with the scaling of ξ mentioned above. This operation means that the scale of temperature is varied by keeping T_c unchanged and, therefore, the temperature is transformed to T' from T and its difference turns out to be

$$T' - T = T_c(t' - t) .$$

This shows that as long as t is small, or in other words, at temperatures close to T_c, T is regarded as invariant. Consequently, the same scaling as (4.6.46) can be applied and we have

$$M' = s^{-m} M, \quad H' = s^{+h} H . \tag{4.6.50}$$

This transformation is physically equivalent to looking at a system by means of a microscope of large magnification and ignoring the details. In fact, ξ becomes shorter by a factor $1/s$ which implies going away from the critical point and magnifying the phenomena close to the critical point, thereby disregarding the irrelevant details in the case $s > 1$. This is the significance of this method of scaling which can derive the relations among the critical exponents and thus proves to be an indispensable tool for the understanding of critical phenomena. However, it fails to give the numerical value of each exponent. To do this, a more powerful method, the renormalization group method, is required.

4.7 Renormalization Group Method

4.7.1 Renormalization Group

Consider a d-dimensional cube of volume L^d with length L as in the case of the Ginzburg-Landau Hamiltonian. We call this system an L-system. The Ls-system with length Ls $(s > 1)$ contains s^d times as many particles as the L-system does and has as many degrees of freedom. For example, the number of Fourier components is $(L/a)^d$ in d-dimensions because it is determined, under the periodic condition, by

$$k_i = \frac{2\pi}{L} n, \quad n = 1, 2, \ldots, n_0 = L/a \qquad (4.7.1)$$

$$i = 1, 2, \ldots, d ,$$

where a is the lattice constant in a regular lattice. The condition $n \leq n_0 = L/a$ is equivalent to the condition $|k| < \Lambda$ in (4.6.32). In the Ls-system, the upper limit of n is Ls/a. If we impose a restriction on k (or n) by $|k| < \Lambda/s$ $(n \leqq n_0/s)$ instead of $|k| < \Lambda$ $(n \leqq n_0)$, the number of components is the same as in the L-system. Alternatively, consider the transformation depicted in Fig. 4.23. Figures 4.23a, 4.23b are L- and $Ls(s = 3)$-systems, respectively. If we disregard the lattices of broken lines, the number of degrees of freedom is unchanged. This is just what has been referred to above as the neglect of the details in the magnified view of a microscope. A mathematical procedure for doing this is to integrate the Fourier components ϕ_k for $\Lambda/s < k < \Lambda$. The new Hamiltonian $\mathscr{H}_{Ls}$ of an Ls-system is defined as

$$\exp(-\mathscr{H}_{Ls}/kT) = \int \cdots \int \exp(-\mathscr{H}_L/kT) \prod_{\Lambda/s<k<\Lambda} d\sigma_k , \qquad (4.7.2)$$

where $\mathscr{H}_{Ls}$ is a function of ϕ_k $(k < \Lambda/s)$.

In the case of Fig. 4.23, where each unit cell is assumed to contain one spin, the neglect of details means to replace s^d spins in a cell of length s, by their average defined by

$$\sigma_i = s^{-d} \sum_{\text{cell}} \sigma_{ci} , \qquad (4.7.3)$$

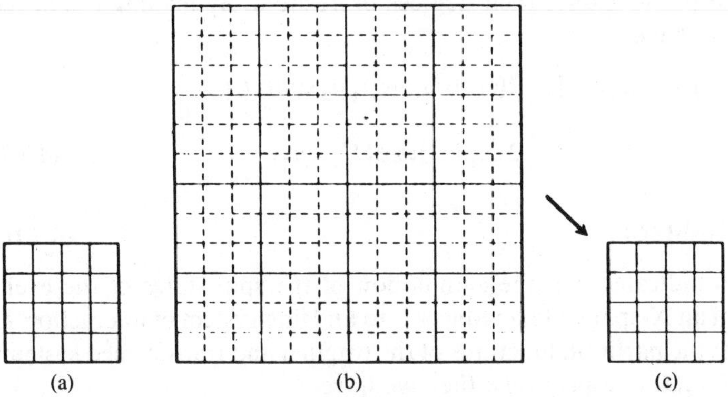

Fig. 4.23. **(a)** → **(b)** Kadanoff transformation, **(a)** → **(b)** → **(c)** renormalization

where σ_{ci} are the spins in the ith cell, and put

$$\exp(-\mathscr{H}_{Ls}/kT) = \sum_{\{\sigma_{ci}\}} \cdots \sum \exp(-\mathscr{H}_L/kT)\delta\left(\sigma_i - s^{-d}\sum \sigma_{ci}\right), \tag{4.7.4}$$

where the δ-function is the n-dimensional one when the spin has n components. These transformations (4.7.2, 4) can be written formally as

$$\mathscr{H}_{Ls} = K_s \mathscr{H}_L \tag{4.7.5}$$

which is called the *Kadanoff transformation*.

The length of a cell is multiplied by s in a Kadanoff transformation. If the length is again shrunk to the original one [the transformation from (b) to (c) in Fig. 4.23]

$$l' = l/s, \quad \text{or} \quad l \to sl' ,$$

the spin density will be transformed to

$$\sigma = \lambda_s \sigma' , \tag{4.7.6}$$

where λ_s can be expressed as s^m, as seen in (4.6.50).

The Kadanoff transformation combined with the scaling (4.7.6) is designated as R_s, and the resulting Hamiltonian $\mathscr{H}_s$ will be written as

$$\mathscr{H}_s = R_s \mathscr{H} . \tag{4.7.7}$$

The transformation R_s is usually nonlinear, but the relation

$$R_{ss'} = R_s R_{s'} \tag{4.7.8}$$

holds. R_s is called the renormalization group, because R_s forms a semigroup lacking inverse elements. The word renormalization comes from the procedure to fold into the new Hamiltonian $\mathscr{H}_s$ the interactions among the irrelevant degrees of freedom which disappear in $\mathscr{H}_s$. An example of a renormalization transformation will be given below. Consider a one-dimensional Ising model discussed in Sect. 4.4.1. Put $J/2kT = L$ and sum over the spin states at the even-number site in order to reduce their degrees of freedom. By making use of the identity (4.4.4) we have

$$\begin{aligned}\sum_{\mu_2 = \pm 1} \exp[L(\mu_1 + \mu_3)\mu_2] &= 2(\cosh^2 L + \mu_1\mu_3 \sinh^2 L) \\ &= 2\exp(L')\exp(L'\mu_1\mu_3) ,\end{aligned} \tag{4.7.9}$$

where

$$\exp(2L') = \cosh(2L) . \tag{4.7.10}$$

Equation (4.7.9) indicates that the elimination of the spin states of the even-numbered site of an N-spin system reduces it to an Ising system of interaction L' of $N/2$ spins. Let the partition functions of the original and transformed systems be $Z_N(L)$ and $Z_{N/2}(L')$, respectively; then we have

$$Z_N(L) = 2^{N/2}\exp(L')^{N/2} Z_{N/2}(L') . \tag{4.7.11}$$

For large N, we can put $Z_N(L) = z(L)^N$ and thus we have

$$z(L)^2 = 2\exp(L')z(L') . \tag{4.7.12}$$

On the other hand, we have from (4.7.10)

$$(2\cosh L)^2 = 2\exp(L')(2\cosh L') ,$$

which is to be compared with (4.7.12), to yield the result

$$Z_N(L) = z(L)^N = (2\cosh L)^N . \tag{4.7.13}$$

This is just (4.4.6).

At the critical points there will be an invariant L under the renormalization transformation. This L is to be determined by

$$\exp(2L) = \cosh 2L \tag{4.7.14}$$

which is satisfied only by $L \to \infty\ (T \to 0)$ or $L \to 0\ (T \to \infty)$, which implies that there is no phase transition in one dimension.

4.7.2 Fixed Point

In a one-dimensional Ising system we have a transformed Hamiltonian of the same functional form. This is not generally true. The relation (4.7.8) implies that repeating R_s many times is equivalent to taking $s \to \infty$. Let a Hamiltonian tend to $\mathscr{H}^*$ by letting $s \to \infty$,

$$\lim_{s \to \infty} R_s \mathscr{H} = \mathscr{H}^* \tag{4.7.15}$$

which is called an invariant Hamiltonian since it does not change by a further application of R_s. At the critical point we can suppose that the invariant Hamiltonian $\mathscr{H}^*$ exists. The way how this $\mathscr{H}^*$ is approached will reveal all the properties inherent and universal to the critical point, because the renormalization procedure consists in magnifying the phenomena near the critical point by eliminating the microscopic details. Consider the Ginzburg-Landau Hamiltonian which has three parameters a_2, a_4 and c in the absence of a magnetic field. In general, these parameters are denoted as a vector μ, which is put as μ^* in the invariant Hamiltonian; μ is transformed to μ' by R_s and μ^* is a fixed point under R_s, as is mentioned in connection with (4.7.15):

$$\mu' = R_s \mu . \tag{4.7.16}$$

For simplicity, let us consider the case of one parameter, i.e., scalar μ. Then the relationship between μ' and μ is represented as shown in Fig. 4.24, where the intersection with a straight line of gradient 1 through the origin is the fixed point μ^*. The fixed point is stable in case (a), but unstable in case (b). The gradient α of the curve is smaller than 1 in case (a) but $\alpha > 1$ in case (b). In the present case with vector μ the same holds. To see this, let μ be close to μ^* and put

$$\mu - \mu^* = \delta\mu . \tag{4.7.17}$$

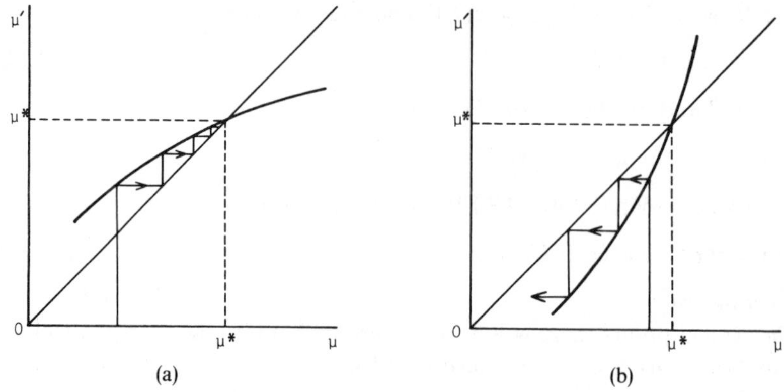

Fig. 4.24. Stable (**a**) and unstable (**b**) fixed points

By linearizing (4.7.16) for small $\delta\mu$ we have

$$\delta\mu' = R_s^L \delta\mu \,, \qquad (4.7.18)$$

where R_s^L is a 3 by 3 matrix in the Ginzburg-Landau model. Let $\rho_i(s)$ and e_i $(i = 1, 2, 3)$ be the eigenvalues and eigenvectors of this matrix. Then $\delta\mu$ can be represented as

$$\delta\mu = \sum_i t_i e_i \qquad (4.7.19)$$

and

$$\delta\mu' = R_s^L t_i e_i = \sum_i \rho_i t_i e_i \,. \qquad (4.7.20)$$

On the other hand, we have the relation similar to (4.7.8)

$$R_s^L R_{s'}^L = R_{ss'}^L \qquad (4.7.21)$$

which implies that R_s^L and $R_{s'}^L$ are commutable, and therefore we have

$$\rho_i(s)\rho_i(s') = \rho_i(ss') \qquad (4.7.22)$$

and can write $\rho_i(s) = s^{a_i}$. When $s > 1$ is assumed, the difference $\delta\mu$ in the direction of i is multiplied by s^{a_i}: it is magnified for $a_i > 0$ $(\alpha = s^{a_i} > 1)$, but contracted for $a_i < 0$ $(\alpha < 1)$ and invariant for $a_i = 0$. Therefore, if $\delta\mu$ lies in a space spanned by the eigenvectors of $a_i < 0$, the fixed-point μ^* can be obtained by letting $s \to \infty$, while if $\delta\mu$ lies in a space spanned by the eigenvectors of $a_i > 0$, μ deviates from μ^*. The critical exponents can be obtained through the process of this deviation.

Let $a_1 > 0 > a_2 > a_3$. Then (4.7.20) becomes

$$\delta\mu' = s^{a_1} t_1 e_1 + O(s^{a_2}) \,. \qquad (4.7.23)$$

Since $\delta\mu' = 0$ at the critical point, t_1 is a function of temperature which vanishes at $T = T_c$;

$$t_1(T) = A(T - T_c) + B(T - T_c)^2 + \cdots, \tag{4.7.24}$$

where $A > 0$ is assumed for simplicity and

$$\mu' = \mu^* + A(T - T_c)s^{a_1}e_1 + \cdots. \tag{4.7.25}$$

By means of the relationship between ξ and $(T - T_c)$, $(T - T_c)s^{a_1}$ is written as $\xi^{-1/\nu}s^{a_1}$ which should be a function of $\xi' = (\xi/s)$. Therefore, the exponent ν can be determined by a_1, an eigenvalue of R_s^L:

$$\nu = 1/a_1 . \tag{4.7.26}$$

This is the success of the renormalization group theory, in contrast to the simple scaling theory.

In the above discussion, we have assumed that one eigenvalue is larger than unity ($a_1 > 0 > a_2 > a_3$). If two eigenvalues larger than unity exist, then we have a tricritical point, where three coexistence phases have the same properties. This tricritical point is observed in ^{4}He and ^{3}He solutions.

In the Gaussian approximation which ignores the quartic terms of ϕ_k, the renormalization procedures can be performed through the well-known integration of the Gaussian function, and the results mentioned already in Sect. 4.6.2 are obtained, $\eta = 0$, $\nu = 1/2$, $\beta = (d - 2)/4$ and $\delta = (d + 2)/(d - 2)$. In the case of the ϕ^4 model, we have to use various methods developed for many-body problems, such as diagram techniques, in order to obtain concrete results. In the Ginzburg-Landau model for $d > 4$, the Gaussian approximation is sufficient. Thus we put $\varepsilon = 4 - d$ and we have the method of ε-expansion. When the number n of components of spin becomes large, Ising models are reduced to spherical models (Sect. 4.4.4) which allows exact treatment. Therefore, we then have the method of $1/n$ expansion.

However, we do not enter into these details, leaving the reader the relevant references at the end of this volume.

4.7.3 Coherent Anomaly Method

In the mean field approximations, we expect better results when we take larger clusters inside of which the fluctuation of spins is taken into account. In practice this is partially the case, as shown in Fig. 4.25. But the critical indices still remain unchanged; for example, 1/2 for magnetization and -1 for susceptibility, see Sect. 4.5.2. The size of the cluster must be increased systematically, and the various fields to be applied to the boundary spins are introduced according to geometrical relations. They are chosen so as to yield average magnitudes of boundary spins equal to that of the central spin. By this process the critical temperature $T_c(l)$ of a cluster of linear dimension l will approach the exact one T_c^* in the limit $l \to 0$. This has been verified by *Suzuki* for a two-dimensional Ising lattice where the exact T_c^* is known. However, the susceptibility, for

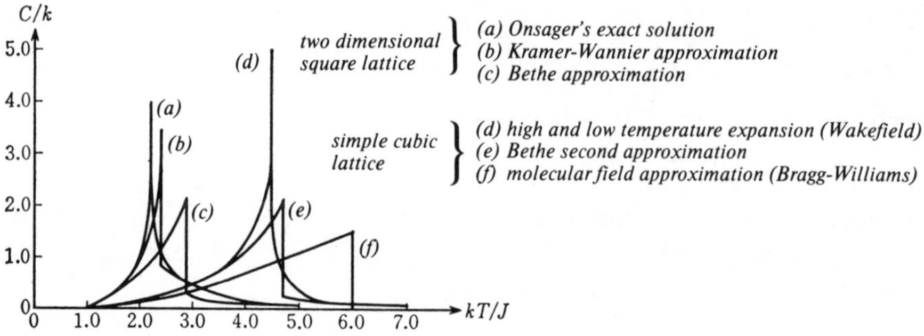

Fig. 4.25. Relationships between specific heat and temperature by various methods

example, has the form with critical index 1:

$$\chi_l(T) = \frac{\bar{\chi}(T_c(l))}{(T - T_c(l))/T_c(l)} \,. \tag{4.7.27}$$

The coefficient $\bar{\chi}(T_c)$ usually increases with l. Consequently, *Suzuki* [4.31] surmised that the exact relation

$$\bar{\chi}(T) \propto (T - T_c^*)^{-\gamma} \tag{4.7.28}$$

will be obtained, if

$$\bar{\chi}(T_c(l)) \propto (T_c^{(l)} - T_c^*)^{-\psi} \tag{4.7.29}$$

holds for large l. This behavior is called a coherent anomaly. The index γ is equal to $\psi + 1$.

The values of $\bar{\chi}(T_c(l))$ are obtained for a systematic series of cluster size l. The indices ψ can also be estimated, and it has been shown that this coherent anomaly method can give critical indices almost correctly when the results for a series of several l's are extrapolated.

5. Ergodic Problems

In the preceding chapters, we have described the general methods of statistical mechanics mostly on the basis of quantum mechanics. But in this chapter, we shall describe the ergodic problems based on classical and quantum mechanics, the reason for which will first be made clear.

The fundamental basis of statistical mechanics is the postulate of equal a priori probabilities, not only in the theory of equilibrium states, but also in dealing with irreversible processes in an isolated system. It is quite reasonable to introduce the concept of probability, but there are two main ways of doing this. Usually only one is described in many textbooks.

One of them is that it is really impossible to consider an actual system to be isolated, as explained in Chap. 1. The textbook of *Landau* and *Lifshitz* [5.1] adopts this viewpoint. Since the system is always more or less subjected to disturbances from the external world, it is permissible to assume that the system exhibits probabilistic behavior in the long run even if it behaves only approximately as an isolated one. In this case, the problem of how to introduce the probability measure without conflicting with the laws of mechanics is intimately connected with the ergodic hypothesis.

The second way is based on this hypothesis, briefly touched on in Chap. 2. The textbook of *Khinchin* is written from this point of view [5.6]. The ergodic hypothesis states that the time average of a mechanical quantity in an isolated system is equal to the ensemble average. In physics, one usually idealizes the objects to be discussed. Thus in statistical mechanics, the systems can be idealized so as to have no disturbance from the surroundings or from the walls of a container consisting of perfectly smooth geometric planes. It is possible that something important might be missing in this idealization. Whether or not the idealization is physically allowable cannot be judged by any other means except the results of the theory. If we can prove the validity of the ergodic hypothesis on idealized isolated systems, the statistical mechanics becomes a systematic theory governed by the logic of mechanics.

The purpose of the present chapter on ergodic problems in idealized systems is to survey the studies done in this field and to suggest novel approaches for the future development of statistical mechanics. The ergodic problem still remains unsolved as a physical problem and has been studied chiefly in classical mechanics. Consequently, this chapter is, in a sense, complementary to the discussions in the introduction (Chap. 1) and is described mainly within classical mechanics, except in the last section. References [5.1–10] serve as general references.

5.1 Some Results from Classical Mechanics

This section is devoted to a short review of classical mechanics, of conservative systems, exclusively.

5.1.1 The Liouville Theorem

Although this theorem was discussed in Chap. 1, we refer to it again.

A dynamical state can be represented by a point in phase space. The state density in phase space is kept constant with the time evolution of the motion. In other words, the volume (Lebesgue measure of a set of points) is constant in the course of time.

5.1.2 The Canonical Transformation

The volume of phase space is invariant under a canonical transformation. In other words, we may say that the absolute value of the Jacobian of the transformation is equal to 1. Because the time evolution of a dynamical system is a canonical transformation, the Liouville theorem follows from this theorem. Both facts make it reasonable to introduce the concept of the equal a priori probability (or measure) in phase space. The canonical transformation is a measure-preserving one.

5.1.3 Action and Angle Variables

Let the system have canonically conjugate variables x_k, y_k $(k = 1, 2, \ldots, n)$, so that the Hamiltonian $\mathscr{H}$ is a function of y_k $(k = 1, 2, \ldots, n)$ only. Namely, x_k are cyclic coordinates. Then the canonical equations of motion become

$$\dot{y}_k = 0 \tag{5.1.1}$$

$$\dot{x}_k = \frac{\partial \mathscr{H}}{\partial y_k} \equiv v_k(y_1, y_2, \ldots, y_n) \tag{5.1.2}$$

and consequently, y_k and v_k are constant. Thus, we can solve the problem

$$x_k = v_k t + \beta_k \,, \tag{5.1.3}$$

where β_k are constants.

Particularly for periodic motions, we can add the value of the period to x_k, say 1, without changing the state of the system. We write this

$$x_k = v_k t + \beta_k \pmod 1 . \tag{5.1.4}$$

Expressed in this way, x_k are called angle variables and y_k action variables. In order to get a precise expression for these quantities, let us consider a two-variable system with the Hamiltonian

$$\mathscr{H} = \frac{1}{2m} p^2 + U(q) \tag{5.1.5}$$

and the canonical equations

$$\dot{p} = -\frac{\partial \mathscr{H}}{\partial q}, \quad \dot{q} = \frac{\partial \mathscr{H}}{\partial p} . \tag{5.1.6}$$

Now suppose that we could have a cyclic coordinate x by a canonical transformation from p, q to y, x; then $\mathscr{H}$ is a function of only y. Let the generating function be $W(q, y)$,

$$p = \frac{\partial W}{\partial q}, \quad x = \frac{\partial W}{\partial y} \tag{5.1.7}$$

and after the transformation, the canonical equations are given by

$$\dot{y} = -\frac{\partial \mathscr{H}}{\partial x} = 0, \quad \dot{x} = \frac{\partial \mathscr{H}}{\partial y} = v(y) , \tag{5.1.8}$$

therefore,

$$y = \text{const.}, \quad x = v(y)t + \text{const.} \tag{5.1.9}$$

Meanwhile, for a periodic motion with the period T and assuming x to change by 1 during this period,

$$\frac{1}{v} = T = \oint dt = \oint \frac{\partial p}{\partial E} dq = \frac{\partial}{\partial E} \oint p dq \tag{5.1.10}$$

holds from the second equation of (5.1.6), where E is the value of $\mathscr{H}$. Now substituting the first equation of (5.1.7) for p in the integral

$$J = \oint p dq , \tag{5.1.11}$$

we get

$$J = \oint \frac{\partial W(q, y)}{\partial q} dq , \tag{5.1.12}$$

therefore, J is found to be a constant because it is a function of y only and it depends on the total energy E through y. We have

$$\frac{1}{v} = T = \frac{\partial J}{\partial E} \tag{5.1.13}$$

from (5.1.10). So far, we have not specified y, but since $\mathscr{H} = \mathscr{H}(y)$, we can take the energy itself as y ($y = E$). Then from the second equation of (5.1.8), the variable conjugate to y is t and the generating function of the transformation $W(q, E)$ becomes, from (5.1.7),

$$\frac{\partial W}{\partial E} = t \,. \tag{5.1.14}$$

Alternatively, we may take J as y because J given by (5.1.12) is a function of y, and then φ is, from (5.1.13),

$$\dot{\varphi} = \frac{\partial \mathscr{H}}{\partial J} = \frac{1}{\partial J/\partial E} = \nu \,, \tag{5.1.15}$$

i.e.,

$$\varphi = \nu t + \text{const.} \tag{5.1.16}$$

Therefore, it turns out that J defined by (5.1.11) is the action variable and φ the angle variable. The above argument can be immediately applied to the system with many degrees of freedom, provided that they are separable in the form

$$\mathscr{H} = \sum_{i=1}^{f} \mathscr{H}_i(p_i, q_i), \quad \mathscr{H}_i = \frac{1}{2m_i} p_i^2 + U_i(q_i) \,. \tag{5.1.17}$$

In this case, the action variables should be taken as

$$J_k = \oint p_k dq_k \,. \tag{5.1.18}$$

As the action variables are expressed by integrals such as (5.1.11) or (5.1.18), they are also called the action integrals. The adiabatic theorem of the action integral or phase volume was mentioned in Sect. 2.3. For a system of n harmonic oscillators

$$\mathscr{H} = \sum_{i=1}^{n} \left(\frac{p_i^2}{2m_i} + \frac{m_i \omega_i^2}{2} q_i^2 \right) . \tag{5.1.19}$$

The transformation

$$P_i = \frac{1}{\sqrt{m_i \omega_i}} p_i, \quad Q_i = \sqrt{m_i \omega_i} q_i \tag{5.1.20}$$

gives

$$\mathscr{H} = \sum_{i=1}^{n} \frac{\omega_i}{2} (P_i^2 + Q_i^2) \,. \tag{5.1.21}$$

Furthermore, putting

$$P_i = \sqrt{\frac{J_i}{\pi}} \cos 2\pi\varphi_i, \quad Q_i = \sqrt{\frac{J_i}{\pi}} \sin 2\pi\varphi_i \,, \tag{5.1.22}$$

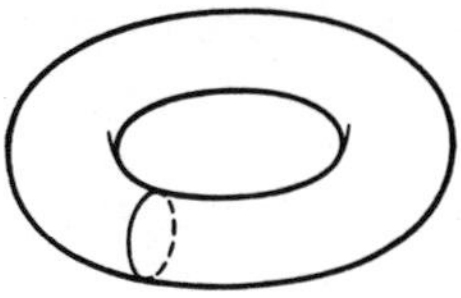

Fig. 5.1. Two-dimensional torus

we obtain

$$\mathscr{H} = \sum_{i=1}^{n} \nu_i J_i\,, \quad \nu_i = \frac{\omega_i}{2\pi} \tag{5.1.23}$$

$$J_i = \frac{1}{2}(P_i^2 + Q_i^2)\,. \tag{5.1.24}$$

J_i's and φ_i's are just the action variables and the angle variables. J_i's are constant and at the time when φ_i's change by 1, P_i, Q_i or p_i, q_i return to their initial values. Accordingly, (5.1.23) expresses an n-dimensional torus (two-dimensional case is shown in Fig. 5.1).

5.1.4 Integrable Systems

A system with n degrees of freedom has $2n$ variables p, q in the equations of motion. But if n integrals $F_1, F_2, \ldots, F_n$ are known and their Poisson bracket vanishes, that is,

$$(F_i, F_j) = 0, \quad (i, j = 1, 2, \ldots, n) \tag{5.1.25}$$

the equations of motion are defined as integrable. F_i, F_j are said to be in involution if they have the relation (5.1.25). In the language of quantum mechanics, they may be said to be commutative. Generally, the system with n integrals in involution is called the integrable one. This theorem was proved by Liouville. The essential point of the proof is to define appropriate action variables and then find a generating function for this transformation. To do this, F_i is taken as a new variable P_i and then its conjugate variable Q_i is a cyclic coordinate:

$$\frac{dQ_i}{dt} = \frac{\partial \mathscr{H}}{\partial P_i}\,, \quad \frac{dP_i}{dt} = -\frac{\partial \mathscr{H}}{\partial Q_i}\,. \tag{5.1.26}$$

Since $P_i = F_i$ is an integral of motion,

$$\frac{dP_i}{dt} = 0\,. \tag{5.1.27}$$

$\mathscr{H}$ does not contain Q_i. Therefore,

$$Q_i = \left(\frac{\partial \mathscr{H}}{\partial P_i}\right) t + \text{const.} \tag{5.1.28}$$

and the problem can be solved, provided that Q_i are known as functions of q and P.

Now, let the integrals of motion be

$$F_i(p_1, p_2, \ldots, p_n, q_1, q_2, \ldots, q_n) = a_i \quad (i = 1, 2, \ldots, n) \tag{5.1.29}$$

and consider the following differential form

$$\sum_{i=1}^{n} p_i \, dq_i \,, \tag{5.1.30}$$

where p_i is given by

$$p_i(q_1, \ldots, q_n, a_1, \ldots, a_n) \tag{5.1.31}$$

as the solution of (5.1.29). Differentiating by q_r the identities obtained by substitution of (5.1.31) into (5.1.29), we obtain

$$\frac{\partial F_i}{\partial q_r} + \sum_k \frac{\partial F_i}{\partial p_k} \frac{\partial p_k}{\partial q_r} = 0 \,. \tag{5.1.32}$$

The condition that the equations have solutions for $\partial p_k/\partial q_r$ $(k = 1, 2, \ldots, n)$ is

$$\det\left(\frac{\partial F_i}{\partial p_k}\right) \neq 0 \,. \tag{5.1.33}$$

Multiplying (5.1.32) by $\partial F_j/\partial p_r$, summing over r, and, furthermore, substracting from it the equation with i and j exchanged in the sum, we arrive at

$$\sum_r \sum_k \partial F_j/\partial p_r \cdot \partial F_i/\partial p_k (\partial p_k/\partial q_r - \partial p_r/\partial q_k) = 0 \tag{5.1.34}$$

by virtue of the relations

$$\sum_r (\partial F_j/\partial p_r \cdot \partial F_i/\partial q_r - \partial F_i/\partial p_r \cdot \partial F_j/\partial q_r) = (F_j, F_i) = 0 \,. \tag{5.1.35}$$

Now, since we have the relations (5.1.33), it must be that

$$\frac{\partial p_k}{\partial q_r} - \frac{\partial p_r}{\partial q_k} = 0 \,. \tag{5.1.36}$$

This means that (5.1.30) is a total differential. Consequently, we can write (5.1.30) as

$$dW = \sum_{i=1}^{n} p_i \, dq_i \tag{5.1.37}$$

and then we have

$$\begin{aligned} W &= W(q_1, q_2, \ldots, q_n, a_1, a_2, \ldots, a_n) \\ &= W(q_1, q_2, \ldots, q_n, P_1, P_2, \ldots, P_n) \end{aligned} \tag{5.1.38}$$

which gives the generating function we wanted.

Since $P_i = \text{const.}$ on the surface given by $F_i = \text{const.}$ ($i = 1, 2, \ldots, n$), they form an n-dimensional torus, provided that this surface is a connected and compact smooth manifold. This theorem was proved by *Arnold*. The simplest example of such integrable systems is a system of harmonic lattice oscillators. The motion on the torus is usually ergodic (Sect. 5.3.2). However, almost all nonlinear Hamiltonian systems are not integrable, and these systems exhibit chaotic behaviors, as will be seen below.

5.1.5 Geodesics

The motion of a dynamical system is derived from the following variational principle:

$$\delta \int_A^B L(q, \dot{q}, t)\,dt = 0\,, \tag{5.1.39}$$

where L is the Lagrangean and q represents the set $(q_1, q_2, \ldots, q_n)$. Equation (5.1.39) means that among all paths connecting two points $A(q(t_1), t_1)$ and $B(q(t_2), t_2)$ in q, t-space, the actual motion is the one corresponding to the minimum of the integral of L along the path. Since the energy is conserved in actual motions, we may limit ourselves to the paths with constant energy. Let K be the kinetic energy

$$L = K - V = 2K - \mathscr{H} \tag{5.1.40}$$

and then the variational principle can be expressed in the following form

$$\delta \int_A^B 2K\,dt = 0 \tag{5.1.41}$$

because of $\delta\mathscr{H} = 0$. Expressing K as $K = (\mathscr{H} - V)^{1/2} K^{1/2}$ and using the general expression of the kinetic energy

$$K = \frac{1}{2} \sum_{i,j=1}^{n} k_{ij}(q)\dot{q}_i\dot{q}_j\,, \tag{5.1.42}$$

we can write

$$\delta \int_A^B 2K\,dt = \delta \int_A^B \sqrt{2}(\mathscr{H} - V)^{1/2}\left(\sum k_{ij}\dot{q}_i\dot{q}_j\right)^{1/2} dt = \delta \int_A^B \sqrt{2}\,ds \tag{5.1.43}$$

$$ds^2 = (\mathscr{H} - V) \sum_{i,j=1}^{n} k_{ij}\,dq_i\,dq_j\,. \tag{5.1.44}$$

Since $\mathscr{H}$ is constant and V is a function of q, the coefficients in (5.1.44) are functions of q alone. Consequently, when we consider a Riemannian space with the metric defined by (5.1.44) in q-space, the equation of motion in dynamics can be represented by the geodesic line in the space. This is a geometric description of Newtonian dynamics.

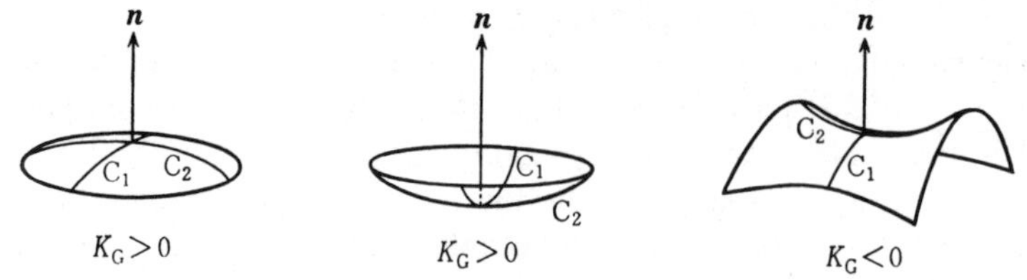

Fig. 5.2. Surfaces with positive or negative Gaussian curvature

The dynamical system of negative curvature is of some interest. It is appropriate to add a short comment on Gaussian curvature.

Consider a curved surface in 3-dimensional Euclidean space, and let the normal at a point P on the surface and a tangential vector through P be denoted by $\boldsymbol{n}$ and $\boldsymbol{t}$, respectively. The intersection of the surface and the plane which contains the vectors $\boldsymbol{n}$ and $\boldsymbol{t}$ forms a curve C. As the direction of tangent $\boldsymbol{t}$ changes, the curvature of C takes a different value. The tangents corresponding to the maximum and minimum curvatures, say K_1 and K_2 and C_1 and C_2 for the corresponding curves, respectively, are perpendicular to each other. The product $K_G = K_1 K_2$ is called the Gaussian curvature and the average of the two $K = (K_1 + K_2)/2$ is the mean curvature. The vector $\boldsymbol{n}$ is the normal to the curve C, but the choice of positive direction is arbitrary and so the signs of K_1 and K_2 are not uniquely determined. The Gaussian curvature, however, has a definite sign, depending on whether C_1 and C_2 are both in the same side of the tangential plane at P (in this case, $K_G > 0$) or in the sides opposite each other (in this case, $K_G < 0$). The surface with $K_G > 0$ ($K_G < 0$) is said to have positive (negative) curvature (Fig. 5.2).

The tangent plane of a surface in n-dimensional space is of $(n-1)$-dimension. Let the independent fundamental vectors of the tangential plane be $\boldsymbol{e}_i$ $(i = 1, 2, \ldots, n-1)$ and consider the 3-dimensional space constructed by any arbitrary two vectors $\boldsymbol{e}_i$, $\boldsymbol{e}_j$ and the normal $\boldsymbol{n}$ of the surface. The intersection of the $(n-1)$-dimensional surface with the 3-dimensional space $(\boldsymbol{e}_i, \boldsymbol{e}_j, \boldsymbol{n})$ gives a curve in 3-dimensional space. If these curves have always positive or negative curvatures, the surface in the n-dimensional space is said to be of positive or negative curvature.

In Riemannian geometry, we can express the curvature in terms of a metric tensor. Thus, we may discuss dynamical systems as motions in the Riemannian space defined by (5.1.44).

5.2 Ergodic Theorems (I)

When we want to measure a physical quantity, we usually need a certain time interval τ. In particular, for the measurement of a thermodynamic quantity, τ is usually taken to be longer than τ_0 and shorter than τ_m. Here τ_0 is the time necessary for the vanishing of the correlation of molecular motion and τ_m is the maximum time during which a macroscopic property does not change. For any τ in between τ_0 and τ_m, the observed value is not supposed to depend on τ. Further, if the system is in thermal equilibrium, the specification of τ_m is not necessary but can be extended to infinity.

A dynamical state of the system is represented by a point P in phase space which moves according to the law of dynamics in a definite manner, if the system is isolated. Let us write P_t for the position at time t of the initial point P and ϕ_t for the transformation from P to P_t. The transformation preserves the measure according to the Liouville theorem. Let $f(P_t)$ be a phase function that is defined by a function of P_t. Hence we can write

$$P_t = \phi_t P \quad \text{and} \quad f(P_t) = f(P, t) \,. \tag{5.2.1}$$

Our measurement[1] thus involves, as mentioned above, the time average

$$\bar{f}_\tau(P) = \frac{1}{\tau}\int_0^\tau f(P, t)\,dt \tag{5.2.2}$$

or

$$\bar{f}(P) = \lim_{\tau\to\infty} \frac{1}{\tau}\int_0^\tau f(P, t)\,dt \,. \tag{5.2.3}$$

If we now consider the isolated system of constant energy, this dynamical system defines a trajectory through P in phase space. *P.* and *T. Ehrenfest* [5.9] gave it the name of a *G*-path. Equation (5.2.2) or (5.2.3) is the time average along a *G*-path from $t = 0$ to $t = \tau$ or ∞. Since for a thermodynamic description of a system one needs only a few thermodynamic variables, the system with the same thermodynamic state, i.e., those with the same values of thermodynamic quantities, are usually in different dynamic states. This means that if $f(P, t)$ is a phase function corresponding to a thermodynamic quantity, (5.2.2 or 3) should not depend on its initial state P. Moreover, according to our experience, $\bar{f}_\tau$ equals $\bar{f}$ for $\tau > \tau_0$ in thermal equilibrium.

Consequently, in order for thermodynamics to hold, the following requirements must be satisfied. First of all, it is necessary that (i) two times τ_0, τ_m ($\tau_0 \ll \tau_m$) exist and the average $\bar{f}_\tau$ be independent of τ for $\tau_0 < \tau < \tau_m$. From the empirical fact that an isolated system finally reaches equilibrium after a long enough time, we could take $\bar{f}_\tau$ as equal to $\bar{f}$ for large enough τ_m. Even if the initial condition is not in equilibrium, $\bar{f}$ is always equal to the value in equilibrium for $\tau \to \infty$. Consequently, instead of (i), we can relax the condition by accepting that (i′) $\bar{f}$ exists (Birkhoff's first theorem). If (i) holds, (i′) does too, but the converse is not always true. Besides this it is also necessary that (ii) $\bar{f}_\tau(P)$

[1] Whether actual measurements are like this or not is a difficult problem in the theory of measurement of classical systems.

We may say the measurement defined by (5.2.2, 3) is an ideal measurement, which is our main concern in this chapter.

or $\bar{f}(\mathrm{P})$ has a definite value independent of the initial condition P. Now we define the phase average of f:

$$\langle f\rangle = \frac{1}{\Omega}\int_{\Gamma} f(\mathrm{P}, t)\, d\Gamma$$
$$\Omega = \int_{\Gamma} d\Gamma\,, \tag{5.2.4}$$

where the integral is taken over the accessible phase space of volume Ω, and $d\Gamma$ is its volume element. Since the integral region is confined to the surface of constant energy, we have

$$\int_{\Gamma} \cdots\, d\Gamma = \int_{\Sigma} \frac{\cdots\, d\Sigma}{|\mathrm{grad}\, E|}\,,$$

where $d\Sigma$ is the surface element. The Jacobian for the canonical transformation is 1 and the integral value does not depend on the canonical variables chosen. $\langle f\rangle$ may seem to depend on t from (5.2.4), but it does not because of (5.2.1) and the measure preserving transformation of ϕ_t. Therefore, $\langle f\rangle$ equals its time average $\overline{\langle f\rangle}$. The time average is also commutative with the phase average, except in special cases. Then

$$\langle f\rangle = \overline{\langle f\rangle} = \langle \bar{f}\rangle\,. \tag{5.2.5}$$

Therefore, from the property (ii) that $\bar{f}$ is independent of P and constant over all phase space, the phase average of $\bar{f}$ is equal to $\bar{f}$, $\langle \bar{f}\rangle = \bar{f}$ and we finally have

$$\langle f\rangle = \bar{f}\,. \tag{5.2.6}$$

This property justifies the use of the phase average, or in other words the ensemble average, instead of the time average. This is the fundamental concept in statistical mechanics.

It can easily be proved from (i′) that the value of $\bar{f}(\mathrm{P})$ is the same for any P on the same G-path (Birkhoff's second theorem). Hence the problem is whether (ii) is satisfied for P on different G-paths or not.

Historically, *L. Boltzmann* first suggested that for satisfying (5.2.6), a G-path should be assumed to pass through all points on the energy surface. Since the G-path then goes everywhere on the surface sooner or later, the time average is equal to the phase average. Therefore, Boltzmann called the above assumption an ergodic hypothesis (*ξργον* means work, energy, and *ὁδός* means path). In the following, this property of the G-path will be called ergodicity in Boltzmann's sense. A dynamical trajectory is, however, a set of one-dimensional continuous points and never intersects itself on a many-dimensional surface of constant energy. Because it is impossible to make a continuous one-to-one correspondence of two or greater dimensional space to one-dimensional space[2], ergodicity

[2] The Peano curve can cover many-dimensional space with a one-dimensional continuous curve, but this correspondence is not one-to-one. We can also find a one-to-one correspondence of many-dimensional space to a one-dimensional one in set theory, but it is not continuous.

in Boltzmann's sense cannot hold. Therefore, the existence of (5.2.6) is now called ergodicity and the hypothesis requiring this relation is the ergodic hypothesis. Thus ergodicity does not always imply ergodicity in the sense of Boltzmann.

5.2.1 Birkhoff's Theorem

The condition (i′) mentioned above is expressed in the following. Let the volume V of a subspace in phase space be invariant for the transformation ϕ_t (this is called the invariant subspace, expressed by the equation $\phi_t V = V$) and $f(\mathrm{P})$ ($\mathrm{P} \in V$) be L_1-integrable in V; then, for almost all P in V, the long-time average of f exists (Birkhoff's first theorem)

$$\bar{f} = \lim_{T\to\infty} \frac{1}{T} \int_0^T f(\mathrm{P}_t)\, dt\,. \tag{5.2.7}$$

This proof was given by *Birkhoff* [5.11]. An elementary proof of this theorem was given by *A. Kolmogorov* which was reviewed in the book by *Khinchin* [5.6]. According to the theorem, f may depend on P at $t = 0$ but the second theorem states that it has the same value for all P on the G-path. From (5.2.7)

$$\lim_{T\to\infty} \frac{1}{T+t} \int_0^{T+t} f(\mathrm{P}, t')\, dt' = \bar{f}(\mathrm{P})\,. \tag{5.2.8}$$

Meanwhile,

$$\frac{1}{T} \int_0^{T+t} f(\mathrm{P}, t')dt' - \frac{1}{T+t} \int_0^{T+t} f(\mathrm{P}, t')dt' = \frac{t}{T(T+t)} \int_0^{T+t} f(\mathrm{P}, t')\, dt' \tag{5.2.9}$$

and then the right-hand side of (5.2.9) tends to zero as $T \to \infty$ when (5.2.8) holds. Therefore,

$$\lim_{T\to\infty} \frac{1}{T} \int_0^{T+t} f(\mathrm{P}, t')\, dt' = \bar{f}(\mathrm{P})\,. \tag{5.2.10}$$

Now,

$$\begin{aligned} \frac{1}{T} \int_0^T f(\mathrm{P}, t_0 + t')\, dt' &= \frac{1}{T} \int_{t_0}^{T+t_0} f(\mathrm{P}, t')\, dt' \\ &= \frac{1}{T} \int_0^{T+t_0} f(\mathrm{P}, t')\, dt' - \frac{1}{T} \int_0^{t_0} f(\mathrm{P}, t')\, dt'\,. \end{aligned} \tag{5.2.11}$$

As $T \to \infty$, the first term on the right-hand side of (5.2.11) tends to $\bar{f}(\mathrm{P})$ by (5.2.10) and the second term goes to zero. The left-hand side may be rewritten as

$$\lim_{T\to\infty} \frac{1}{T} \int_0^T f(\mathrm{P}, t_0 + t)\, dt = \lim_{T\to\infty} \frac{1}{T} \int_0^T f(\mathrm{P}_{t_0}, t)\, dt = \bar{f}(\mathrm{P}_{t_0})$$

and we have

$$\bar{f}(\mathrm{P}_{t_0}) = \bar{f}(\mathrm{P}) \tag{5.2.12}$$

which proves the second theorem. Consequently, we can have the *fundamental theorem*:

The necessary and sufficient condition that the time evolution ϕ_t is ergodic is that all the invariant functions[3] f are constant.

It is necessary because, if f is an invariant and ergodic function, $f(\mathrm{P}_t) = f(\mathrm{P}) = \bar{f}$ is equal to $\langle f \rangle$ which is a constant independent of P. It is also sufficient because, as mentioned in connection with (5.2.6), when $\bar{f}$ is constant independent of P, it is ergodic.

The above Birkhoff's theorem is called an individual ergodic theorem. That is because the theorem is related to each G-path separately as defined by (5.2.7). On the other hand, in 1929, *J. von Neumann* proved the mean ergodic theorem in quantum mechanics. Stimulated by this, *Birkhoff* [5.11] obtained the ergodic theorem in classical mechanics in 1931.

5.2.2 Mean Ergodic Theorem

Koopman [5.12] noticed that the transformation can be represented by a linear operator in functional space. That is, we define the operator U_t by

$$U_t f(\mathrm{P}) = f(\phi_t \mathrm{P}) . \tag{5.2.13}$$

The operator U_t is called the operator induced by ϕ_t, since U_t is represented on a Hilbert space constructed on Lebesgue L_2-integrable functions.

Now, if the variable t is continuous, as we consider here, the transformation ϕ_t is called a flow. We can introduce the induced operator U for an abstract dynamical system to be mentioned later (Sect. 5.3). Since P is a point in phase space in Hamiltonian systems, its function $f(\mathrm{P})$ changes according to the equation expressed in Poisson bracket as

$$\frac{df}{dt} = (f, \mathscr{H}) = -\mathrm{i}\mathscr{L} f .$$

$\mathscr{L}$ defined by this equation is a self-adjoint operator and called a Liouville operator. Therefore, U_t in (5.2.13) can be written in the form

$$U_t = \exp(-\mathrm{i}\mathscr{L} t) .$$

Clearly U_t is a unitary operator.

von Neumann proved that there exists a certain function f^* such that

$$\left\| \frac{1}{\tau} \int_0^\tau f(\mathrm{P}, t) dt - f^*(\mathrm{P}) \right\| \to 0 \tag{5.2.14}$$

with $\tau \to \infty$, where $\|g\|$ is the distance in the Hilbert space given by

$$\|g\|^2 = \int_V |g|^2 dV . \tag{5.2.15}$$

[3] f is called a ϕ_t-invariant function or simply an invariant function, if $f(\mathrm{P}_t) \equiv f(\phi_t \mathrm{P}) = f(\mathrm{P})$.

Convergency defined by (5.2.14) is called the average or strong convergence. Average convergence means that the standard deviation of the difference between the average over τ and $f^*(\mathrm{P})$ tends to zero with $\tau \to \infty$.

von Neumann's average convergence is sufficient for statistical mechanics, since averages are its main concern. Birkhoff's individual ergodic theorems give stronger results and are more important in getting the dynamics. However, von Neumann's theorem has an advantage in the possible application of the notion of spectral resolution used in quantum mechanics.

Let g_λ be the eigenfunction of the eigenvalue λ of U_t, then

$$U_t g_\lambda = \lambda g_\lambda \, . \tag{5.2.16}$$

Since U_t is unitary, $|\lambda| = 1$ [therefore we write it in the form $\lambda = \exp(2\pi i \mu t)$ and sometimes also call μ the eigenvalue]. The above relation means $g_\lambda(\phi(x)) = \lambda g_\lambda(x)$ and then

$$|g_\lambda(\phi(x))| = |g_\lambda(x)| \, .$$

That is, the absolute value of the eigenfunction is an invariant quantity. If the system is ergodic, almost all invariant quantities are constant almost everywhere (*fundamental theorem*, Sect. 5.2.1), and thus the absolute value of the eigenfunction of U_t is constant almost everywhere too. Also, if another eigenfunction h_λ other than g_λ exists,

$$U_t\left(\frac{h_\lambda}{g_\lambda}\right) = \frac{\lambda h_\lambda}{\lambda g_\lambda} = \frac{h_\lambda}{g_\lambda} \, , \tag{5.2.17}$$

then h_λ / g_λ is invariant and constant from the *fundamental theorem*. Consequently, the eigenfunction is simple (not degenerate). Thus we have the following theorem:

The necessary and sufficient condition for a system to be ergodic is that U_t has 1 as its simple eigenvalue. Proof: let f be an invariant function of U_t,

$$U_t f = f \, , \tag{5.2.18}$$

i.e., f is an eigenfunction of eigenvalue 1. A constant is also an eigenfunction of eigenvalue 1 of U_t. Since λ is simple, the eigenfunction of $\lambda = 1$ is unique and hence the invariant function f must be a constant. Thus, this theorem holds from the *fundamental theorem*.

An invariant quantity of U_t is also called a spectral invariant. Spectral invariants are, of course, ϕ-invariants.

5.2.3 Hopf's Theorem

While von Neumann's theorem is a theory of strong convergence of the time average of a phase function f, *Hopf* [5.7] proved the following theorem on the inner product (f, g) of a pair of functions f and g belonging to the L_2 function.

A function f^* exists, satisfying

$$\lim_{\tau\to\infty} \frac{1}{\tau} \int_0^\tau |(U_t f, g) - (f^*, g)|^2 \, dt = 0 \qquad (5.2.19)$$

for any phase function f, where g is an arbitrary function. We can examine the convergency of a distribution function ρ in phase space on the basis of this theorem. In Hamiltonian systems, ρ changes according to

$$\frac{\partial \rho}{\partial t} = \mathrm{i}\mathscr{L}\rho \qquad (5.2.20)$$

with the Liouville operator. Therefore,

$$\rho_t = U_{-t}\rho(0) \, . \qquad (5.2.21)$$

In general dynamical systems, the time evolution of a phase function is represented by U_t, and the time evolution of the distribution function is by U_{-t}. Then we have for the average of $f(P)$ with respect to $\rho_t = \rho(\mathrm{P}, t)$,

$$\int f(\mathrm{P})\rho_t dV = (f, \rho_t) = (f, U_{-t}\rho) = (U_t f, \rho) \qquad (5.2.22)$$

and if the last form has a limit for $t \to \infty$ in the sense of Hopf, the limit ρ^* of $U_{-t}\rho$ too. The limit for $t \to \infty$ of (5.2.22) will be discussed in Sect. 5.8.1.

The theorems of Birkhoff, von Neumann and Hopf mentioned above are concerned with the convergence of a phase function or distribution function. They cannot prove ergodicity by themselves. Besides, condition (ii) explained in Sect. 5.2 is required. The property that assures the condition is a metrical transitivity to be discussed next.

5.2.4 Metrical Transitivity

Let V denote an invariant subspace of phase space Γ for the transformation ϕ_t defined in (5.2.1). When V cannot be decomposed in two invariant subspaces V_1, V_2 with positive measure, V is said to be metrically transitive or metrically indecomposable. We assume the measure of V to be 1 [$\mu(V) = 1$]. When V is metrically transitive, an arbitrary measurable invariant set in V has the measure of either 1 or 0. The concept of metrical transitivity was introduced by *Birkhoff* [5.13] and *Smith* [5.14]. Then the following theorem holds.

If a measurable invariant space V is metrically transitive, the time mean (5.2.7) of any measurable phase function is equal to the phase mean (5.2.4) for almost all points P on V. The converse is true, too. In other words, metrical transitivity is equivalent to ergodicity. Proof: If V is metrically transitive, the long-time average of a phase function is a constant on almost all G-paths, because otherwise we can assume that V can be decomposed into V_1 and V_2 of positive measure such that $\bar{f}(\mathrm{P}) > \alpha$ on the G-path from a point on V_1 and $\bar{f}(\mathrm{P}) \leq \alpha$ from a point on V_2. The G-path through V_1 or V_2 cannot go out from V_1 or V_2, respectively, and hence V_1 and V_2 are invariant subspaces, which contradicts the assumption of metrical transitivity of V. Since the long-time average of any phase function is constant on V, ergodicity follows from the *fundamental theorem*. Conversely, assume that V can be decomposed into two invariant subspaces V_1 and V_2 of positive measure. Consider a phase function f such that it takes the value 1 on V_1 and 0 on V_2, then $\bar{f}$ becomes 1 on V_1 and 0 on V_2, and the phase average takes a value between 0 and 1, which is different from the time average and hence nonergodicity follows.

5.2.5 Mixing

Owing to Birkhoff's work, the study of ergodicity of a dynamical system can be reduced to that of metrical transitivity. One of the properties that are responsible for metrical transitivity is mixing. This is defined as follows. Consider the measurable subspaces A, B in the space V. A is subjected to the transformation ϕ_t as a dynamical system. A dynamical system is called mixing when we have, for any pair of A, B,

$$\lim_{t\to\infty} \mu[(\phi_t A)\cap B] = \mu(A)\mu(B) \tag{5.2.22}$$

with normalization $\mu(V) = 1$, where μ denotes the measure and the intersection $(\phi_t A)\cap B$ means the common part of $\phi_t A$ with B. $\mu[\phi_t(A)\cap B]/\mu[B]$ is the fraction of A in B after the elapse of time t, and (5.2.22) says that after a sufficiently long time, it is equal to the fraction (measure) of A as a whole.

Theorem: A dynamical system is ergodic if it is mixing.

Proof: Let A be an invariant measurable set and take $B = A$.
Then we get

$$(\phi_t A)\cap A = A \tag{5.2.23}$$

and from (5.2.22), $\mu(A) = 0$ or 1, which is a condition for metrical transitivity.

5.2.6 Khinchin's Theorem

Khinchin [5.2] proposed another method for studying ergodicity without recourse to metrical transitivity. Since the system under consideration usually consists of very many particles, a phase function $f(\mathrm{P})$ that contains the coordinates of only one (or a few) particles among them will have no correlation with the initial value after sufficient time has elapsed. Then we define the correlation coefficient

$$R(u) = \frac{1}{Df}\langle f(\mathrm{P},t)f(\mathrm{P},t+u)\rangle\,, \tag{5.2.24}$$

where $\langle\ \rangle$ denotes the phase average. Because the average is taken over the energy surface, it can be written as

$$\langle\cdots\rangle = \frac{1}{\Omega(\Sigma)}\int_\Sigma \cdots \frac{d\Sigma}{|\mathrm{grad}\, E|}\,. \tag{5.2.25}$$

Further, we put

$$Df = \langle f^2(\mathrm{P}) \rangle \, . \tag{5.2.26}$$

Without loss of generality, the average of f can be taken as zero:

$$\langle f \rangle = 0 \, . \tag{5.2.27}$$

Let the upper bound of $|f|$ on Σ be M. The correlation coefficient $R(u)$ is always smaller than 1 (Schwarz inequality) and it should be proportional to $\langle f \rangle^2$ if correlation disappears for $u \to \infty$. Since $\langle f \rangle = 0$ is assumed, we shall have $R(u) \to 0$ for $u \to \infty$. *Khinchin* [5.2] proved the following:

Theorem: The phase function $f(\mathrm{P})$ is ergodic if $R(u) \to 0$ with $u \to \infty$.

Proof: Put

$$\lim_{\tau \to \infty} \frac{1}{\tau} \int_0^\tau f(\mathrm{P}, t) dt = \bar{f}(\mathrm{P}) \, . \tag{5.2.28}$$

Then we can write

$$\begin{aligned}\langle \bar{f}(\mathrm{P})^2 \rangle &= \frac{1}{\Omega(\Sigma)} \int_\Sigma \frac{d\Sigma}{|\mathrm{grad}\, E|} \bar{f}(\mathrm{P})^2 \\ &= \frac{1}{\Omega(\Sigma)} \int_\Sigma \frac{d\Sigma}{|\mathrm{grad}\, E|} \left[\bar{f}(\mathrm{P})^2 - \frac{1}{\tau^2} \int_0^\tau \int_0^\tau f(\mathrm{P}, u) f(\mathrm{P}, v) \, du \, dv \right] \\ &\quad + \frac{1}{\Omega(\Sigma)} \int_\Sigma \frac{d\Sigma}{|\mathrm{grad}\, E|} \frac{1}{\tau^2} \int_0^\tau \int_0^\tau f(\mathrm{P}, u) f(\mathrm{P}, v) du \, dv \, . \end{aligned} \tag{5.2.29}$$

Now we define

$$Q = \bar{f}(\mathrm{P})^2 - \left[\frac{1}{\tau} \int_0^\tau f(\mathrm{P}, t) \, dt \right]^2 \tag{5.2.30}$$

and split the integral region of the first term containing Q in the rhs of (5.2.29) into G_ε where $|Q| < \varepsilon$ and the remaining G'_ε. Since $Q \to 0$ for almost all P on the surface Σ, the measure $\mu(G'_\varepsilon)$ becomes smaller with $\tau \to \infty$ and we can get

$$\frac{\mu(G'_\varepsilon)}{\Omega(\Sigma)} = \frac{1}{\Omega(\Sigma)} \int_{G'_\varepsilon} \frac{d\Sigma}{|\mathrm{grad}\, E|} < \varepsilon \, . \tag{5.2.31}$$

Then (5.2.29) is

$$\begin{aligned}\langle \bar{f}(\mathrm{P})^2 \rangle &= \frac{1}{\Omega(\Sigma)} \frac{1}{\tau^2} \int_0^\tau \int_0^\tau du \, dv \int_\Sigma \frac{f(\mathrm{P}, u) f(\mathrm{P}, v)}{|\mathrm{grad}\, E|} d\Sigma \\ &\quad + \frac{1}{\Omega(\Sigma)} \int_{G_\varepsilon} \frac{Q \, d\Sigma}{|\mathrm{grad}\, E|} + \frac{1}{\Omega(\Sigma)} \int_{G'_\varepsilon} \frac{Q \, d\Sigma}{|\mathrm{grad}\, E|} \, . \end{aligned} \tag{5.2.32}$$

The first term is represented by $R(u - v)$ and, as mentioned above, the second term is smaller than ε and the third term is smaller than $M^2 \varepsilon$. Therefore,

$$\langle \bar{f}(\mathrm{P})^2 \rangle \leqq \left| \frac{Df}{\tau^2} \int_0^\tau \int_0^\tau R(u - v) \, du \, dv \right| + \varepsilon + M^2 \varepsilon \, .$$

According to the assumption of $R(u) \to 0 (u \to \infty)$, we can take $|R(u)| < \varepsilon$ if $|u| > u_0$ and

$$\langle \bar{f}(\mathrm{P})^2 \rangle \leqq \frac{Df}{\tau^2} \int_0^\tau du \int_{\max(0, u-u_0)}^{\min(\tau, u+u_0)} |R(u-v)|\, dv + \frac{\varepsilon Df}{\tau^2} \int_0^\tau \int_0^\tau du\, dv + \varepsilon + M^2 \varepsilon$$

$$\leqq \frac{2u_0 Df}{\tau} + \varepsilon(Df + 1 + M^2) \,. \qquad (5.2.33)$$

Consequently, for sufficiently large τ and sufficiently small ε, we have

$$\langle \bar{f}(\mathrm{P})^2 \rangle = 0$$

or, almost everywhere on the energy surface

$$\bar{f}(\mathrm{P}) = 0 = \langle f(\mathrm{P}) \rangle \,. \qquad (5.2.34)$$

Therefore, $f(\mathrm{P})$ is ergodic.

Now, ergodicity in the sense of Boltzmann can never hold, as mentioned already, but it may be possible that if G-paths are assumed to pass through any vicinity (however small it is) of every point, a phase function that is a continuous function of P may satisfy condition (ii), or in other words, $\bar{f}(\mathrm{P})$ has a constant value independent of the initial state P. This is called a quasi-ergodic hypothesis. *ter Haar* [5.3] said that if the system is quasi-ergodic, the energy surface is metrically indecomposable because the trajectory will always cross any region A of positive measure on an energy surface. In opposition to this view, *Van Hove* [5.15] noted that if the region A is an open set, the argument is valid but otherwise it is not the case. For example, take A as a complementary set of a trajectory. If the trajectory is quasi-ergodic, the complementary set of the trajectory has no inner point anywhere and is not an open set. Then it is necessary for ergodicity that the measure of the complementary set is zero.

5.3 Abstract Dynamical Systems

5.3.1 Bernoulli Schemes and Baker's Transformation

Ergodicity is initially discussed as a global property of a Hamiltonian system, but it is often generalized to systems that are defined by one-parameter automorphisms[4] ϕ in measurable space.

---

[4] Consider a measurable space (M, μ), where μ is a measure defined on a manifold M. (M, μ) is transformed to (M', μ') in one-to-one correspondence by a transformation ϕ_t of one-parameter t, except for the region of measure 0. Put $\phi_t A = A'$, and if $\mu(A) = \mu'(A')$ (measure preserving), ϕ_t is an isomorphism, and moreover if $(M, \mu) = (M', \mu')$, it is an automorphism.

We call this system with automorphic ϕ_t an abstract dynamical system (M, μ, ϕ_t). If the transformation is continuous and differentiable any number of times, this system is called diffeomorphism, or a classical dynamical system.

Sometimes a transformation ϕ may take a discrete parameter. For example, the transformation defined by

$$\phi\begin{bmatrix} x \\ y \end{bmatrix} = \begin{cases} \begin{bmatrix} 2x \\ y/2 \end{bmatrix} \pmod 1 & \left(0 \leqq x < \dfrac{1}{2}\right) \\ \begin{bmatrix} 2x \\ (y+1)/2 \end{bmatrix} \pmod 1 & \left(\dfrac{1}{2} \leqq x < 1\right) \end{cases} \tag{5.3.1}$$

on the two-dimensional torus $\{(x, y) \bmod 1\}$ is called a baker's transformation (remember the process of making a piecrust). This is illustrated in Fig. 5.3. ϕ^{-1} is the operator extending twice in the longitudinal direction, contracting in half in the transversal direction and moving the above half to the right side. Then $\phi^{-1}\phi = 1$. Obviously it is measure preserving and automorphic. But it is not diffeomorphic because of the discontinuity at $x = 1/2$.

On the other hand, consider the operation of the throwing of a die. Here, the number of the faces of a die is generalized to n, and the set of the numbers of spots is denoted as $Z_n = \{1, 2, \ldots, n\}$. Let $M = Z_n^z$ be the direct product of countably many Z_n and the number of spots at the jth throw be a_j. The sequence m formed by a_j ($j = -\infty, \ldots, \infty$)

$$m = m(\ldots, a_{-1}, a_0, a_1, \ldots) \tag{5.3.2}$$

is an element of M. The probability (measure) of a particular number i irrespective of the order of sequence is given by p_i. The measure $\mu(m)$ for a sequence m is given by the product of the probabilities of an individual throw, then $\sum_{i=1}^{n} p_i = 1$ leads to $\mu(M) = 1$. Now define ϕ by

$$\phi m(\ldots, a_j, \ldots) = m'(\ldots, a'_j, \ldots), \tag{5.3.3}$$

where $a'_j = a_{j-1}$ for every j. This means a shift of the sequence to the right by one. Clearly the transformation ϕ is measure-preserving. This dynamical system is called a Bernoulli scheme and is denoted by $B(p_1, p_2, \ldots, p_n)$. The game

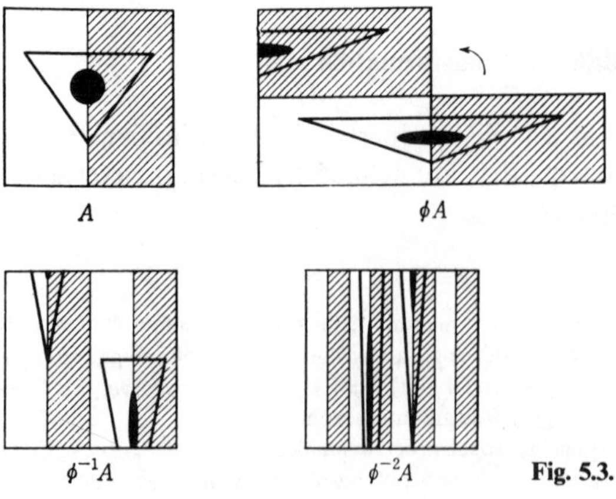

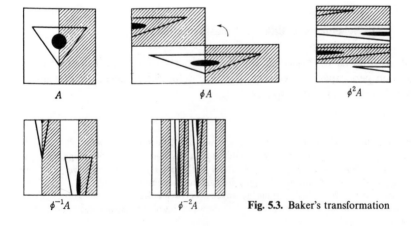

Fig. 5.3. Baker's transformation

of heads or tails by tossing a coin is represented by $B(p_1, p_2)$ and usually $p_1 = p_2 = 1/2$. $B(\frac{1}{2}, \frac{1}{2})$ is proved to be isomorphic to the baker's transformation.

Let U and U' be the induced operators of the dynamical systems (M, μ, ϕ) and (M', μ', ϕ'), respectively. Then U and U' are called equivalent if they are related by an isometric operator[5] F:

$$U' = FUF^{-1} . \tag{5.3.4}$$

Two dynamical systems (M, μ, ϕ) and (M', μ', ϕ') are called isomorphic if the transformation $f: M \to M'$ is isomorphic. Since ϕ is automorphic in M,

$$x_2 = \phi x_1, \quad x_1, x_2 \in M$$

and similarly

$$x_2' = \phi' x_1', \quad x_1', x_2' \in M' ,$$

where

$$x_2' = f x_2, \quad x_1' = f x_1 ,$$

and we have

$$\phi' = f \phi f^{-1} . \tag{5.3.5}$$

On the other hand, in a similar manner to the induced operator U in (5.2.13), we can introduce an induced operator F by the relation

$$Fg(x) = g(fx) .$$

F is an isometric operator, since f is isomorphic. Consequently, the lhs and the rhs of the relation

$$U'g(x') = g(\phi' x')$$

can be rewritten as

$$U'g(fx) = U'Fg(x)$$

and

$$g(\phi' x') = g(f\phi x) = FUg(x) ,$$

respectively, whence

$$U' = FUF^{-1} ,$$

i.e., U and U' are equivalent. When U and U' are equivalent, ϕ and ϕ' are said to be spectrally isomorphic or to have the same spectral structure.

Therefore isomorphism leads to spectral isomorphism. Conversely, does spectral isomorphism imply isomorphism? von Neumann showed that if the

---

[5] Let $L_2(M, \mu)$ be a Hilbert space constructed by L_2 functions on M and similarly $L_2(M', \mu')$; then the isometric operator F transforms a function of $L_2(M, \mu)$ to a function $L_2'(M', \mu')$, where the length defined in (5.2.15) is unchanged, i.e., $\| g \|^2 = \int |g|^2 \, d\mu = \int |Fg|^2 \, d\mu' = \| Fg \|^2$.

spectrum of U is discrete and simple, the converse is valid. But in general it is not: a counter-example is given by the Bernoulli systems. They have the same spectral structure, but they are not always isomorphic (see below).

Many studies have been done on various abstract dynamical systems. To clarify whether or not they can be regarded as isomorphic, the study of invariants under isomorphic transformation has been a major subject. The spectrum of U is an invariant, since equivalent U's have the same structure (i.e., all the eigenvalues are the same). To study the isomorphism in the Bernoulli systems, Kolmogorov introduced the entropy as a new invariant and showed that innumerably many Bernoulli schemes exist which are nonisomorphic. Further, *Ornstein* [5.16] recently showed that Bernoulli schemes with the same entropy are isomorphic.

5.3.2 Ergodicity on the Torus

In classical dynamical systems, there is a simple mathematical model which is ergodic. Consider a circle whose circumference length is 1. Let the length of arc from a definite point be x (mod 1). ϕ denotes the translation $x \to x + \omega$ (mod 1) on the circle, where ω is real. Then each orbit of ϕ is dense everywhere on the circle and ergodic if, and only if, ω is irrational (Fig. 5.4).

Proof: If ω is rational and can be represented by the ratio of integers p, q $(q > 0)$, ϕ^q is an identity transformation and every orbit is closed and consists of a finite set of points. Thus it is not dense. Here consider $f(x) = \exp(2\pi i q x)$. It is invariant for the translation ϕ and the long time average is also $f(x)$. Since the value depends on x and is not constant, the translation could not be ergodic. Next, assume ω to be irrational. Then $\phi^n x$ $(n = 0, 1, 2, \ldots)$ takes infinitely many distinct points between 0 and 1. Therefore, there is always a limit point. Consequently, for any $\varepsilon > 0$ there are two distinct integers n and m such that

$$|\phi^n x - \phi^m x| < \varepsilon . \tag{5.3.6}$$

The translation ϕ being length-preserving; setting $|n - m| = p$, we have

$$|\phi^p x - x| < \varepsilon . \tag{5.3.7}$$

Therefore, $\phi^p x, \phi^{2p} x, \phi^{3p} x, \ldots$ partition the section (0, 1) into segments of length less than ε. Thus the orbit is dense. Then let $f(x)$ be an arbitrary

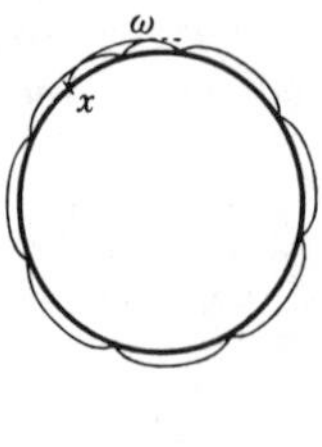

Fig. 5.4. Translation on a circle

measurable phase function and define the Fourier coefficients

$$a_k = \int_0^1 \exp(-2\pi i k x) f(x)\,dx$$
$$f(x) = \sum_{k=0}^{\infty} a_k \exp(2\pi i k x)\,. \tag{5.3.8}$$

The Fourier coefficients of $\phi f(x) = f(\phi x)$ become

$$b_k = \int_0^1 \exp(-2\pi i k x) f(\phi x)\,dx = \int_0^1 \exp[-2\pi i k(x-\omega)] f(x)\,dx$$
$$= \exp(2\pi i k \omega) a_k\,. \tag{5.3.9}$$

Consequently, we can generally write

$$\phi^p f(x) = \sum_{k=0}^{\infty} \exp(2\pi i k p \omega) a_k \exp(2\pi i k x)\,. \tag{5.3.10}$$

The long-time average of $f(x)$ by ϕ is

$$\bar{f} = \lim_{n\to\infty} \frac{1}{n} \sum_{p=0}^{n-1} \phi^p f(x) = \lim_{n\to\infty} \frac{1}{n} \sum_{k=1}^{\infty} \frac{1-\exp(2\pi i k n \omega)}{1-\exp(2\pi i k \omega)} a_k \exp(2\pi i k x)\,. \tag{5.3.11}$$

Since ω is irrational, all terms except a_0 on the rhs vanish for $n \to \infty$. Thus

$$\bar{f} = a_0 = \int_0^1 f(x)\,dx = \langle f \rangle\,. \tag{5.3.12}$$

Alternatively, assume an invariant function f and use it in (5.3.9). In order that (5.3.9) holds for the invariant function f, the relation $b_k = a_k$ must hold, but since $\exp(2\pi i k\omega) \neq 1 (k \neq 0)$, $a_k = 0 (k \neq 0)$. Consequently, $f \doteq a_0$ and the invariant function is always a constant. Since all the invariant functions are constant, the orbit is ergodic from the *fundamental theorem*. The above proofs were done for the one-dimensional torus, but the extension for the n-dimensional one is obvious. That is, x, ω, k(integer) should be considered as vectors. Ergodicity follows, if and only if the scalar product $k \cdot \omega$ for any k is not equal to an integer, $\exp(2\pi i k\omega) \neq 1$. ϕ can also be extended to the continuous one. Then the translation on the torus becomes $\phi_t\colon x \to x + \omega t$ (t: real).

5.3.3 *K*-Systems (Kolmogorov Transformation)

To explain K-systems let us begin with an example. We shall consider the baker's transformation ϕ^n ($n = -\infty, \ldots, -1, 0, 1, \ldots, \infty$) mentioned in Sect. 5.3.1 through partitions of the square. A partition of the set M is a collection of nonempty nonintersecting measurable subsets of M that cover M, ϕ^0 is a partition leaving everything undivided and is denoted by v. ϕ^1 is a transverse partition into two parts and ϕ^{-1} a longitudinal partition. We denote the partition by the transformation ϕ^n by a_n ($n = -\infty, \ldots, \infty$). Now we are given two partitions ξ, ζ. If ζ is a refinement of ξ, we write this as

$$\zeta \geq \xi$$

and say that ζ is finer than ξ or ξ is coarser than ζ. Moreover, the sum of ξ and $\zeta (\xi \vee \zeta)$ is defined as the coarsest partition among the common partitions of ξ and ζ. The product of ξ and $\zeta (\xi \wedge \zeta)$ is the finest partition among the many partitions that are coarser than ξ or ζ. For example, $a_{-1} \vee a_1$ is the partition that consists of four parts formed by vertical and horizontal bisectors and $a_{-1} \wedge a_1$ is the partition a_0 $(= v)$. Now consider the sum of a_n $(n = -\infty, \ldots, \infty)$

$$\bigvee_{n=-\infty}^{\infty} a_n$$

which means the partition into individual points. This is denoted by ε. The product

$$\bigwedge_{n=-\infty}^{\infty} a_n$$

is the partition v. Generally, if for a partition ξ, $\phi\xi$ is the refinement of ξ and $\bigvee_{n=-\infty}^{\infty} \phi^n \xi$ is ε and $\bigwedge_{n=-\infty}^{\infty} \phi^n \xi$ is v, the system is defined as a K-system. Because the baker's transformation satisfies the above three conditions by taking ξ as $\bigvee_{n=-\infty}^{0} a_n$, it is a K-system. By definition, a dynamical system isomorphic to a K-system is also a K-system. Consequently, the Bernoulli scheme $B(\frac{1}{2}, \frac{1}{2})$ is a K-system too.

Next, the entropy of the partition α is defined by using the function

$$z(t) = \begin{cases} -t\log_2 t & (0 < t \leqq 1) \\ 0 & (t = 0) \end{cases} \tag{5.3.13}$$

as

$$h(\alpha) = -\sum_{A_i \in \alpha} \mu(A_i)\log_2 \mu(A_i) = \sum_i z[\mu(A_i)] , \tag{5.3.14}$$

where A_i is an element of α and $\mu\{A_i \cap A_j\} = 0$ $(i \neq j)$.

On the other hand, let α be a partition, ϕ an automorphic transformation and n a positive integer and define

$$h(\alpha, \phi) = \lim_{n \to \infty} \frac{h(\alpha \vee \phi\alpha \vee \cdots \vee \phi^{n-1}\alpha)}{n} \tag{5.3.15}$$

which is given the name of the entropy of α with respect to ϕ. It can be shown that the rhs of (5.3.15) exists and converges. Moreover, we can write the supremum of $h(\alpha, \phi)$ over all partitions α as $h(\phi)$, which is called the entropy of ϕ:

$$h(\phi) = \sup h(\alpha, \phi) . \tag{5.3.16}$$

Hence the following theorem: $h(\phi)$ is an invariant of the dynamical system

(M, μ, ϕ). Proof: Let (M', μ', ϕ') be a dynamical system isomorphic to (M, μ, ϕ). Then an isomorphism $f: M \to M'$ exists such that $\phi' = f\phi f^{-1}$. Further, if α is a partition of M, $f\alpha$ is a partition of M'. Hence

$$\begin{aligned} h(f\alpha, \phi') = h(f\alpha, f\phi f^{-1}) &= \lim_{n\to\infty} \frac{h(f\alpha \vee \cdots \vee f\phi^{n-1}f^{-1}f\alpha)}{n} \\ &= \lim_{n\to\infty} \frac{h[f(\alpha \vee \cdots \vee \phi^{n-1}\alpha)]}{n} \\ &= \lim_{n\to\infty} \frac{h(\alpha \vee \cdots \vee \phi^{n-1}\alpha)}{n} = h(\alpha, \phi) . \end{aligned}$$

When α runs over all the partitions of M, $f\alpha$ runs over all the partitions of M'. Consequently, we get $h(\phi) = h(\phi')$.

Take $\{[0, 1) \times [0, \frac{1}{2}), [0, 1) \times [\frac{1}{2}, 1)\}$ as α in the baker's transformation. Then ϕ gives

$$\begin{aligned} \alpha \vee \phi\alpha \vee \cdots \vee \phi^{n-1}\alpha &= \phi^{n-1}\alpha \\ &= \left\{[0, 1) \times \left[\frac{k-1}{2^n}, \frac{k}{2^n}\right), k = 1, 2, \ldots, 2^n\right\} , \end{aligned}$$

therefore,

$$h(\alpha, \phi) = \lim_{n\to\infty} \frac{1}{n}\left(-\frac{1}{2^n}\log_2\frac{1}{2^n}\right) \times 2^n = \log_2 2 . \tag{5.3.17}$$

It can be shown that $h(\alpha, \phi)$ does not exceed $\log_2 2$ whatever partition α takes. Therefore, we obtain

$$h(\phi) = \log_2 2 .$$

Consequently, the entropy of the Bernoulli scheme $B(\frac{1}{2}, \frac{1}{2})$ is $\log_2 2$, too.

It is clear from the definition that the entropy does not generally become negative. Kolmogorov proved that the entropy of a K-system is positive and finite, such as the Bernoulli scheme $B(\frac{1}{2}, \frac{1}{2})$.

It is surmised that the baker's transformation $\phi^n (n = -\infty, \ldots, -1, 0, 1, \ldots, \infty)$ is mixing and ergodic because the sum of its partition becomes the partition to each point. Actually, *Kolmogorov* and *Sinai* proved that the K-system is mixing [5.8].

The partition of a space corresponds to the observations of a trajectory in the space by dividing it into cells of small size. If a system is ergodic, the entropy is the same wherever the orbit starts.

5.3.4 *C*-Systems

Consider the following automorphic and measure-preserving transformation ϕ:

$$\phi\begin{bmatrix} x \\ y \end{bmatrix} = \begin{bmatrix} 1 & 1 \\ 1 & 2 \end{bmatrix}\begin{bmatrix} x \\ y \end{bmatrix} \quad (\text{mod}\, 1) . \tag{5.3.18}$$

Since the Jacobian, that is, the determinant of the matrix on the rhs is 1, it is a measure-preserving transformation. More generally, when the point (x, y) is transformed to (x_1, y_1) by ϕ, we have

$$x_1 = \phi_1(x, y), \quad y_1 = \phi_2(x, y)$$

and

$$dx_1 = \frac{\partial \phi_1}{\partial x} dx + \frac{\partial \phi_1}{\partial y} dy, \quad dy_1 = \frac{\partial \phi_2}{\partial x} dx + \frac{\partial \phi_2}{\partial y} dy \tag{5.3.19}$$

then we write them as

$$\begin{bmatrix} dx_1 \\ dy_1 \end{bmatrix} = \phi^* \begin{bmatrix} dx \\ dy \end{bmatrix} \tag{5.3.20}$$

and ϕ^* is said to be the differential of ϕ. When one direction (dx, dy) is changed to the direction (dx_1, dy_1) by ϕ, ϕ^* gives the quantitative relation between them.

Again we return to the example (5.3.18). Let the eigenvalues of the matrix of the equation be λ_1, λ_2, and since their product is 1, we can set

$$0 < \lambda_2 < 1 < \lambda_1 \, .$$

Let the eigenvectors corresponding to λ_1, λ_2 be $\boldsymbol{u}$, $\boldsymbol{v}$. If (dx, dy) is parallel to the direction $\boldsymbol{u}$, the transformation ϕ^* multiplies it by λ_1 in the same direction. When this is repeated n times, the length is increased by $\lambda_1^n \gg 1$. In other words, the distance between two points which have a small separation in the direction of $\boldsymbol{u}$ increases exponentially with the repetition of the transformation ϕ. In contrast to this, in the direction of $\boldsymbol{v}$, the separation contracts and converges to a stable solution. Since a vector connecting two points generally has a component in the direction of $\boldsymbol{u}$, the length of the vector increases exponentially after repeated operations of ϕ. Thus, the C-system is defined as a system in which the vectors indicating the difference of two starting points have two components of dilation and contraction. Ergodicity of the transformation of (5.3.18) is proven as follows. Remove the condition of mod 1 and let $\boldsymbol{x}$ be an arbitrary point. Since $\boldsymbol{x}$ can be written as $\boldsymbol{x} = a\boldsymbol{u} + b\boldsymbol{v}$, i.e., the combination of $\boldsymbol{u}$ and $\boldsymbol{v}$, after applying n times the transformation ϕ, $\boldsymbol{x}$ is transformed to $\boldsymbol{x}(n)$:

$$\boldsymbol{x}(n) \equiv \phi^n \boldsymbol{x} = a\lambda_1^n \boldsymbol{u} + b\lambda_2^n \boldsymbol{v} \, . \tag{5.3.21}$$

Since $\lambda_2^n \to 0$ with $n \to \infty$, $\phi^n \boldsymbol{x} \to a\lambda_1^n \boldsymbol{u}$ (if $a \neq 0$) and then the $\boldsymbol{x}(n)$'s lie in a straight line defined by the vector $\boldsymbol{x}$ after a long time. With the condition of mod 1, the straight line becomes the one on the two-dimensional torus. The ratio of the direction cosines of $\boldsymbol{u}$ is $\lambda_1 - 1$ and irrational, and therefore $\boldsymbol{x}(n)$ distributes densely over the torus and is ergodic (as explained in Sect. 5.3.2). The proof of ergodicity for general C-systems was given by *Anosov* [5.15]. Moreover, Sinai proved that those C-systems that are C-diffeomorphous are K-systems [5.14].

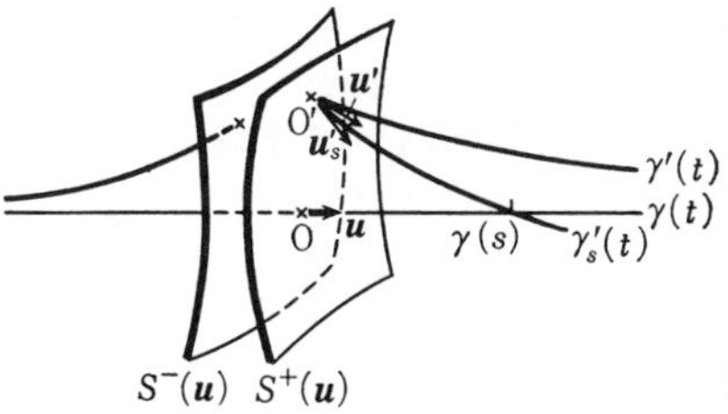

Fig. 5.5. Geodesics on a surface of negative curvature

The motion along geodesics on a surface with negative curvature is an example of a C-system. In Fig. 5.5, the geodesics starting from a point O in the direction of $\boldsymbol{u}$ is shown by $\gamma(t)$, where t is the length along the geodesics measured from O. Let $\gamma_s'(t)$ be the geodesics passing through a point $\gamma(s)$ on $\gamma(t)$ emanating from a point O′ different from O, and $\boldsymbol{u}_s'$ be the tangential direction at O′. Let $s \to \infty$ with O fixed. Then the following statement holds: if the surface is of negative curvature, $\gamma_s'(t)$ converges to a geodesics $\gamma'(t)$ and $\boldsymbol{u}_s'$ does so to a constant direction $\boldsymbol{u}'$. The proof is omitted, but we can easily understand that it does not hold for surfaces with non-negative curvature, such as a spherical surface. $\gamma'(t)$ is called a positive asymptote. Similarly, we can obtain a negative asymptote by taking t or s to be negative. Let us denote by $S^+(\boldsymbol{u})$ the positive horosphere which passes through O and is orthogonal to all positive asymptotes, and $S^-(u)$ the negative horosphere orthogonal to the negative asymptotes. Since we consider the surface with negative Gaussian curvature, we can put its upper limit as $-k^2$. Then, if $t > 0$, the distance between a geodesic emanating orthogonally from the positive (negative) horosphere $S^+(S^-)$ and the positive (negative) asymptote converges as $\exp(-kt)$ (diverges as $\exp(kt)$) for $t \to \infty$. The properties agree with the conditions of a C-system as mentioned before.

In a Hamiltonian system whose motions are limited in a finite space, the curvature cannot be negative over the whole space if the potential is analytic. In systems which have negative curvature in parts of the space, computer experiments show that unstable regions are of a wider range than regions with negative curvature [5.17].

The phenomenon that a small difference at initial states enlarges with time is found in collisions between two spheres, or between a sphere and a wall. In fact, *Sinai* [5.18] proved the ergodicity of this system. The problem becomes a little simpler by considering one of the two spheres as fixed, but then it is almost intuitively clear that orbits enlarge their difference by collisions with the fixed sphere and by reflections from walls.

5.4 The Poincaré and Fermi Theorems

The dynamical system under consideration is a conservative one and has an energy integral, but other integrals can exist in some cases. Thus the orbits do not densely cover the energy surface. For ergodic problems it is important whether there are constants of motion other than the energy integral or not. Here we mention the classical theorems on this problem [5.19–28].

5.4.1 Bruns' Theorem

Bruns [5.29], *Poincaré* [5.30] and others reached important conclusions on integrals of dynamical systems, originating from the so-called three-body problem. For a system which consists of three mass points interacting through central forces of a Newtonian potential, there are 18 independent variables in total but 10 integrals exist among them: three for the momentum conservation of translational motion of the center of gravity, three obtained by integrating them once, three for the conservation of angular momentum around the center of gravity and one for the total energy. All can be written in algebraic forms. Bruns proved the nonexistence of algebraic integrals other than these.

5.4.2 Poincaré-Fermi's Theorem

Poincaré extended Bruns' result to the restricted three-body problem [5.30]. The theorem is not limited to the restricted three-body problem and can be applied to canonical equations with a Hamiltonian of the following form. That is, let $\mathscr{H}$ be a function of $y_1, y_2, \ldots, y_n, x_1, x_2, \ldots, x_n$ and a parameter μ and be able to be expanded in the form

$$\mathscr{H} = \mathscr{H}_0 + \mu \mathscr{H}_1 + \mu^2 \mathscr{H}_2 + \cdots . \tag{5.4.1}$$

The equations of motion are

$$\frac{dx_i}{dt} = \frac{\partial \mathscr{H}}{\partial y_i}, \quad \frac{dy_i}{dt} = -\frac{\partial \mathscr{H}}{\partial x_i} . \tag{5.4.2}$$

We assume that $\mathscr{H}$ is a periodic function having the period 2π with respect to x and $\mathscr{H}_0$ is a function of y alone. This is sometimes called a canonical normal system. Moreover, we assume that its Hessian is not 0:

$$\det\left|\frac{\partial^2 \mathscr{H}_0}{\partial y_i \partial y_k}\right| \neq 0 . \tag{5.4.3}$$

Putting $\partial \mathscr{H}_0/\partial y_i = \Omega_i$, (5.4.3) is the Jacobian between Ω and y.

Poincaré's theorem claims the nonexistence of integrals of the form $\Phi = \text{const.}$ other than $\mathscr{H}$ in a system with the above conditions, where Φ is assumed to be a one-valued analytic function with respect to sufficiently small μ for all values of x and for y in a certain region D (which Poincaré called a "fonction analytique uniforme"). In short, Φ is such a function that can be expanded in the form

$$\Phi = \Phi_0 + \mu\Phi_1 + \mu^2\Phi_2 + \cdots . \tag{5.4.4}$$

When we interpret y as a momentum and x as a coordinate according to (5.4.2), it is natural to consider that $\mathscr{H}_0$ is a function of momenta alone, but Poincaré considered the case $\mathscr{H}_0$ as a function of coordinates alone and exchanged x for y. Now we consider systems of nonlinear lattice vibrations. q_i is the displacement of the lattice point i from its equilibrium position and p_i the momentum. The Hamiltonian is written as

$$\mathscr{H} = \sum_{i=1}^{n}\left(\frac{1}{2m_i}p_i^2 + \frac{m_i\omega_i^2}{2}q_i^2\right) + (\text{higher terms of } q)\,, \tag{5.4.5}$$

where, if we introduce action variables J_i and angular variables φ_i according to (5.1.20, 22), we can write it as

$$\mathscr{H} = \sum_{i=1}^{n} \nu_i J_i + (\text{higher terms of } J \text{ and } \varphi)\,. \tag{5.4.6}$$

Consequently, systems of nonlinear lattice vibrations are canonical normal systems. But if we take the first term in (5.4.6) as $\mathscr{H}_0$, the Hessian becomes 0 and Poincaré's condition is not satisfied. We must put a part of higher nonlinear terms into $\mathscr{H}_0$.

The surface of $\Phi = \text{const.}$ forms a family by changing the constant. Poincaré's theorem asserts that such a family does not exist unless it is the family of the surface of energy. *Fermi*, extending Poincaré's theorem further, proved that systems with more degrees of freedom than 2 ($n > 2$ degrees of freedom) do not have even the isolated surface of $\Phi = 0$ [5.31]. We do not give the proof of Poincaré's and Fermi's theorems here. But Fermi asserted from his theorem that canonical normal systems are quasi-ergodic.

Now consider a small region σ on the surface of $\mathscr{H} = \text{const.} = E$. If σ' is the region on the surface attainable from initial points in σ, then σ' will cover over the whole surface (in the sense of quasi-ergodic) or a part of it only. Since the former case is quasi-ergodic, we consider the latter one. Let σ'' be the part excluding σ' on the surface and S the boundary between them. A trajectory connecting a point of σ' and a point of σ'' does not exist. Now, take a point P on S and points P′ and P″ close to P in the region σ' and σ'', respectively. Then the trajectory passing through P′ or P″ is in the region σ' or σ'', respectively, and one through P is always on S. Thus, S is an isolated surface of an integral. Fermi asserted that the system is quasi-ergodic from the nonexistence of S different from $\mathscr{H} = E$.

However, the surface S considered by Fermi is analytic. The boundary between σ' and σ'' is not necessarily analytic. Hence, Fermi's consideration does not prove that canonical normal systems are quasi-ergodic. Furthermore, according to Van Hove's remark mentioned previously (Sect. 5.2.6) even if a system is quasi-ergodic, we cannot conclude the metrical indecomposability.

5.5 Fermi-Pasta-Ulam's Problem

5.5.1 Nonlinear Lattice Vibration

The lattice specific heat of solids can be satisfactorily explained on the basis of the quantum mechanical calculation of lattice vibration in the harmonic approximation. At high temperatures, classical treatments hold true. This fact means that the energies of normal modes are redistributed among the modes as the temperature of the system changes. On the other hand, as we have already mentioned, any normal mode in harmonic vibrations is independent of the others and the energy of each mode is an integral of motion and is a constant. In order to explain the specific heat, we must consider that the magnitude of the energy itself can be expressed quite well by the harmonic approximation, but the existence of a small nonlinear term or anharmonicity can give rise to an exchange of energy among the normal modes. Even if the anharmonic term is very small and the contribution to the total energy is negligible, its existence must fulfill the important function of the energy exchange among the normal modes.

It was a long standing belief of many physicists that the existence of small nonlinear terms would be sufficient to bring about ergodicity or irreversibility. *E. Fermi*, *J. Pasta* and *S. Ulam* had tried to investigate this problem by numerical calculations on a computer around 1950 [5.32]. It was just about the time when high-speed electronic computers had been put to practical use and they took up the ergodicity of nonlinear oscillator systems as an important problem in physics which could be investigated by the use of a computer.

Now, as shown in Fig. 5.6, we consider a linear chain of $N + 1$ identical particles of mass m connected with identical springs. For simplicity, the 0th and Nth particles are assumed to be fixed. Let p_k and q_k be the momentum of the kth particle and its displacement from the equilibrium position, respectively. In the harmonic approximation of Hooke's spring of elastic constant κ, the Hamiltonian is written as

$$\mathscr{H} = \sum_{k=1}^{N-1} \frac{1}{2m} p_k^2 + \frac{\kappa}{2} \sum_{k=1}^{N-1} (q_{k+1} - q_k)^2, \quad q_0 = q_N = 0 . \tag{5.5.1}$$

Transforming to the normal coordinates (y_j, x_j) by

$$q_k = \left(\frac{2}{N}\right)^{1/2} \sum_{j=1}^{N-1} x_j \sin\frac{jk\pi}{N}, \quad m\dot{x}_j = y_j \tag{5.5.2}$$

we have

$$\mathscr{H} = \sum_{j=1}^{N} \frac{1}{2m} (y_j^2 + m^2 \omega_j^2 x_j^2) \tag{5.5.3}$$

Fig. 5.6. One-dimensional oscillator system

$$\omega_j = 2\left(\frac{\kappa}{m}\right)^{1/2} \sin\frac{j\pi}{2N} . \tag{5.5.4}$$

If nonlinear interactions are included in the original Hamiltonian, which is written, for example, as

$$\mathscr{H} = \sum_k \frac{1}{2m} p_k^2 + \frac{\kappa}{2}\sum_k (q_{k+1} - q_k)^2 + \frac{\lambda}{s}\sum_s (q_{k+1} - q_k)^s , \tag{5.5.5}$$

λ is the nonlinear coupling constant and $s = 3, 4$ represent cubic and quartic nonlinear potentials, respectively. Generally, a potential will be expanded in powers of distance. Then the equation of motion is written as

$$m\ddot{q}_k = \kappa(q_{k+1} - 2q_k + q_{k-1}) + \lambda[(q_{k+1} - q_k)^{s-1} - (q_k - q_{k-1})^{s-1}] \tag{5.5.6}$$

or, in a normal coordinate,

$$\ddot{x}_j = -\omega_j^2 x_j + \lambda F_j(x) , \tag{5.5.7}$$

where $F_j(x)$ are nonlinear forces. In particular, if $N = 3$, the system has 2 degrees of freedom with two movable particles and the equations of motion become

$$\begin{aligned} \ddot{x}_1 &= -\omega_1^2 x_1 - \sqrt{2}\lambda x_1 x_2 \\ \ddot{x}_2 &= -\omega_2^2 x_2 - \frac{\lambda}{\sqrt{2}}(x_1^2 - 3x_2^2) . \end{aligned} \tag{5.5.8}$$

The long-time behavior of the system (5.5.6) or (5.5.7) was studied by computer solutions [5.32]. The nonlinear forces the authors used are quadratic, cubic and linear ones. They let the frequencies of the normal vibrations (modes) in the harmonic approximation be $\omega_1, \omega_2, \ldots$, beginning with the lowest, and the energy of the kth mode be

$$E_k = \frac{1}{2}m(\dot{x}_k^2 + \omega_k^2 x_k^2) . \tag{5.5.9}$$

They investigated how the energies $E_1, E_2, \ldots$ change with time when an amount of energy is given initially to the lowest mode. Their conjecture was that the energy E_1 would be gradually transferred to the other modes and in the end almost equal amounts of energy would be alloted to every mode. However, this was not the case. Figure 5.7 shows the energy sharing among normal modes as time proceeds for the cubic potential and $N = 32$. The unit of time is $2\pi/\omega_1$, the period of the lowest mode, and $\kappa = m = 1$, $\lambda = 1/4$. The initial velocities of all the particles are set to zero, and the initial displacements are given by a sine curve corresponding to the lowest mode. As is seen in this figure, the energy comes back to the 1st mode at about 158 periods so that the behavior seems to be almost periodic and the energy is exchanged more or less among the lower modes but scarcely with the higher modes. These features are almost the same for the case of a quartic potential. In this case, when only the first mode is initially excited as described above, the energy is not transferred to the even

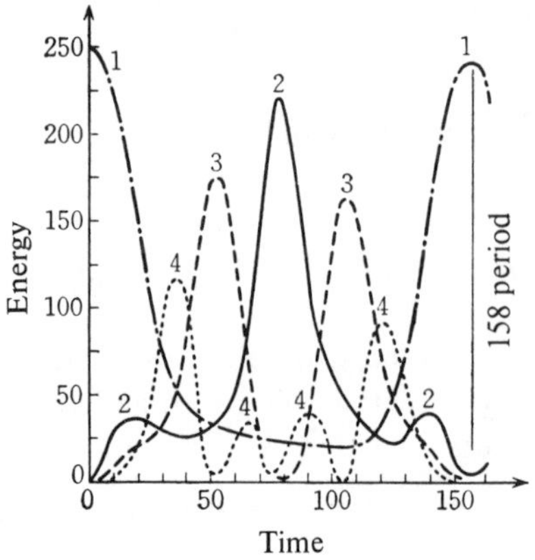

Fig. 5.7. Recurrence property of the mode energy. Numerals denote the mode numbers [5.32]

modes of 2, 4, 6, ... because of the symmetry properties of the potential function. But this case also shows the recurrence property of the normal mode energy. Thus it is revealed that the mere presence of anharmonicity does not result in ergodicity as far as the present simple system is concerned.

5.5.2 Resonance Conditions

If the frequencies $\omega_1, \omega_2, \ldots$ can satisfy the relation

$$\omega(n) = \sum_j n_j \omega_j = 0 \tag{5.5.10}$$

for integers $n_1, n_2, \ldots$, not all 0, they are said to be commensurate or to have the resonance condition among them. *Hemmer* showed that the condition where the frequencies given by (5.5.4) do not give rise to resonance occurs only when the particle number N is a prime number or a power of 2 [5.33]. The experiment displayed in Fig. 5.7 was performed for $N = 32$, a nonresonant case. However, this is not necessarily essential, because, even if the resonance condition is satisfied, the degree of energy-sharing among ω's satisfying it may happen to be small or when the resonance condition is satisfied not rigorously but approximately, energy exchange may take place. Nonlinear terms generally loosen the resonance condition as will be seen later. Therefore, energy exchange will occur in the presence of nonlinear terms provided that the resonance condition is approximately satisfied. The occurrence of energy exchange among the lower modes in Fermi's experiments can be understood from the resonance condition $\omega_j = j\omega_1$ satisfied approximately for small j/N.

Now let us examine the relationship between the resonance condition and the energy exchange by means of the perturbation theory of nonlinear vibration. We consider (5.5.8), where ω_1, ω_2 are regarded as parameters chosen arbitrarily. This is the case where two oscillators with frequencies ω_1, ω_2 are connected by the same nonlinear springs. Assuming the solution to be expressed in the form of a series expansion of λ, the result obtained by *Ford* and *Waters* [5.34] is as follows:

$$x_1 = A_1\cos\tau_1 - \lambda\left(\frac{A_1A_2\cos(\tau_1+\tau_2)}{\sqrt{2}[\omega_1^2-(\Omega_1+\Omega_2)^2]} + \frac{A_1A_2\cos(\tau_1-\tau_2)}{\sqrt{2}[\omega_1^2-(\Omega_1-\Omega_2)^2]}\right)$$
$$+\lambda^2\left(\frac{A_1^3\cos3\tau_1}{4(\omega_2^2-4\Omega_1^2)(\omega_1^2-9\Omega_1^2)} - \frac{3A_1A_2^2\cos(\tau_1+2\tau_2)}{4(\omega_2^2-4\Omega_2^2)[\omega_1^2-(\Omega_1+\Omega_2)^2]}\right.$$
$$-\frac{3A_1A_2^2\cos(\tau_1-2\tau_2)}{4(\omega_2^2-4\Omega_2^2)[\omega_1^2-(\Omega_1-2\Omega_2)^2]}$$
$$+\frac{A_1A_1^2\cos(\tau_1+2\tau_2)}{2[\omega_1^2-(\Omega_1+\Omega_2)^2][\omega_1^2-(\Omega_1+2\Omega_2)_2]}$$
$$\left.+\frac{A_1A_2^2\cos(\tau_1-2\tau_2)}{2[\omega_1^2-(\Omega_1-\Omega_2)^2][\omega_1^2-(\Omega_1-2\Omega_2)^2]}\right.+\lambda^3(\cdots)+\cdots \tag{5.5.11}$$

$$x_2 = A_2\cos\tau_2 - \lambda\left(\frac{A_1^2-3A_2^2}{2\sqrt{2}\omega_2^2} + \frac{A_1^2\cos2\tau_1}{2\sqrt{2}(\omega_2^2-4\Omega_1^2)} + \frac{3A_2^2\cos2\tau_2}{2\sqrt{2}(\omega_2^2-4\Omega_2^2)}\right)$$
$$+\lambda^2\left(\frac{A_1^2A_2\cos(2\tau_1+\tau_2)}{2[\omega_1^2-(\Omega_1+\Omega_2)^2][\omega_2^2-(2\Omega_1+\Omega_2)^2]}+\cdots\right.$$
$$\cdots+\frac{A_1^2A_2\cos(2\tau_1-\tau_2)}{2[\omega_1^2-(\Omega_1-\Omega_2)^2][\omega_2^2-(2\Omega_1-\Omega_2)^2]}$$
$$-\frac{3A_1^2A_2\cos(2\tau_1+\tau_2)}{4(\omega_2^2-4\Omega_1^2)[\omega_2^2-(2\Omega_1+\Omega_2)^2]}$$
$$-\frac{3A_1^2A_2\cos(2\tau_1-\tau_2)}{4(\omega_2^2-4\Omega_1^2)[\omega_2^2-(2\Omega_1-\Omega_2)^2]}$$
$$\left.+\frac{9A_2^3\cos3\tau_2}{4(\omega_2^2-4\Omega_2^2)(\omega_2^2-9\Omega_2^2)}\right)+\lambda^3(\cdots)+\cdots \tag{5.5.12}$$

$$\tau_1 = \Omega_1t+\theta_1,\quad \tau_2=\Omega_2t+\theta_2$$

$$\Omega_1^2 = \omega_1^2 - \lambda^2\left(\frac{A_1^2-3A_2^2}{2\omega_2^2} + \frac{A_1^2}{4(\omega_2^2-4\Omega_1^2)} + \frac{A_2^2}{2[\omega_1^2-(\Omega_1+\Omega_2)^2]}\right.$$
$$\left.+\frac{A_2^2}{2[\omega_1^2-(\Omega_1-\Omega_2)^2]}\right)+\cdots \tag{5.5.13}$$

$$\Omega_2^2 = \omega_2^2 - \lambda^2\left(\frac{A_1^2}{2[\omega_1^2-(\Omega_1+\Omega_2)^2]} + \frac{A_1^2}{2[\omega_1^2-(\Omega_1-\Omega_2)^2]}\right.$$
$$\left.-\frac{3(A_1^2-3A_2^2)}{2\omega_2^2} + \frac{9A_2^2}{4(\omega_2^2-4\Omega_2^2)}\right)+\cdots. \tag{5.5.14}$$

Here A_1, A_2, θ_1, and θ_2 are constants determined by initial conditions. Putting $\Omega_1 = \omega_1$, $\Omega_2 = \omega_2$ approximately for small λ from (5.5.13, 14) and substituting them into (5.5.11, 12) we have approximately $x_1 = A_1 \cos\tau_1$ and $x_2 = A_2 \cos\tau_2$, provided that none of the denominators is zero or of the order of λ. The oscillators move as if in the absence of nonlinear terms and no appreciable energy-sharing occurs. In other words, energy-sharing occurs, in general, at

$$n_1\omega_1 + n_2\omega_2 \approx \lambda\,, \tag{5.5.15}$$

where n_1 and n_2 are small integers. Substituting $\Omega_1 = \omega_1$ and $\Omega_2 = \omega_2$ in the denominators on the rhs of (5.5.13), we have

$$\Omega_1^2 = \omega_1^2 - \lambda^2\left(\frac{A_1^2 - 3A_2^2}{2\omega_2^2} + \frac{A_1^2}{4(\omega_2^2 - 4\omega_1^2)} + \frac{A_2^2}{2[\omega_1^2 - (\omega_1 + \omega_2)^2]}\right.$$
$$\left. + \frac{A_2^2}{2[\omega_1^2 - (\omega_1 - \omega_2)^2]}\right) + \cdots\,. \tag{5.5.16}$$

We would have the small denominator of λ when

$$2\omega_1 - \omega_2 \approx \lambda\,. \tag{5.5.17}$$

The same occurs in (5.5.14). Thus, if $2\omega_1 - \omega_2 \approx \lambda$, (5.5.13, 14) become

$$\Omega_1 = \omega_1 + \kappa_1\lambda + \cdots\,,\quad \Omega_2 = \omega_2 + \kappa_2\lambda + \cdots$$

and the second term of order λ in (5.5.11) may be written as follows:

$$-\frac{\lambda A_1 A_2 \cos(\tau_1 - \tau_2)}{\sqrt{2}[\omega_1^2 - (\Omega_1 - \Omega_2)^2]} = K A_1 A_2 \cos(\tau_1 - \tau_2)\,,$$

where K is a constant. Consequently, one of the first-order terms of λ in (5.5.11) becomes a zeroth-order term. This means that energy-sharing can really occur. If (5.5.17) does not hold, we find that $\Omega_1 = \omega_1 + \kappa_3\lambda^2$ and $\Omega_2 = \omega_2 + \kappa_4\lambda^2$. However, if $\omega_1 - \omega_2 \approx \lambda^2$, then some of the terms of order λ^2 in (5.5.11) may go over into terms of zeroth order in λ. This is seen from the third term of order λ^2. Thus we find that up to the order λ^2 in (5.5.11), the conditions $2\omega_1 \simeq \omega_2$ and $\omega_1 \simeq \omega_2$ are effective for energy-sharing. For terms of higher orders of λ to grow, we would need different relationships between ω_1 and ω_2. The results examined on a computer show that the amount of energy-sharing for the case $2\omega_1 \simeq \omega_2$ is much bigger than that for the case $\omega_1 \simeq \omega_2$. These results are shown in Figs. 5.8, 9. Here E_1, E_2 are the energies of the modes defined by (5.5.9).

Extending the results mentioned above to the case of many oscillators, we see that energy-sharing occurs generally provided

$$\sum_j n_j\omega_j = O(\lambda)\,. \tag{5.5.18}$$

The rhs of this equation is of the first order of λ or higher.

Consequently, in the case of large nonlinearity, the system will be able to share energy even if the resonance condition $\sum n_j\omega_j = 0$ is not satisfied rigor-

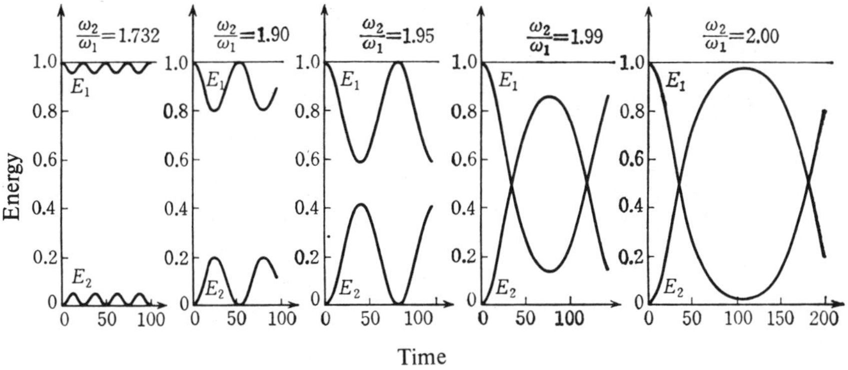

Fig. 5.8. Energy-sharing between two oscillators. The ratio of two frequencies at the left end corresponds to that calculated by (5.5.4) [5.34]

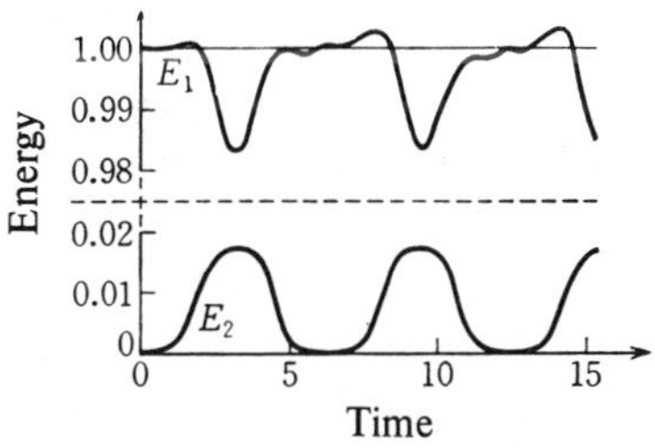

Fig. 5.9. Energy-sharing for $\omega_1 = \omega_2$ [5.34]

ously, but approximately, because the frequencies themselves are modulated like (5.5.13). Furthermore it would be interesting to investigate how the energy-sharing changes as λ is varied. Next we shall describe the results of computer experiments.

5.5.3 Induction Phenomenon

We have shown that for the occurrence of energy-sharing among normal modes in the harmonic approximation, an approximate resonance condition (5.5.18) is required. In one-dimensional lattices, the frequencies of higher modes are close to each other and a great amount of energy exchange will occur rather easily when one of the higher modes is initially excited. Figure 5.10 shows the energy-sharing among normal modes when the 11th mode ($k = 11$) is initially excited in the lattice, which has quartic potentials ($s = 4$) as nonlinear terms and parameters $\lambda = 0.5$, $m = \kappa = 1$, $N = 16$ with fixed ends ($q_0 = q_N = 0$) [5.35]. The energies of modes other than the 11th increase gradually, and, after an elapse of a certain amount of time, abrupt energy transfer takes place owing to the increase of the energy of the 9th mode in particular. The abrupt drop of the 11th

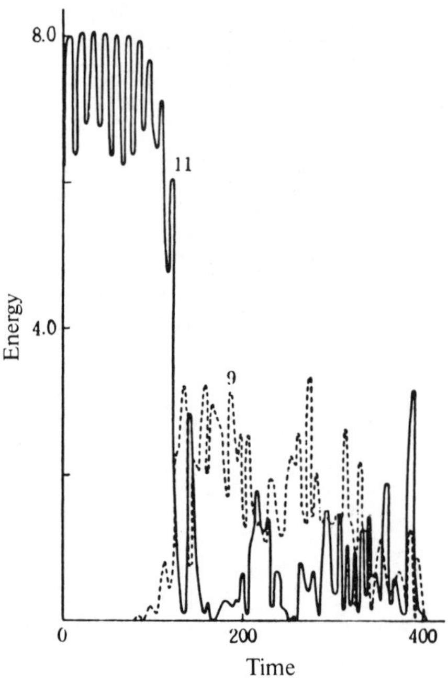

Fig. 5.10. Induction phenomenon when only the 11th mode is excited. Here $N = 16$, $m = \kappa = 1$ and $\lambda = 0.5$

mode energy is brought about by the increase of other mode energies up to that time. This is the so-called induction phenomenon. When λ is too small, the energy of modes other than the 11th mode do not increase appreciably and the induction phenomenon does not occur.

Now we define complex normal coordinates, which are a little different from the normal coordinates introduced by (5.5.2), as follows,

$$q_k = i \sum_{j=-N}^{N} \left(\frac{a_j}{\omega_j}\right) \exp\left(-i\frac{\pi jk}{N}\right) \tag{5.5.19}$$

$$a_0 = 0, \quad a_j = a_{-j}, \quad \omega_{-j} = -\omega_j \,,$$

where ω_j is given by (5.5.4). a_j is related to x_j in (5.5.2) through the relation

$$a_j = \frac{1}{\sqrt{2N}} \omega_{|j|} x_{|j|} \,. \tag{5.5.20}$$

The equation of motion for a_k is written as

$$\ddot{a}_k = -\omega_k^2 a_k - \lambda\omega_k^2 \sum_{k'} \sum_{k''=-N}^{N} \sum_{k'''} a_{k'} a_{k''} a_{k'''} D(k + k' + k'' + k''') \tag{5.5.21}$$

$$(k = -N, \ldots, N) \,,$$

where one defines for integers $k, -4N \leqq k \leqq 4N$,

$$D(k) = \frac{1}{N}\sum_{j=1}^{N} \cos\left[\left(j - \frac{1}{2}\right)\frac{k\pi}{N}\right]$$

which takes values of $\pm 1, 0$, as follows:

$$D(0) = 1, \quad D(\pm 2N) = -1, \quad D(\pm 4N) = 1$$
$$\mathrm{D} = 0 \quad \text{for others} . \tag{5.5.22}$$

Equation (5.5.22) provides the selection rule of the energy transfer. If the 11th mode is excited, the mode which first receives energy from it is $k = 1$. In order that the kth mode can get excited in accordance with (5.5.22) when initially the 11th mode is excited [in other words, under the initial conditions $a_{11} \neq 0, a_k = 0$ (for $k \neq 11$), $\dot{a}_k = 0$ (for all k)] the sum $k + k' + k'' + k'''$ must be equal to zero, $\pm 2N$ or $\pm 4N$, with $|k'| = |k''| = |k'''| = 11$. This is possible only for $k = 1$. When the 1st and 11th modes have been excited, the 3rd, 9th and 13th modes can be excited next. In this case, the even modes are never excited.

However, unless the modes excited in this way continue to increase their energies further, the induction phenomenon does not occur. Now assume that only the k_0th mode is initially excited. At first the a_k of the modes other than k_0 is small and, as far as the nonlinearity is small, we can put

$$a_{k_0} = a\cos(\omega_{k_0} t) . \tag{5.5.23}$$

Then we have, from (5.5.21)

$$\ddot{a}_k + (\omega_k^2 + 6\lambda\omega_k^2 a_{k_0}^2) a_k = f_k(t) \tag{5.5.24}$$

in a linear approximation for a_k ($k \neq k_0$), taking account of (5.5.22). $f_k(t)$ is an inhomogeneous term which contains a_{k_0} and $a_{k'}$'s ($k' \neq k$). The existence of this term gives rise to the energy transfer to the kth mode.

Making use of the approximation (5.5.23) for a_{k_0}, the lhs of (5.5.24) is Mathieu's equation. Now define new variables by

$$\tau = \omega_{k_0} t$$
$$\alpha_k = (\omega_k/\omega_{k_0})^2 (1 + 3\lambda a^2) \tag{5.5.25}$$
$$h_k^2 = (\omega_k/\omega_{k_0})^2\, 3\lambda a^2/2$$

and the homogeneous part of (5.5.24) becomes

$$\frac{d^2 a_k}{d\tau^2} + (\alpha + 2h^2 \cos 2\tau) a_k = 0 . \tag{5.5.26}$$

A Mathieu's equation has stable or unstable solutions depending on the set of parameters α and h^2 (Fig. 5.11). Since we have

$$\frac{\alpha_k}{h_k^2} = 2 + \frac{2}{3\lambda a^2} \tag{5.5.27}$$

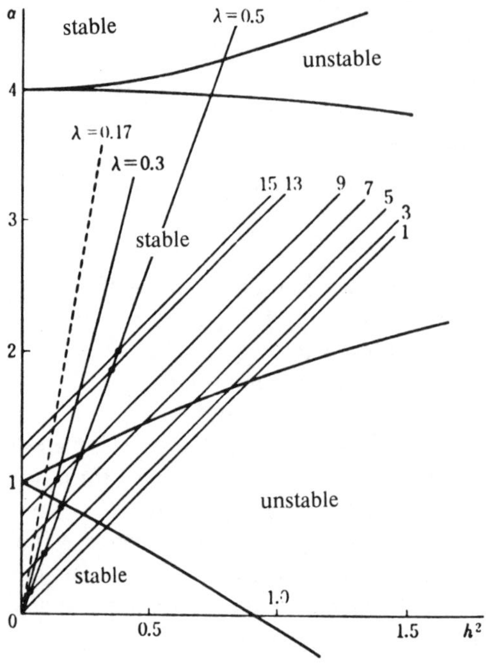

Fig. 5.11. Stable and unstable regions for the solutions of Mathieu's equation

and

$$\alpha_k = 2h_k^2 + (\omega_k/\omega_{k_0})^2 \tag{5.5.28}$$

from (5.5.25), whether the kth mode is unstable or not is determined by whether the intersection of the two straight lines (5.5.27, 28) lies in the unstable region or not. Equation (5.5.28) represents a straight line with slope 2 on the (α, h^2) plane and which is translated parallel to itself as k varies. On the other hand, the line (5.5.27) goes through the origin and its slope becomes steeper as λ becomes smaller. Figure 5.11 shows the stable and unstable regions for the solutions of Mathieu's equation and we find that, for the example shown in Fig. 5.10, i.e., for $N = 16$, $k_0 = 11$ and $\lambda = 0.5$, the mode $k = 9$ is in an unstable region. When λ is smaller than about 0.17, we see that all the modes are stable and the induction phenomenon does not occur.

When the energy-sharing becomes large, the approximation of (5.5.23) is not appropriate and thus our approximation by Mathieu's equation becomes poor. But, after the occurrence of the induction phenomenon, we can surmise that a thermal equilibrium state can be realized. In a C-system, as mentioned in Sect. 5.3.4, the distance of two trajectories initially close to each other grows exponentially. To confirm this on a computer, the distances of two trajectories, one with the same initial conditions as those of Fig. 5.10 and the other with initial conditions close to them are calculated. Figure 5.12 thus obtained shows a remarkable exponential separation after the elapse of the induction period.

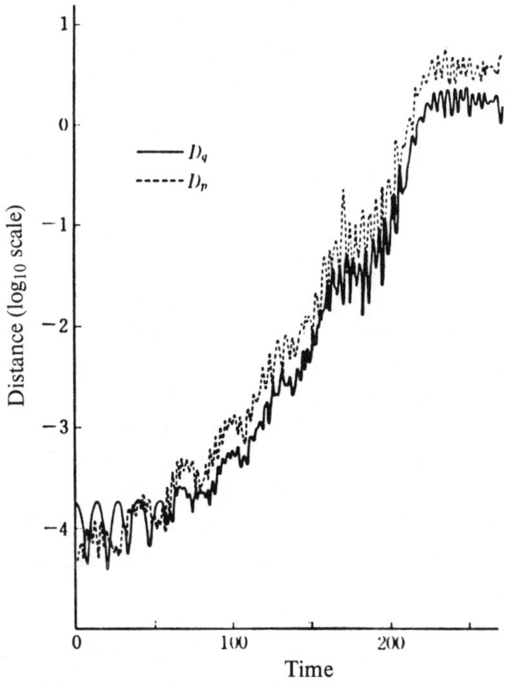

Fig. 5.12. Orbital separation D_p and D_q for $N = 16$, $k_0 = 11$, $\lambda = 0.5$, $D_p^2 \equiv \sum_i (\Delta p_i)^2$, $D_q^2 \equiv \sum_i (\Delta q_i)^2$. Δp_i and Δq_i are the differences of the coordinates of the ith particle

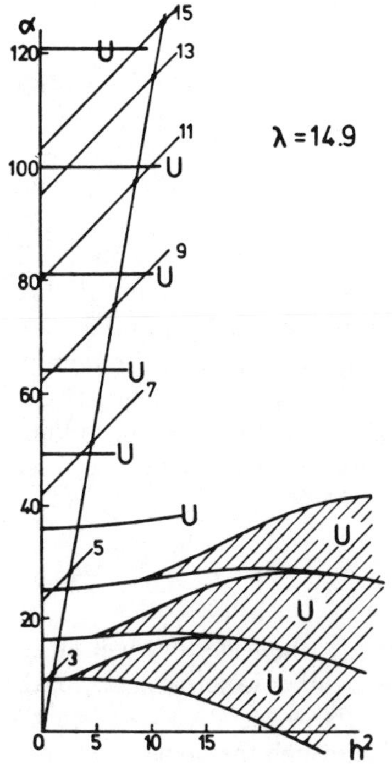

Fig. 5.13. Stability of the normal modes for $N = 16$, $k_0 = 1$, $\lambda = 14.9$. The regions designated as U are unstable regions

If one excites the lowest mode initially, the trajectories are usually stable and thus quasi-periodic as shown in Fig. 5.13: the chance of unstable sets of parameters (α, h^2) is quite small. This is the case of *Fermi* et al. described earlier (Sect. 5.5.1). It can be explained in terms of solitons and is now understood.

5.6 Third Integrals

Provided that the equations of motion of a dynamical system with n degrees of freedom are solved, n coordinates and n momenta are given as functions of time. The most familiar relations obtained by eliminating the time among them are the energy and momentum integrals (if any). They are conservative, analytic and have no singular points, and are generally called isolating integrals. If the hypersurface representing an integral of motion is folded innumerably and fills up the space (e.g., Lissajous' figures with an irrational ratio of frequencies), the integral is not isolating. An integral of this kind is an orbit itself and not useful in solving the problem. Isolating integrals other than those of energy and momentum are called third integrals. When we note, the theorems of Poincaré and Fermi, does the third integral exist?

In stellar dynamics, this has often been under discussion. It is sometimes an advantage to examine the existence of integrals by computer experiments. *Hénon* and *Heiles* [5.36] considered the system whose Hamiltonian or energy integral I_1 is given by

$$I_1 = \frac{1}{2}(\dot{x}^2 + \dot{y}^2) + U(x, y) \tag{5.6.1}$$

$$U(x, y) = \frac{1}{2}(x^2 + y^2) + \alpha\left(x^2 y - \frac{1}{3}y^3\right). \tag{5.6.2}$$

This equation is proposed for describing the motion of a star in an axially symmetric gravitational potential. It applies also to the motions of three identical particles on a ring connected by springs of harmonic and cubic potentials by reducing the problem to a system of two degrees of freedom. In addition, the system of two degrees of freedom is tractable and intuitive as shown below. By virtue of the energy integral $E(=I_1)$, it is sufficient to know three coordinates such as $x, y, \dot{y}$. The trajectory is confined to the three-dimensional space $(x, y, \dot{y})$ which satisfies

$$U(x, y) + \frac{1}{2}\dot{y}^2 \leqq E \tag{5.6.3}$$

since $\dot{x}^2 \geqq 0$. If there is no isolating integral but I_1, the trajectory will fill the space and the system will be ergodic. If there is another isolating integral, the trajectory will lie on a surface of the domain. Now let $P_1, P_2, \ldots$ be the successive intersections of the trajectory passing through the plane $x = 0$ from

$x < 0$ to $x > 0$, that is, the points on the trajectory which lie in the $(y, \dot{y})$ plane and satisfy $x = 0$, $\dot{x} > 0$. If we follow the trajectory for an infinitely long time, there will be, in general, an infinite sequence of points $P_1, P_2, \ldots$. When we regard the point P_n as the point mapped from the preceding point P_{n-1} (the Poincaré mapping), the mapping can be proved to be area-preserving (the invariant of Poincaré-Cartan). If there is no isolating integral other than the energy, these points will densely fill the surface of the section which is the intersection of the volume (5.6.3) with the plane $x = 0$. But if another integral exists, the points P_i will lie on a curve (which is called the curve of section or invariant curve).

Figure 5.14 is the result of a Poincaré mapping of the system (5.6.1) for $E = 0.08333$, where the sequence $P_1, P_2, \ldots$ was obtained one after another by using a computer. Different curves correspond to different initial points $y, \dot{y}$. As shown by the sign $\times$ in the figure, there are at least four stable periodic orbits (periodic solutions). The figure shows the existence of an isolating integral (or rather quasi-integral).

Figure 5.15 corresponds to a somewhat higher value of E. Here we have several closed curves, but besides them, a trajectory exists whose intersections on the $(y, \dot{y})$ plane do not lie on a regular curve. All the points which are not on regular curves in Fig. 5.15b belong to a trajectory with the same initial condition. They seem to be distributed at random over an area left free between the closed curves of the section. The way in which the points of a trajectory do not lie on a curve and diffuse in disorder is, however, extremely complicated and tangled. As seen in Fig. 5.15, when a stochastic region (or sea) is formed around a closed curve, several islands of small closed curves are found in the stochastic sea. Around these islands other ones are found.

At every center of an island lies a periodic solution. In fact, Poincaré showed the existence of infinitely many periodic solutions in a system with two degrees of freedom (the Poincaré-Birkhoff fixed point theorem to be discussed later).

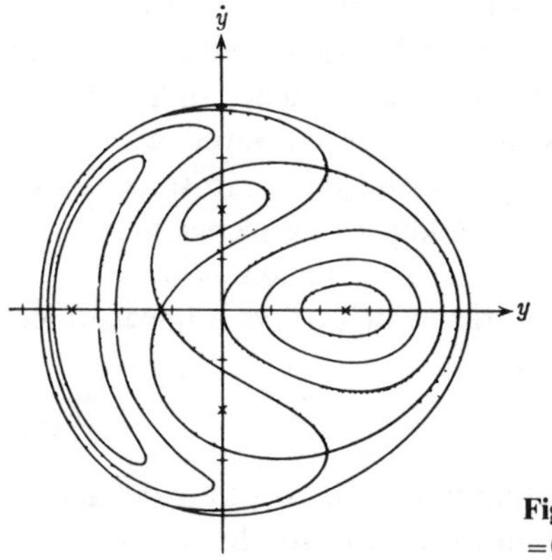

Fig. 5.14. The curve of section for $\alpha = 1$ and $E = 0.08333$. $\times$ shows the stable periodic solutions [5.36]

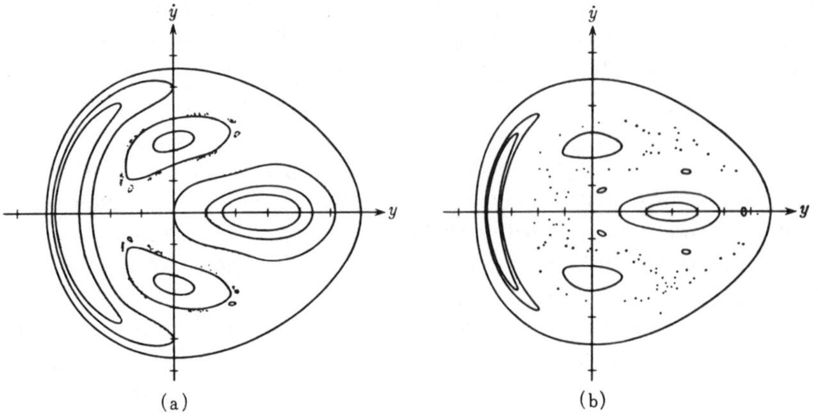

Fig. 5.1.5 a, b. The curve of section $\alpha = 1$. (**a**) For $E = 0.10629105$ and (**b**) for $E = 0.12500$ [5.36]

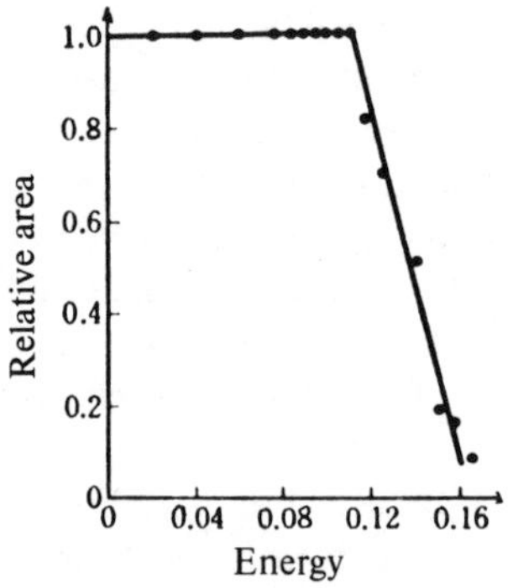

Fig. 5.16. As E increases, a stochastic region appears rather abruptly beyond a certain value, but the transition is diffuse on a magnified scale

Now we compute, as a function of energy E, the fraction of the total area which is covered by the invariant curves with the result shown in Fig. 5.16. When the value of E grows beyond a certain value, the stochastic region comes out, no sharp transition exists, as shown in (Sect. 5.7). The closed curve of the section is an isolating integral or quasi-integral. The existence of the closed curve of section means that this system is not ergodic. However, we cannot always say that a stochastic region is ergodic because there are some whose stochastic regions are composed of several ergodic components which are metrically intransitive. Our problem in the stochastic region is to see how rapidly the difference of two adjacent trajectories which obeys a linear differential equation of the form (see also Fig. 5.12)

$$\dot{x} = A(t)x \,. \tag{5.6.4}$$

grows. The relevant quantity is just the Lyapunov number which is defined as follows: Let $\phi(t)$ be a function whose asymptotic behavior for $t \to \infty$ is in question. Now we compare the behavior of $\phi(t)$ with $\exp(t)$. If a real number λ exists, for $\varepsilon > 0$, such that

$$\lim_{t\to\infty} \sup|\phi(t)|\exp(\lambda + \varepsilon)t = \infty$$

and

$$\lim_{t\to\infty} |\phi(t)|\exp(\lambda - \varepsilon)t = 0 \,,$$

λ is called the Lyapunov number of $\phi(t)$. If (5.6.4) has n independent solutions, n Lyapunov numbers exist.

In the Hénon-Heiles systems, the Lyapunov numbers are found to be all zero for the regular curve and (−0.05, 0, 0, 0.05) for the stochastic region in Fig. 5.15b.

As stated above, the Poincaré mapping of a Hamiltonian dynamical system with two degrees of freedom is measure-preserving. Therefore, the properties of the mapping will be clarified by studying a general measure-preserving transformation on a plane [5.37]:

$$\begin{aligned} x_1 &= f(x_0, y_0) \\ y_1 &= g(x_0, y_0) \end{aligned} \quad \text{or} \quad \begin{pmatrix} x_1 \\ y_1 \end{pmatrix} = T\begin{pmatrix} x_0 \\ y_0 \end{pmatrix} \tag{5.6.5}$$

with Jacobian 1:

$$J = \frac{\partial(f, g)}{\partial(x, y)} = 1 \,. \tag{5.6.6}$$

The fixed points of the transformation correspond to the periodic orbits of the original dynamical system and are given by

$$\begin{pmatrix} x_f \\ y_f \end{pmatrix} = T\begin{pmatrix} x_f \\ y_f \end{pmatrix} . \tag{5.6.7}$$

The transformation may be linearized in the vicinity of the fixed point as

$$\begin{pmatrix} \Delta x_1 \\ \Delta y_1 \end{pmatrix} = M\begin{pmatrix} \Delta x_0 \\ \Delta y_0 \end{pmatrix}, \quad M = \begin{pmatrix} a+d & c+b \\ c-b & a-d \end{pmatrix}, \tag{5.6.8}$$

where $\Delta x = x - x_f$, $\Delta y = y - y_f$ and

$$\det M = a^2 + b^2 - c^2 - d^2 = 1 \tag{5.6.9}$$

in accordance with (5.6.6). Now consider the quadratic form

$$\psi = (b - c)\Delta x^2 + 2d\Delta x\Delta y + (b + c)\Delta y^2 \tag{5.6.10}$$

which is invariant under the transformation M. Since the discriminant is $b^2 - c^2 - d^2 = 1 - a^2$ by virtue of (5.6.9), the quadratic curve (5.6.10) represents an ellipse, parallel lines, or a hyperbola according to $a^2 \lesseqgtr 1$. Define R by

$$R = \frac{1}{2} - \frac{1}{4}\mathrm{tr}\{M\} = \frac{1}{2}(1 - a) \,, \tag{5.6.11}$$

Table 5.1. Classification on invariant curves

R a	$R<0$ $a>1$	$R=0$ $a=1$	$0<R<1$ $1>a>-1$	$R=1$ $a=-1$	$1<R$ $a<-1$
λ	both positive	1	complex	-1	both negative
ψ	hyperbola (ordinary)	line	ellipse	line	hyperbola (reflection)
index	-1	0	1	1	1

where $a = 1/2\,\mathrm{tr}\{M\}$ is used and the eigenvalues λ of M are written as

$$\lambda = 1 - 2R \pm 2\sqrt{R(1-R)}\,. \tag{5.6.12}$$

R is called the residue of the fixed point. The curve of ψ is classified by the value of R as given in Table 5.1. In accordance with this, we can call the fixed point elliptic or hyperbolic.

For classification of the fixed point, it is convenient to use a quantity called an index. Consider a vector **PQ** joining a point P to its image Q on a closed invariant curve. We define the index of the curve by the number of rotations of the vector **PQ** when P makes a turn on the closed curve, and the index of the fixed point is the index of an invariant curve around the fixed point. The index is taken to be positive if the rotational directions of P on the closed curve and of the vector **PQ** are the same. The index of a curve is equal to the sum of indices of the fixed points within the closed curve. The index of a fixed point has the same sign as the residue (Table 5.1). In the case of $0 < R < 1$, the invariant curve of the second order is an ellipse and the eigenvalues of M are complex. This means that the images of the mapping go around on the curve. Now we put $a = \cos 2\pi\omega_0$ and if an integer s exists such that $s\omega_0$ is an integer, then one easily verifies $M^s = 1$ (unit matrix). In other words, for such a, all the points in the invariant ellipses are s-periodic points. Otherwise the images run densely on the ellipses. The closed invariant curve continuously covers the region around the fixed point in the linear approximation. What situation is expected in the presence of a nonlinear part in the transformation T? In an integrable Hamiltonian system of two degrees of freedom, we always have an invariant torus and its Poincaré mapping has a family of invariant curves which covers continuously the region around a fixed point. The transformation (say T_0) is generally quasi-periodic on the invariant curves and eventually periodic on some of the invariant curves. We say s-periodic when the period is s and not s' for $1 \leqslant s' < s$, s and s' being integers. This is different from the case of linear approximation. The residue at the s-period fixed point is zero in the integrable cases including the linear approximation. In the nonintegrable system, where a small perturbation T' is added to an integrable nonlinear Hamiltonian, the existence of the invariant curves around an elliptic fixed point was proved by *Kolmogorov*, *Arnol'd* and *Moser* (KAM) [5.8, 22]. This theorem will be discussed later. Before entering into this subject, we make a short mention of the fixed point theorem proposed first by *Poincaré* and later proved by *Birkhoff* [5.22]. In an integrable system under the transformation T_0, any point on the s-period invariant curve is an s-period fixed point. When a small perturbation T' is added to T_0, then s-period fixed points of numbers $2s$ still exist around an elliptic fixed point under the transformation $T = T_0 + T'$ but not continuously, in contrast to the case of an integrable system. Half of them are elliptic and the others are hyperbolic. Thus we have a hierarchy of many-period fixed points. Now let P be an s-period fixed point. Then $\mathrm{P}' = T(\mathrm{P})$, $\mathrm{P}'' = T^2(\mathrm{P}), \ldots, \mathrm{P}^{(s-1)} = T^{s-1}(\mathrm{P})$ are all s-period fixed points and have the same residue because

$$M(\mathrm{P}') = J(\mathrm{P})M(\mathrm{P})J^{-1}(\mathrm{P}), \tag{5.6.13}$$

where $J(\mathrm{P})$ is the Jacobian matrix of the transformation T and $M(\mathrm{P})$ is the linearized matrix of T', and consequently, we have

$$\mathrm{tr}\{M(\mathrm{P}')\} = \mathrm{tr}\{M(\mathrm{P})\}. \tag{5.6.14}$$

If P is an elliptic fixed point, P′, P″, ... are elliptic fixed points and all have index 1. If P is a hyperbolic fixed point, P′, P″, ... are all hyperbolic fixed points with index -1 because according to the KAM theory, invariant tori of index 1 exist and the total number of indices of the fixed points lying between two invariant tori must be zero. There are equal numbers of elliptic fixed points of index 1 and hyperbolic fixed points of index -1.

5.7 The Kolmogorov, Arnol'd and Moser Theorem

Consider a Hamiltonian in canonical normal form described in terms of action and angle variables:

$$\mathscr{H} = \mathscr{H}_0(J) + \varepsilon\mathscr{H}_1(J, \varphi), \tag{5.7.1}$$

where J and φ are the abbreviations for $(J_1, J_2, \ldots, J_f)$ and $(\varphi_1, \varphi_2, \ldots, \varphi_f)$. ε is a small parameter. Assume that $\mathscr{H}_0$ has a nonzero Hessian

$$\det|\partial^2\mathscr{H}_0/\partial J_i \cdot \partial J_i| \neq 0 \tag{5.7.2}$$

so that $\mathscr{H}_0$ must contain nonlinear terms other than the harmonic part like (5.1.23). The motion determined by the unperturbed system $\mathscr{H}_0(J)$ is given by

$$\begin{gathered} \dot{J}_j = 0 \\ \dot{\varphi}_j = \frac{\partial\mathscr{H}_0}{\partial J_j} \equiv \nu_{0j}(J) \equiv \frac{\omega_{0j}(J)}{2\pi}. \end{gathered} \tag{5.7.3}$$

We put $J_j = \tilde{J}_j$ (= const.) and $\omega_{0j}(\tilde{J}) = \tilde{\omega}_j$. The unperturbed system is integrable and has invariant tori supporting quasi-periodic orbits if ν_{0j}'s are incommensurate[6]; i.e., for any set of integers k_i which are not all zero:

$$(v, k) \equiv \sum v_i k_i \neq 0 .$$

On the other hand, it has periodic orbit if v's are commensurate. When a small perturbation is added, the total Hamiltonian $\mathscr{H}$ has periodic orbits around the fixed points of $\mathscr{H}_0$ in accordance with the Poincaré–Birkhoff fixed-point theorem. Does $\mathscr{H}$ have invariant tori also? The answer to this question was given by *Kolmogorov* in 1954 [5.38], but he published it only in a short note and did not show the concrete calculations. After about a decade, *V.I. Arnol'd* and *J. Moser* [5.8, 39] extended the theorem and gave the proof independently.

If the Hamiltonian (5.7.1) is transformed to

$$\mathscr{H} = \mathscr{H}'_0(J') + \varepsilon^2 \mathscr{H}'_2(J', \varphi') + \cdots \tag{5.7.4}$$

by a canonical transformation generated by the function $W(J', \varphi)$ from (J, φ) to (J', φ'), and this procedure is repeated infinitely many times, then we get a form

$$\mathscr{H} = \mathscr{H}_0^\infty(J^\infty) . \tag{5.7.5}$$

Thus we have the integrals $J^\infty(J, \varphi) = \text{const}$. Such an integral forms an invariant torus. In order to see the validity of this formal procedure, let us examine in detail the generating function W. Putting W as $J'\varphi + \varepsilon S(J', \varphi)$, we have

$$J = \partial W/\partial\varphi = J' + \varepsilon \partial S/\partial\varphi \tag{5.7.6}$$

and

$$\varphi' = \partial W/\partial J' = \varphi + \varepsilon \partial S/\partial J' . \tag{5.7.7}$$

We assume that S is a periodic function of φ with period 1 and expand it in a Fourier series

$$S(J', \varphi) = \sum_k S_k(J') \exp(2\pi i \{k, \varphi\}) , \tag{5.7.8}$$

where $\{k, \varphi\}$ denotes $\sum k_j \varphi_j$ in a system with many degrees of freedom. Furthermore, we expand

$$\mathscr{H}_1 = h_0(J) + \sum_{k \neq 0} h_k(J) \exp(2\pi i \{k, \varphi\}) . \tag{5.7.9}$$

Then the Hamiltonian can be written, by introducing (5.7.6),

$$\begin{aligned} \mathscr{H} &= \mathscr{H}_0(J) + \varepsilon h_0(J) + \varepsilon \sum_{k \neq 0} h_k(J) \exp(2\pi i \{k, \varphi\}) \\ &= \mathscr{H}_0(J') + \varepsilon h_0(J') + \varepsilon [\{\partial \mathscr{H}_0/\partial J', \partial S/\partial\varphi\} \\ &\quad + \sum_k h_k(J') \exp(2\pi i \{k, \varphi\})] + O(\varepsilon^2) + \cdots . \end{aligned} \tag{5.7.10}$$

---

[6] The commensurate v's are dense, but the measure is zero. In other words, incommensurate v's means almost all v's.

Putting $\partial \mathscr{H}_0/\partial J'_i = \nu_i(J')$ and determining S_k so as to make the term in [. . .] vanish:

$$2\pi i\{\nu, k\} S_k(J') + h_k(J') = 0$$

or

$$S_k(J') = i h_k(J')/\{2\pi\nu, k\} . \tag{5.7.11}$$

The generating function thus obtained must be convergent. However, the denominator $\{2\pi\nu, k\}$ can be made as small as possible for a given set of ν_j by choosing appropriate integers k_j. This difficulty is called the problem of small denominators. In particular, ν of unperturbed motion must be incommensurate (a condition of the KAM theory) because otherwise $\{\nu, k\}$ becomes zero for some set of $\boldsymbol{k}$. A region of J of finite measure can be shown to exist where a lower bound for the value $\{2\pi\nu(J'), k\}$ is assured. Consider a system of two degrees of freedom for simplicity. Then for any k_1 and k_2, an α of finite measure exists which satisfies

$$\begin{aligned} &|\alpha - |k_2/k_1|| > K/|k_1|^\mu \\ &\alpha = |\nu_1/\nu_2| \end{aligned} \tag{5.7.12}$$

for appropriate K and μ. To see this, we can put $\alpha = |\nu_1/\nu_2| < 1$ without loss of generality, and we first determine the measure of the regions of α which satisfy $|\alpha - |k_2/k_1|| < K|k_1|^{-\mu}$; we have regions for every set of $(|k_1|, |k_2|)$ in Table 5.2.

The measure I of the sum of these regions in $0 \le \alpha \le 1$ is estimated as

$$I < \sum_{k_1=1}^{\infty} (K/k_1^\mu) k_1 = K \sum_{k_1=1}^{\infty} (k_1^{\mu-1})^{-1} = KC(\mu - 1) , \tag{5.7.13}$$

where $C(x)$ is the ζ function of Riemann which is convergent for $x > 1$. If we take K sufficiently small and $\mu > 2$ (for example, $\mu = 4$), we can let $I < 1$. Therefore, the measure of the region of α which satisfies $|\alpha - |k_2/k_1|| > K/|k_1|^\mu$ for any k_1, k_2 is larger than $1 - I > 0$. This inequality is sufficient to prove the convergence

Table 5.2

$\lvert k_2\rvert$ \ $\lvert k_1\rvert$	1	2	. . .	n
0	$\lvert\alpha\rvert < K$	$\lvert\alpha\rvert < K/2^\mu$	. . .	$\lvert\alpha\rvert < K/n^\mu$
1	$\lvert\alpha - 1\rvert < K$	$\lvert\alpha - 1/2\rvert < K/2^\mu$	. . .	$\lvert\alpha - 1/n\rvert < K/n^\mu$
2		$\lvert\alpha - 1\rvert < K/2^\mu$	. . .	$\lvert\alpha - 2/n\rvert < K/n^\mu$
⋮				⋮
n				$\lvert\alpha - 1\rvert < K/n^\mu$

of the Fourier series (5.7.8):

$$
\begin{aligned}
|S| &\leqslant \sum |h_k(J')/2\pi(v_1 k_1 + v_2 k_2)| \\
&\leqslant \sum |h_k(J')/2\pi v_2 k_1 (|v_1/v_2| - |k_2/k_1|)| \\
&< |2\pi v_2 K^{-1}| \sum |h_k(J')| |k_1|^{\mu-1} ,
\end{aligned}
\tag{5.7.14}
$$

where the sum of the last expression is convergent because the Fourier coefficient $h_k(J')$ of an analytic function becomes exponentially zero for $|k| = |k_1| + |k_2| + |k_3| \to \infty$. The next difficult problem is to show the convergence of the generating function leading to (5.7.5). The final $\bar{W}$ is expressed in powers of ε. However, *Kolmogorov* suggested a series having its nth term of order 2^n for better convergence in an analogous way to Newton's method of approximation for evaluating the zeros of a function [5.38].

In this way it can be shown that invariant tori exist close to the invariant tori of the unperturbed integrable system. The invariant tori do not cover densely the whole space, but only the restricted region satisfying the incommensurability condition. Therefore, the KAM theorem is not inconsistent with the theorem of Poincaré and Fermi.

The KAM theorem refers to a noncontinuous discrete family of invariant tori around an elliptic fixed point, showing nonintegrability, and nonergodicity. Then what will happen around a hyperbolic fixed point? Consider a system of two degrees of freedom and its Poincaré mapping.

In the linear approximation one has two asymptotes through the hyperbolic fixed point and a continuous family of hyperbolic invariant curves around it. In the presence of nonlinear terms, Birkhoff and Moser proved that the situation is the same, i.e., one has two asymptotes through the fixed point and a continuous family of nearly hyperbolic invariant curves, showing integrability around the small region of a hyperbolic fixed point.

One of the asymptotes is an H^+ branch on which image points run towards the hyperbolic fixed point in infinitely many repetitions of mapping, and the other is an H^- branch on which image points run outwards from the fixed point (Fig. 5.17). Let the hyperbolic point under consideration be a $T^s(=T')$ invariant fixed point (s-periodic point), and the neighboring hyperbolic point be $P' = TP$. If the asymptotes H^+ going to P and H^- going out from P′ intersect at a point A_1, then there are infinitely many intersecting points. In fact let $A_2 = T'A_1$ be the image of A_1 on H^-, then A_2 is also on the H^+ curve and thus there are infinitely many intersections $A_1, A_2, \ldots$ towards P. They are called heteroclinic points. (If P and P′ are the same point, $A_1, A_2, \ldots$ are called homoclinic points.). If H^- cuts H^+ at A_1 from the right side of H^+, then H^- cuts H^+ at A_2 in the same direction because H^+ and H^- are separatrices. Consequently, another intersection point B_1 exists between A_1 and A_2. In the same way we have intersections $B_2, B_3, \ldots$. By virtue of the area-preserving property of T, the shaded parts of Fig. 5.17 have an equal area and the unshaded parts also have (a different) equal area. Since $A_1B_1 > A_2B_2 > A_3B_3 > \ldots$, the shapes of the shaded and unshaded parts become elongated. Consider a point P_0

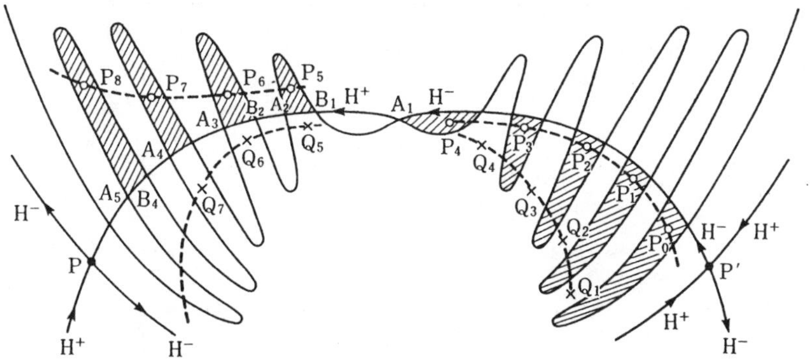

Fig. 5.17. Heteroclinic points around hyperbolic fixed points, and the images of two initial points P_0 and Q_0. *Broken lines* are Moser-Birkhoff invariant curves

in a shaded region near P′. The images $P_1, P_2, \ldots$ marked by an open circle are in consecutive shaded regions. $P_1, P_2, \ldots$ lie on a Moser–Birkhoff invariant curve, but on approaching P, the images marked by the circle lie on another invariant curve.

If one considers a point Q_0 near P_0 but outside the shaded region, the images of Q_0 marked by a cross lie on a different invariant curve on approaching P. A slight displacement of the initial point yields a great difference of the image points. This gives rise to chaotic behavior of the mapping around a hyperbolic fixed point. Figure 5.18 shows an example. The dots around the origin are the image points of a single orbit.

These chaotic regions are usually separated by KAM invariant curves as shown in Fig. 5.19. Thus in a system with two degrees of freedom, the chaotic regions are separated by a KAM invariant torus. Ergodicity cannot be expected in a system of two degrees of freedom.

In a system with n degrees of freedom ($n > 2$), the phase space has $2n$ dimensions, and the energy surface is $2n - 1$ dimensional. Thus a KAM torus of n dimensions can not separate a $(2n - 1)$-dimensional surface for $n > 2$. The image points in a chaotic region are not confined to a restricted region, but can move into another region; this phenomenon is called Arnold diffusion. In thermodynamical systems that are necessarily high dimensional therefore, it is expected that chaotic behavior prevails in the entire space.

Computer experiments can reveal various phenomena occurring in cases where mathematical theories do not yet yield transparent results. Examples are the Fermi-Pasta-Ulam problem (Sect. 5.5), Hénon-Heiles experiment, etc. Generally speaking, in physics progress is made through the interaction between theories and experiments. However, as far as ergodic problems are concerned, except through computers, no experiments are possible, which allow one to control a dynamical system by specifying its initial conditions, and to observe

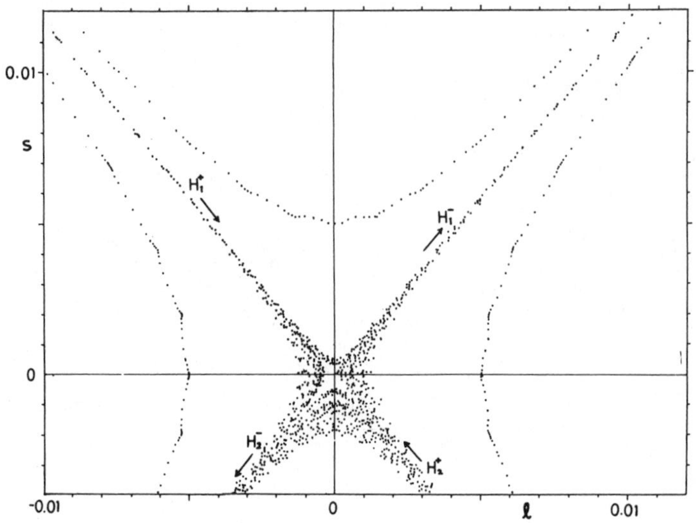

Fig. 5.18. An example of chaos around a hyperbolic fixed point at the origin (from billiard motion in an annulus bounded by non-concentric circles [5.40]

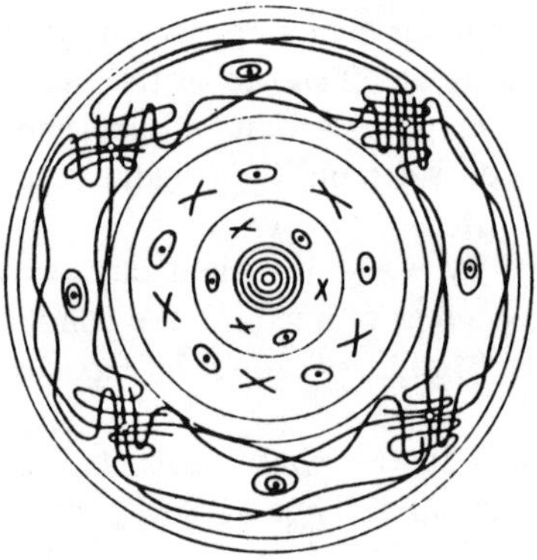

Fig. 5.19. Generic behavior on the surface of section. ⊙ are elliptic fixed points, × are hyperbolic fixed points, around which an intricate network is formed [5.8]

the relevant required quantities. Note that thermodynamical systems have many degrees of freedom, whereas computer experiments are usually restricted to a few degrees of freedom. Nevertheless, we can extend the results obtained there and predict what is expected in thermodynamical systems.

In the next section we again discuss the ergodic theorem for chaotic dynamics. We are concerned with the distribution function or density function $\rho(\mathrm{P}, t)$ in phase space, where P is a point in phase space. Our main interest is its behavior at $t \to \infty$.

5.8 Ergodic Theorems (II)

5.8.1 Weak Convergence

In (5.2.22) a phase function $f(\mathrm{P})$ is usually a continuous function of P almost everywhere, but $\rho(\mathrm{P}, t)$ is a function rapidly changing with respect to P and t. For example, consider the baker's transformation discussed in Sect. 5.3.1. In Fig. 5.3 the shaded (black, $\rho = 2$) regions and unshaded (white, $\rho = 0$) ones are stacked alternately 2^n times after the transformation ϕ^n, and if we look at a certain point P, it is in either a black or a white region irregularly with increasing n. To see this let the coordinates of P be x, y $(0 \leqslant x < 1, 0 \leqslant y < 1)$, and $y = 0, a_1 a_2 \ldots a_n \ldots$ $(a_i = 0$, or $1)$ in binary representation. P is in a white region if $a_n = 0$, and in a black region if $a_n = 1$ after n transformations. Consequently $\rho(\mathrm{P}, n)$ does not tend in a pointwise manner to a definite function for $n \to \infty$. However, in the limit $n \to \infty$, the integral corresponding to (5.2.22)

$$\int f(\mathrm{P})\rho(\mathrm{P}, n)\, d\Gamma \tag{5.8.1}$$

exists and is equal to

$$\int f(\mathrm{P})\rho^{+}(\mathrm{P}, n)\, d\Gamma, \quad \rho^{+}(\mathrm{P}) = 1\ , \tag{5.8.2}$$

where ρ^{+} is a coarse-grained function of $\rho(\mathrm{P})$ over a small region around P for large n [5.41]. This limiting process is sometimes referred to as weak convergence or convergence in law.

The same is expected to hold for Hamiltonian systems of many degrees of freedom. In a chaotic region, the largest Lyapunov number λ is always positive, implying that the separation of two nearby points grows exponentially as $e^{\lambda t}$ for $t > 0$ and as $e^{-\lambda t}$ for $t < 0$. In other words, two neighboring points P_1, P_2 in phase space come from two exponentially separated points P'_1, P'_2 at $t = 0$. Thus the density functions $\rho_t(\mathrm{P}_1)$ and $\rho_t(\mathrm{P}_2)$ are respectively equal to the values of the density functions at P'_1 and P'_2. Even if the density function $\rho_0(\mathrm{P})$ at $t = 0$ is a smoothly changing function, the difference of $\rho_0(\mathrm{P}'_1)$ and $\rho_0(\mathrm{P}'_2)$ is not small, and the gradient of, and consequently the rate of change of, $\rho_t(\mathrm{P})$ is great. In other words, the density function $\rho_t(\mathrm{P})$ varies rapidly and irregularly with respect to P and t [5.42]. This is similar to the baker's transformation. Thus we can say that $U_{-t}\rho_0(\mathrm{P})$, equation (5.2.21), does not tend to a definite function. However, since $f(P)$ is a continuous function or smoothly varying function, the integral of (5.2.22) will exist for $t \to \infty$, or for t large, and we can write

$$\lim_{t \to \infty} \int f(\mathrm{P}) U_{-t}\rho_0(\mathrm{P})\, d\Gamma = \int f(\mathrm{P})\rho^{+}(\mathrm{P})\, d\Gamma\ , \tag{5.8.3}$$

where $\rho^{+}(\mathrm{P})$ is a coarse-grained function of $\lim_{t\to\infty} U_{-t}\rho_0(\mathrm{P})$. This function $\rho^{+}(\mathrm{P})$ is defined irrespective of the function $f(\mathrm{P})$.

Convergence in law, introduced here, is a concept in the theory of probability implying that the law of probability governing a random process can be different from that of $t \to \infty$ which governs a different probability space. In the present case $\rho_t(\mathrm{P})$ obeys Hamiltonian mechanics in phase space, but $\rho^+(\mathrm{P})$ obeys the law of evolution of a thermodynamical state in a space of gross variables of thermodynamical quantities. Now it is usual to use the term of weak convergence in Sect. 5.2.3 we defined ρ^* as a limit $\rho_t(\mathrm{P})$ for $t \to \infty$ in the sense of Hopf. If ρ^+ exists then $\rho^* = \rho^+$.

5.8.2 Ergodicity

Since ergodicity, or metrical transitivity, implies that the phase point representing a state of a dynamical system wanders all over the energy surface of the phase space as time goes on, the distribution function of the phase space is supposed to be uniform on the energy surface (5.2.4). The distribution function $\rho(\mathrm{P}, t)$ in phase space can be expressed as

$$\rho(\mathrm{P}, t) = U_{-t}\rho(\mathrm{P}, 0) = \rho(\mathrm{P}_0, 0)\,, \tag{5.8.4}$$

where P_0 is the point at $t = 0$, $\mathrm{P}_0 = \phi_{-t}\mathrm{P}$. The last equation is just the Liouville theorem. The average of an arbitrary continuous function $g(\mathrm{P})$ is at time t

$$\begin{aligned}\langle g \rangle_t &= \int g(\mathrm{P})\rho(\mathrm{P}, t)\,d\Gamma \\ &= \int g(\mathrm{P})U_{-t}\rho(\mathrm{P}, 0)\,d\Gamma\,.\end{aligned} \tag{5.8.5}$$

Now we assume that the limit

$$\lim_{t\to\infty} \rho(\mathrm{P}, t) = \lim_{t\to\infty} U_{-t}\rho(\mathrm{P}, 0) \quad (= \text{const.}) \tag{5.8.6}$$

exists and tends to a quantity independent of P. The meaning of this limit will be discussed later.

Ergodicity can be proved from the condition (5.8.6). In fact, from (5.8.5) we have

$$\lim_{t\to\infty} \langle g \rangle_t = \langle g \rangle_0\,, \tag{5.8.7}$$

where $\langle\ \rangle_0$ stands for the average in terms of a uniform distribution function. Equation (5.8.7) implies that this limit does not depend on the initial distribution.

On the other hand (5.8.5) can be rewritten as

$$\begin{aligned}\langle g \rangle_t &= \int U_t g(\mathrm{P})\rho(\mathrm{P}, 0)\,d\Gamma \\ &= \int g(\phi_t(\mathrm{P}))\rho(\mathrm{P}, 0)\,d\Gamma \\ &= \int \bar{g}(\mathrm{P})\rho(\mathrm{P}, 0)\,d\Gamma\end{aligned} \tag{5.8.8}$$

if we take $g(\mathrm{P})$ as an invariant function: $g(\mathrm{P}) = g(\mathrm{P}_t) = \bar{g}(\mathrm{P})$. Equation (5.8.8) depends on the initial distribution, contrary to (5.8.7). To reconcile these two, $\bar{g}(\mathrm{P})$ must be independent of P. Therefore, according to the fundamental theorem the system is ergodic.

The meaning of (5.8.6) must be examined more closely. First consider ergodicity on a torus discussed in Sect. 5.3.2. Let ϕ denote the translation with irrational ω,

$$\phi x = x + \omega \ (\mathrm{mod}\ 1)\ , \tag{5.8.9}$$

and $\rho(x, p)$, the density function after p translations, is given by

$$\rho(x, p) = \rho(x - \omega, p - 1)\ . \tag{5.8.10}$$

This is to be compared with a phase function $f(x)$ discussed in Sect. 5.3.2, where $\phi f(x) = f(x + \omega)$. Therefore the same arguments as (5.3.8–12) hold by expanding $\rho(x, p)$ in a Fourier series provided that ω is replaced by $-\omega$. Thus we have

$$\begin{aligned} \rho(x, p) &= \sum_{k=0}^{\infty} \exp(-2\pi i k p \omega)\rho_k \exp(2\pi i k x)\ , \\ \rho_k &= \int_0^1 \rho \exp(-2\pi i k x)\rho(x, 0)\, dx\ , \end{aligned} \tag{5.8.11}$$

and thus $\rho(x, p)$ does not have a limit at $p \to \infty$. Then consider the long-time average of $\rho(x)$

$$\bar{\rho} = \lim_{n\to\infty} \frac{1}{n} \sum_{p=0}^{n-1} \rho(x, p) \tag{5.8.12}$$

$$\begin{aligned} &= \lim \frac{1}{n} \sum_{k=0}^{\infty} \frac{1 - \exp(-2\pi i k n \omega)}{1 - \exp(-2\pi i k \omega)} \rho_k \exp(2\pi i k x) \\ &= \rho_0 \\ &= \int_0^1 \rho(x, 0)\, dx\ . \end{aligned} \tag{5.8.13}$$

In dynamics on a torus the limit (5.8.6) is therefore taken as the limit for the long-time average.

For a baker's transformation, or in a system obeying Hamiltonian dynamics with positive Lyapunov number, it was shown in Sect. 5.8.1 that the limit (5.8.6) should be regarded as the one corresponding to weak convergence. Even though this limit exists, we have to show furthermore that $\rho^+(\mathrm{P})$ is constant on the energy surface, for the system to be ergodic. A baker's transformation can be shown to have this property.

The uniform distribution of the limit (5.8.6) implies metrical transitivity, but does more. The mixing property is also a result of a uniform distribution. Consider two regions A and B where a point in A obeys the law of dynamics, while B is kept unchanged. Then the product of $\phi_t A$ and B can be expressed as

$$\mu[\phi_t A \cap B] = \int_B d\mathrm{Q} \int_A d\mathrm{P}\, \rho(\mathrm{P}_t)\delta(\mathrm{P}_t, \mathrm{Q}) \tag{5.8.14}$$

and

$$\delta(\mathrm{P}_t, \mathrm{Q}) = 1, \quad \text{if } \mathrm{P}_t = \phi_t \mathrm{P} \in B, \quad \mathrm{Q} \in B\,, \qquad = 0, \quad \text{otherwise}\,, \tag{5.8.15}$$

where, by assumption, $\rho(\mathrm{P}_t, t) = U_{-t}\rho(\mathrm{P}_t, 0) = \rho(\mathrm{P}, 0)$ converges weakly to a uniform distribution when P_t is kept unchanged (and thus $\mathrm{P} = \phi_{-t}\mathrm{P}_t$ changes irregularly) for $t \to \infty$. Thus we have the mixing property

$$\lim_{t\to\infty} \mu[\phi_t A \cap B] = \mu[A]\mu[B]\,.$$

5.8.3 Entropy and Irreversibility

Mathematically, the meaning of weak convergence is concerned with finding a limit for $t \to \infty$. However, as long as the function $f(\mathrm{P})$ in (5.8.2) is smooth, as is usual in physical quantities, we can take t as a large finite value. Then the limiting function $\rho^+(\mathrm{P}, t)$ is considered as t-dependent, and the time t in this function is not a microscopic, but a macroscopic variable in the thermodynamic sense, i.e., the small time interval $\delta t = \tau$ is also macroscopic and much larger than the small time interval in a microscopic description. Thus the function $\rho^+(\mathrm{P}, t)$ obeys

$$\rho^+(\mathrm{P}, t+\tau) = [\rho'(\mathrm{P}, t+\tau)]^+\,, \qquad \rho'(\mathrm{P}, t+\tau) = U_{-\tau}\rho^+(\mathrm{P}, t)\,, \tag{5.8.16}$$

$$\frac{\partial \rho^+}{\partial \tau} = \frac{\rho^+(\mathrm{P}, t+\tau) - \rho^+(\mathrm{P}, t)}{\tau} = \left[\left(\frac{U_{-\tau} - 1}{\tau}\right)\rho^+(\mathrm{P}, t)\right]^+ = \mathscr{U}\rho^+(\mathrm{P}, t)\,, \tag{5.8.17}$$

where $(U_\tau - 1)/\tau$ is just equal to $\mathrm{i}L$ (5.2.20) for small τ. In this connection, it should be noted that the process of weak convergence is taken at every macroscopic time.

Entropy is a quantity on a thermodynamic scale, and is defined in terms of $\rho^+(\mathrm{P}, t)$:

$$S(t) = -\int \rho^+ \ln \rho^+ \, d\Gamma\,. \tag{5.8.18}$$

This quantity is shown, in the following, to increase with time, in contrast to the quantity

$$S'(t) = -\int \rho \ln \rho \, d\Gamma\,, \tag{5.8.19}$$

which remains constant because $dS'(t)dt = 0$ by virtue of (5.2.20).

Since weak convergence is a kind of coarse graining in space, we may take from (5.8.16)

$$\rho^+(\mathrm{P}, t+\tau) = [\rho'(\mathrm{P}, t+\tau)]^+ = \int a(\mathrm{P} - \mathrm{P}')\rho'(\mathrm{P}', t+\tau)\, d\Gamma'\,, \tag{5.8.20}$$

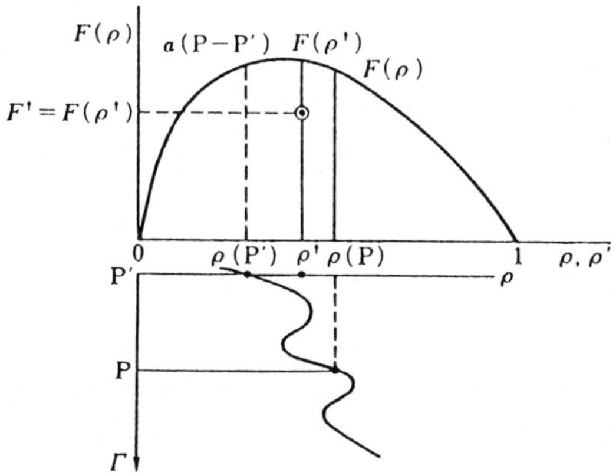

Fig. 5.20. Explanation of the inequality (5.8.23). (ρ^+, F^+) is the center of mass for mass distributed on the curve $F(\rho(\mathrm{P}))$

where the weighting function $a(\mathrm{P}-\mathrm{P}')>0$ can be assumed to tend quickly to zero for large $|\mathrm{P}-\mathrm{P}'|$, and to be normalized as

$$\int a(\mathrm{P})\,d\Gamma = 1\,. \tag{5.8.21}$$

Now consider a convex function

$$F(x) = -x\ln x, \quad 0 \leqslant x \leqslant 1\,. \tag{5.8.22}$$

It can be shown that

$$F(\rho^+(\mathrm{P}, t+\tau)) > F^+(\rho'(\mathrm{P}, t+\tau)) \quad \text{at } \rho' = \rho^+\,, \tag{5.8.23}$$

$$F^+(\rho'(\mathrm{P}, t+\tau)) \equiv \int a(\mathrm{P}-\mathrm{P}')F(\rho'(\mathrm{P}', t+\tau))\,d\Gamma' \tag{5.8.24}$$

because (5.8.20, 24) imply that the point (ρ^+, F^+) in a two-dimensional space is the center of mass of the curve $F(\rho'(\mathrm{P}))$ with mass $a(\mathrm{P}-\mathrm{P}')$ distributed at $\rho'(\mathrm{P}')$, and consequently, the center of mass (ρ^+, F^+) is below the convex function $F(\rho^+)$ (Fig. 5.20).

The entropy defined by (5.8.18) is thus

$$S(t+\tau) = \int F(\rho^+(\mathrm{P}, t+\tau))\,d\Gamma > \int F(\rho'(\mathrm{P}, t+\tau))\,d\Gamma \tag{5.8.25}$$

by virtue of (5.8.21), and the third member of this inequality is equal to $\int F(\rho^+(\mathrm{P}, t))\,d\Gamma = S(t)$, since ρ' evolves according to the operator $U(-\tau)$ of Hamiltonian dynamics (5.8.16).

The increase of entropy is a manifestation of irreversibility. Irreversibility is caused by the change of the time- evolution operator from the Liouville operator L defined in (5.2.20) to $\mathscr{U}$, which is achieved by coarse graining related to weak convergence [5.26].

5.9 Quantum Mechanical Systems

In classical Hamiltonian systems, the foundation of statistical mechanics is thus attributable to their chaotic behavior. This is also expected to be the case in quantum systems.

5.9.1 Theorems in Quantum Mechanical Systems

In quantum mechanical systems, we have results similar to those of classical mechanical systems. Consider a Hermite operator $\hat{A}$ as a mechanical quantity and let the wave function be $\Psi(t)$; the measured value of $\hat{A}$ is defined as the long-time average:

$$\bar{A} = \lim_{T\to\infty} \frac{1}{T} \int_{\tau_0}^{\tau_0+T} A(t)\,dt\,, \tag{5.9.1}$$

where

$$A(t) = (\Psi, \hat{A}\Psi)\,. \tag{5.9.2}$$

The quantum mechanical counterpart of the Birkhoff theorems is that $\bar{A}$ exists and does not depend on the initial condition. Let φ_j be the eigenfunctions of the Hamiltonian $\mathscr{H}$ of the system and E_j the energy eigenvalues, then $\Psi(0)$ and $\Psi(t)$ are represented as

$$\Psi(0) = \sum_{j=1}^{\infty} c_j(0)\varphi_j \tag{5.9.3}$$

$$\Psi(t) = \sum_{j=1}^{\infty} c_j(0)\exp(-\mathrm{i}E_i t/\hbar)\varphi_j = \sum_{j=1}^{\infty} c_j(t)\varphi_j\,. \tag{5.9.4}$$

Then (5.9.2) becomes

$$\begin{aligned} A(t) &= \sum_{i,j} c_i(t)c_j^*(t)A_{ji}, \\ A_{ji} &= (\varphi_j, \hat{A}\varphi_i)\,. \end{aligned} \tag{5.9.5}$$

Taking the time average, we have

$$\frac{1}{\tau}\int_{\tau_0}^{\tau_0+\tau} c_i(t)c_j^*(t)\,dt = c_i(0)c_j^*(0)\frac{1}{\tau}\int_{\tau_0}^{\tau_0+\tau} \exp[-\mathrm{i}(E_i - E_j)t/\hbar]\,dt$$

but, since

$$\lim_{\tau\to\infty}\frac{1}{\tau}\int_{\tau_0}^{\tau_0+\tau} \exp[-\mathrm{i}(E_i - E_j)t/\hbar]\,dt = \begin{cases} 0 & (E_i \neq E_j) \\ 1 & (E_i = E_j)\,, \end{cases} \tag{5.9.6}$$

$\bar{A}$ exists provided that the operations $\sum_{i,j}$ and $\lim_{\tau\to\infty}$ are commutable. In addition, if the energy eigenvalues are not degenerate, we have

$$\bar{A} = \sum_i |c_i(0)|^2 A_{ii} \tag{5.9.7}$$

and since $|c_i(0)|^2 = |c_i(t)|^2$, $\bar{A}$ does not depend on the phase angle of the complex quantity c_i and thus on τ_0. This is the same as Birkhoff's second theorem. If the energy eigenvalues are degenerate, $\bar{A}$ depends on the phase angle of c_i. Then by taking the average of phase angles (random phase assumption), we can obtain the same result as (5.9.7).

On the other hand, the counterpart of Hopf's theorem is given as follows. It is a density matrix $\hat{\rho}$ (defined as $\hat{\rho}_{ij} = c_i c_j^*$) which corresponds to a distribution function in classical mechanics. Its motion is written as

$$\begin{aligned}\hat{\rho}(t) &= \exp\left(-\frac{\mathrm{i}}{\hbar}\mathscr{H}t\right)\hat{\rho}(0)\exp\left(\frac{\mathrm{i}}{\hbar}\mathscr{H}t\right)\\ &\equiv \hat{U}_{-t}\hat{\rho}(0)\end{aligned} \tag{5.9.8}$$

corresponding to (5.2.21). The average of a dynamical quantity $\hat{A}$ is

$$\mathrm{tr}\{\hat{A}\hat{\rho}(t)\} = \mathrm{tr}\{\hat{A}\hat{U}_{-t}\hat{\rho}(0)\}\ , \tag{5.9.9}$$

and then we will examine the existence of $\hat{\rho}^*$, similar to (5.2.19), such that

$$D = \lim_{\tau\to\infty}\frac{1}{\tau}\int_0^\tau |\mathrm{tr}\{\hat{A}\hat{U}_{-t}\hat{\rho}(0)\} - \mathrm{tr}\{\hat{A}\hat{\rho}^*\}|^2\,dt = 0\ . \tag{5.9.10}$$

$\hat{\rho}(0)$ is generally a sum of pure states

$$\hat{\rho}(0) = \sum_\nu w_\nu \hat{\rho}_\nu(0)\ . \tag{5.9.11}$$

The second expression of (5.9.10) is written, provided that ρ^* also has the form of (5.9.11), as

$$D = \lim_{\tau\to\infty}\frac{1}{\tau}\int_0^\tau \left|\sum_\nu w_\nu[\mathrm{tr}\{\hat{A}\hat{U}_{-t}\hat{\rho}_\nu(0)\} - \mathrm{tr}\{\hat{A}\hat{\rho}_\nu^*(0)\}]\right|^2 dt\ . \tag{5.9.12}$$

Using the inequality of Schwartz and then assuming commutability of the time average and $\sum$,

$$\begin{aligned}D &\leq \lim_{\tau\to\infty}\frac{1}{\tau}\int_0^\tau \sum_\nu w_\nu|\mathrm{tr}\{\hat{A}\hat{U}_{-t}\hat{\rho}_\nu(0)\} - \mathrm{tr}\{\hat{A}\hat{\rho}_\nu^*(0)\}|^2\,dt\\ &= \sum_\nu w_\nu\left[\lim_{\tau\to\infty}\frac{1}{\tau}\int_0^\tau |\mathrm{tr}\{\hat{A}\hat{U}_{-t}\hat{\rho}_\nu(0)\} - \mathrm{tr}\{\hat{A}\hat{\rho}_\nu^*(0)\}|^2\,dt\right]\ .\end{aligned} \tag{5.9.13}$$

Thus it is sufficient if (5.9.10) holds for the density matrix $\hat{\rho}_\nu$ of the pure state. Then

$$A(t) = \mathrm{tr}\{\hat{A}\hat{U}_{-t}\hat{\rho}_\nu(0)\} = \sum_{i,j}\exp[\mathrm{i}(E_i - E_j)t/\hbar]A_{ij}\rho_{\nu ji}(0)\ . \tag{5.9.14}$$

Among the terms in $A(t)$, the terms which become constant values independent of time when averaged over a long time are those with $E_i = E_j$ and therefore we can take $\hat{\rho}_\nu^*(0)$ as

$$\sum_i A_{ii}\rho_{\nu ii}(0) \equiv \mathrm{tr}\{\hat{A}\hat{\rho}_\nu^*(0)\} \tag{5.9.15}$$

if E_i are not degenerate. Consequently, we may write

$$D_\nu \equiv \lim_{\tau\to\infty} \frac{1}{\tau}\int_0^\tau |A(t) - \mathrm{tr}\{\hat{A}_\nu \hat{\rho}^*(0)\}|^2\, dt$$

$$= \lim_{\tau\to\infty} \frac{1}{\tau}\int_0^\tau \left|\sum_\omega \exp(\mathrm{i}\omega t) g_\nu(\omega)\right|^2 dt\,, \qquad (5.9.16)$$

where

$$\omega = \frac{E_i - E_j}{\hbar} \qquad (5.9.17)$$

$$g_\nu(\omega) = \begin{cases} 0 & (\text{when } \omega = 0) \\ \sum_{i,j} A_{ij}\rho_{\nu ji}\,, \end{cases} \qquad (5.9.18)$$

where the sum over i, j is taken over all i, j which satisfy (5.9.17). From (5.9.16) we have

$$D_\nu = \sum_\omega |g_\nu(\omega)|^2\,. \qquad (5.9.19)$$

Since this quantity depends quadratically on $\hat{A}$ as seen in (5.9.16), *Ludwig* [5.43] took the ratio

$$\delta = D_\nu/\mathrm{tr}\{\hat{A}^2\rho_\nu^*\} = D_\nu/\sum_{ij}|A_{ij}|^2\rho_{\nu ii}(0) \qquad (5.9.20)$$

and required that this ratio tend to zero. This is equivalent to the consideration that the average of the variation of $A(t)$ around $\mathrm{tr}\{\hat{A}^2\rho_\nu^*\}$ is small compared with $\mathrm{tr}\{\hat{A}\rho_\nu^*\}$. The quantum parallel of Hopf's theorem is to prove that δ is small. Assuming that the resonance conditions among energy levels are not satisfied, i.e., for any pairs (i, j) and (i', j'), $E_i - E_j \neq E_{i'} - E_{j'}$ unless i and j are equal to i' and j', respectively (nondegeneracy of energy differences), Ludwig showed that δ is small when there are many nonzero nondiagonal elements A.

In quantum mechanics, however, there is the difficulty of establishing whether the long-time average (5.9.7) of $\hat{A}$ is equal to the phase average. If the system is a mixed state of several pure states of the same energy, the phase average of $\hat{A}$ is $(1/n)\sum_i A_{ii}$ where the sum is taken over the degenerate states of the energy and n is the number of them. On the other hand, since all the degenerate states have the same time evolution, the long-time average of $\hat{A}$ is given by the same form of $\bar{A}$ of (5.9.7) under the random phase assumption where the sum over i is the sum of the degenerate n states. Consequently, the long-time average is not equal to the phase average unless $|c_i(0)|^2$ is given by a value independent of i. To avoid this difficulty, *von Neumann* introduced the concept of coarse observables which are called macro-observables. The mathematical method of *von Neumann* was modified by *van Kampen* [5.44] so as to be physical. Completely isolated systems cannot exist and in the case of quantum mechanical systems, an observation is always accompanied with uncertainty. Furthermore, an accurate measurement of energy takes an infinitely long time.

Therefore, we always need to take more or less an amount of error into account. To do this, let the energy eigenvalues obtained by diagonalization of the Hamiltonian $\mathscr{H}$ be arranged in the order of magnitude and grouped into cells with a finite energy interval where all the energy eigenvalues of a cell are considered the same. The Hamiltonian defined in such a way is called the macroscopic Hamiltonian $\hat{\mathscr{H}}_{\mathrm{m}}$. The macroscopic observation should be related to $\hat{\mathscr{H}}_{\mathrm{m}}$. Further, for any mechanical quantity $\hat{A}$, the commutation relation of $\hat{\mathscr{H}}$ and $\hat{A}$ is

$$(\hat{\mathscr{H}}\hat{A} - \hat{A}\hat{\mathscr{H}})_{kl} = (E_k - E_l)A_{kl} \tag{5.9.21}$$

in the representation diagonalizing $\hat{\mathscr{H}}$. It is either when $A_{kl} = 0$ or when $E_k = E_l$ that $\hat{\mathscr{H}}$ and $\hat{A}$ can be commutable. Since in the macroscopic observation both $\mathscr{H}$ and $\hat{A}$ are observable at the same time, A_{kl} will be zero for a large difference between k and l. Therefore, taking $\mathscr{H}_{\mathrm{m}}$ instead of $\mathscr{H}$ and making the energies degenerate like $E_k = E_l$ for $A_{kl} \neq 0$, $\mathscr{H}_{\mathrm{m}}$ and $\hat{A}$ can commute and simultaneously be diagonalized. For a second quantity $\hat{B}$, however, it is not always possible to make $\mathscr{H}_{\mathrm{m}}$, $\hat{A}$ and $\hat{B}$ simultaneously diagonal. The macroscropic operators $\hat{A}_{\mathrm{m}}$ and $\hat{B}_{\mathrm{m}}$, constructed by letting the diagonal elements of $\hat{A}$, $\hat{B}$ be degenerate, commute with $\mathscr{H}_{\mathrm{m}}$ and between themselves. Having thus obtained the complete system, we can divide the Hilbert space into a set of cells by such macroscopic operators. *von Neumann* assumed that the cells obtained in this way are still sufficiently large and have a great number of microstates, and that the cell energies $E_1, E_2, \ldots$ and their differences are not degenerate, and furthermore, that microstates in the same cell take the same a priori probability and are averaged over the microcanonical ensemble. As previously shown, these conditions are necessary for the time average to be independent of the phase angles of the initial state.

Now let a state in the αth cell be expressed by j_α and let its eigenvector be ω_{j_α}. Here j takes the value $1, 2, \ldots, S_\alpha$. Then we can express the wave function of the system $\Psi(t)$ as

$$\Psi(t) = \sum_{\alpha=1}^{N} \sum_{j=1}^{S_\alpha} c_{j\alpha}(t)\omega_{j\alpha}\,, \tag{5.9.22}$$

where N is the number of cells. The probability of finding the system in the state α by a microscopic observation is

$$u_\alpha(t) = \sum_{j=1}^{S_\alpha} |(\Psi(t), \omega_{j\alpha})|^2 = \sum_{j=1}^{S_\alpha} |c_{j\alpha}(t)|^2\,. \tag{5.9.23}$$

Assuming $\Psi(t)$ to be normalized, we have

$$|\Psi(t)|^2 = 1 = \sum_{\alpha=1}^{N} \sum_{j=1}^{S_\alpha} |c_{j\alpha}(t)|^2\,. \tag{5.9.24}$$

Since from the same a priori probability of eigenvectors in a cell, all $\langle |c_{j\alpha}(t)|^2 \rangle$

have the same value $\bar{W}_\alpha$ regardless of j, we have

$$\langle u_\alpha(t)\rangle = s_\alpha \bar{W}_\alpha, \quad \sum_{\alpha=1}^{N} \bar{W}_\alpha = 1 . \tag{5.9.25}$$

If we observe a macroscopic operator $\hat{A}_{\mathrm{m}}$, we have, putting a^α as the eigenvalue for the cell α,

$$\begin{aligned} \hat{A}_{\mathrm{m}}\Psi(t) &= \sum_{\alpha, j} a^\alpha c_{j\alpha}(t)\omega_{j\alpha} \\ A(t) &= (\Psi(t), \hat{A}_{\mathrm{m}}\Psi(t)) = \sum_\alpha a^\alpha \sum_j |c_{j\alpha}(t)|^2 \end{aligned} \tag{5.9.26}$$

and therefore, by taking an average over the microcanonical ensemble for each cell,

$$\langle A(t)\rangle = \sum_\alpha a^\alpha s_\alpha \bar{W}_\alpha . \tag{5.9.27}$$

von Neumann's objective was to prove the theorem of Hopf convergence:

$$\lim_{T\to\infty} \frac{1}{T}\int_0^T |A(t) - \langle A(t)\rangle|^2\, dt = 0 . \tag{5.9.28}$$

The ergodic theorems in quantum mechanics were discussed, among others, in the context of von Neumann's theorem and aimed to criticize and to improve it at the same time. The nondegeneracy of the energy spectrum and the absence of resonance frequencies seem to play an analogous role as the metric indecomposibility in classical dynamical systems, but they are not the same. The following is an example of *Fierz* [5.45]. A fluid which rotates in a cylindrical container should not be ergodic because of the conservation of its angular momentum. On the other hand, however, we may understand the absence of resonance conditions between the energy levels of the system if appropriate intermolecular forces are taken into account. According to von Neumann's theorem, the system results in being ergodic. *Fierz* and *Farquhar* and *Landsberg* [5.46] showed the possibility of establishing ergodicity without the condition of nondegeneracy in energies and their differences.

The method by von Neumann and van Kampen of introducing macro-observables is based on the process of observations, but *Ludwig* [5.43] considered that they should arise from the more intrinsic properties of a system. For this purpose, he introduced the concepts of "discernibility" and defined a distance between two different ensembles by

$$d(\hat{\rho}_1, \hat{\rho}_2) = \frac{1}{\sqrt{2}}\|\sqrt{\hat{\rho}_1} - \sqrt{\hat{\rho}_2}\| , \tag{5.9.29}$$

where $\hat{\rho}_1, \hat{\rho}_2$ represent the density matrices for two ensembles, respectively, and $\| \ \|$ is defined by

$$\|\hat{A}\|^2 = \mathrm{tr}\{\hat{A}^+\hat{A}\} \quad (\hat{A}^+ \text{ is the transposed matrix of } \hat{A}) .$$

Since the difference of the average values of $\hat{A}$ over these two ensembles satisfies the inequality

$$|\langle A_1\rangle - \langle A_2\rangle| = |\mathrm{tr}\{\hat{\rho}_1\hat{A}\} - \mathrm{tr}\{\hat{\rho}_2\hat{A}\}|$$
$$\leq 2\sqrt{2}|\hat{A}|d(\hat{\rho}_1, \hat{\rho}_2)\,, \tag{5.9.30}$$

the differences of $\langle A\rangle$ become small if $d(\hat{\rho}_1, \hat{\rho}_2)$ is small. This treatment refers to a microscopic observable, but *Ludwig* attempted to derive a definition of macroscopic observables by introducing the definition of the macroscopic measure of discernibility. However, it has not been completely established how to do it from the Hamiltonian of the system and therefore his procedure is not yet certain of being independent of the method of observations.

On the other hand, since the method of von Neumann yields results that are too general, Italian physicists [5.47] attempted to prove ergodicity by taking averages over initial states.

Among the works of studying ergodicity from a point of view different from that of von Neumann, one should mention of the one by *Golden* and *Longuet-Higgins* [5.48]. They started with an infinite system whose energy spectrum is continuous and tried to relate this property to ergodicity. There has also been an attempt to apply the method of *Khinchin* in classical systems to the case of quantum systems [5.2].

5.9.2 Chaotic Behavior in Quantum Systems

This section discusses some characteristic properties of quantum chaos with emphasis on the classical mechanical correspondence and ergodic properties. Quantum mechanics is usually concerned with energy levels and wave functions. Thus, the chaos in quantum mechanics will be revealed in their behavior or in the Hamiltonian matrix itself. Overviews of quantum chaos can be found in *Eckhardt*, and many papers in Les Houches Lectures [5.49].

a) *Level Spacing*

A system of one degree of freedom is usually solvable and its energy levels and eigenfunctions are regular, in parallel with the regular behavior of the corresponding classical counterparts.

Now consider the case of two degrees of freedom. As an example of an integrable system, we discuss a particle of mass μ in a rectangular box with sides of length a and b. The word "integrability" is used here in its classical meaning for simplicity. Classical integrability and quantal integrability are considered not necessarily to coincide. We will not enter into this sophisticated subject. The energy levels are given by

$$E(l,m) = \frac{h^2}{8\mu}\left(\frac{l^2}{a^2} + \frac{m^2}{b^2}\right), \qquad l \geqslant 1, m \geqslant 1\,,$$
$$= \frac{h^2}{8\mu a^2}(l^2 + km^2), \qquad k = \frac{a^2}{b^2}\,. \tag{5.9.31}$$

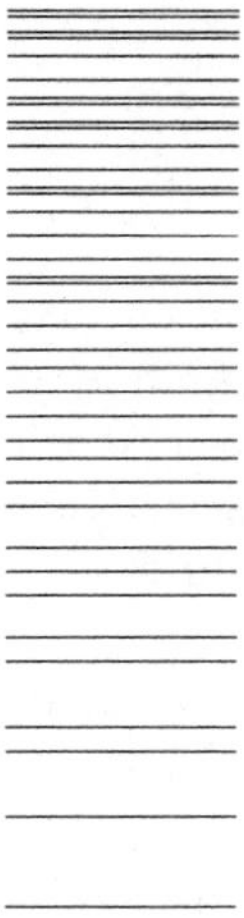

Fig. 5.21. Energy levels $l^2 + \sqrt{3}m^2$

Each term or its spacing is regular with respect to l or m, but when the levels (5.9.31) are rearranged in order of magnitude the sequence and the level spacing are not regular, as shown in Fig. 5.21. When the ratio $k = a^2/b^2$ is rational, some degenerate levels exist. However, in the case of irrational k we can expect the existence of some neighboring levels with arbitrary small spacing. Thus the distribution $\mathrm{P}(s)$ for level spacing s tends to a finite value for $s \to 0$.

On the other hand, in nonintegrable systems, a single quantum number is assigned to each energy level, in contrast to the case of (5.9.31) with two quantum numbers. The levels do not intersect when a system parameter is changed, except for the case of accidental degeneracy, provided that energy eigenvalues belong to the same symmetry class. Thus we can expect $\mathrm{P}(s) \to 0$ for $s \to 0$. The above considerations suggest that the distribution of level spacing, rather than the energy sequence itself, is appropriate to characterize the difference between integrable and nonintegrable systems. In fact, for integrable systems the Poisson distribution

$$\mathrm{P}(s) = \mathrm{e}^{-s} \tag{5.9.32}$$

with the property $\mathrm{P} \neq 0$ for $s \to 0$, and for nonintegrable systems with fully chaotic behavior in the classical counterpart the Wigner distribution

$$\mathrm{P}(s) = \frac{\pi}{2} s \exp\left(\frac{-\pi}{4} s^2\right) \tag{5.9.33}$$

with the property $\mathrm{P}(s) \to 0$ for $s \to 0$ are proposed. They are shown to hold, at least approximately, for integrable and nonintegrable systems respectively. The validity of the distribution (5.9.32) was discussed theoretically for some integrable system (5.9.31) by *Berry* and *Tabor* [5.50]. The Wigner distribution was proposed for nuclear energy levels of heavy atoms by a random matrix theory under the assumption of a Gaussian orthogonal ensemble for their matrix elements [5.51]. Heavy atoms are considered as systems composed of many nucleons and supposed to be in chaotic states. The distribution (5.9.33), however, is shown from numerical calculations still to be valid for systems of a few degrees of freedom which are classically in fully developed chaos.

In a partially chaotic state, usually observed in a classical dynamical system of two degrees of freedom discussed in the above sections, the distribution function $\rho(s)$ of energy spacing is represented by an interpolation between (5.9.32) and (5.9.33): for example,

$$\begin{aligned} &\mathrm{P}_\beta(s) = A s^\beta \exp(-\alpha s^{1+\beta}), \\ &A = (1+\beta)\alpha, \quad \alpha = \left[\Gamma\left(\frac{2+\beta}{1+\beta}\right)\right]^{1+\beta}, \end{aligned} \tag{5.9.34}$$

which is called the Brody distribution function with free parameter β [5.52]. Equations (5.9.32, 33) are for $\beta = 0$ and $\beta = 1$, respectively. It is to be noted that the distribution functions (5.9.32–34) are all normalized and have mean spacing equal to unity. This specification is necessary to compare various level spacing distributions because in a sequence of level spacings the mean spacing is not always unity. The mean spacing should be defined for an infinite sequence, but in an actual calculation one has to be satisfied with a finite number n of members. Then it happens that the mean spacing, and consequently rescaling of the energy levels, depends on n if linear scaling is adopted. To avoid this difficulty, unfolding of energy levels E_i to new ε_i can be done according to the nonlinear scaling

$$\varepsilon_i = \bar{N}(E_i)\,. \tag{5.9.35}$$

In (5.9.35) $N(E)$ is the cumulative distribution function of energy level E_i (the number of levels with energy less than E), and thus it is a step-wise increasing discontinuous function, and $\bar{N}$ is its smoothed one. By this procedure the mean spacing turns out to be unity on the average, irrespective of n, and the method of unfolding does not depend on n.

The distributions (5.9.32–34) are shown to be valid numerically in various systems. For example, consider a system with the Hamiltonian

$$\mathscr{H} = \frac{1}{2}(p_1^2 + p_2^2) + \frac{\alpha}{4}(2q_1^4 + q_2^4 + 2Cq_1^2 q_2^2), \quad \alpha = 0.088\,, \tag{5.9.36}$$

which is integrable when $C = 0$. The distribution functions P(s) for several values of C are shown in Fig. 5.22. The classical behavior corresponding to the same values of C is shown in Fig. 5.23. In the case $C = 0$, the Poincaré mapping in the classical system is regular (Fig. 5.23a) and the level spacings in the quantum system have a Poisson distribution (Fig. 5.22a). In the case $C = 10$, the Poincaré mapping is irregular (Fig. 5.23d) and the level spacings have a Wigner distribution (Fig. 5.22d). At the intermediate values of C ($C = 1, 4$), the Poincaré mapping is partly regular and partly chaotic (Fig. 5.23b,c) and the level-spacing distribution lies between the Poisson and Wigner distributions (Fig. 5.22b,c).

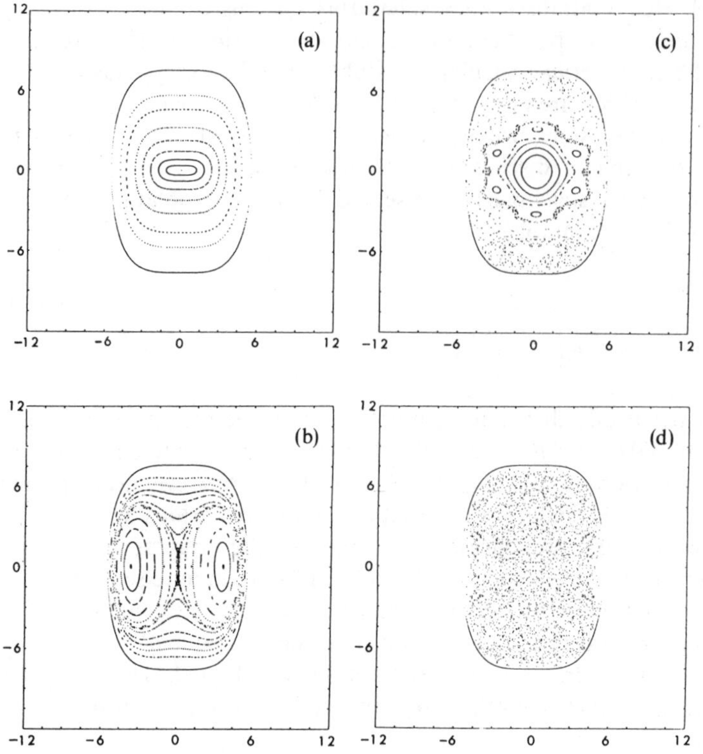

Fig. 5.22 a–d. Poincaré mapping (q_2–p_2 plane). (**a**) $C = 0$, (**b**) $C = 1$, (**c**) $C = 4$, (**d**) $C = 10$

One sees that the classical and quantum behavior is parallel from $C = 0$ (integrable) to $C = 10$ (fully chaotic) [5.53].

An avoided crossing is also another indication of nonintegrability and thus must have some connection with chaotic behavior. Figures 5.24a,b show the energy levels for the ranges $0 < C \leqslant 2$ and $9 \leqslant C \leqslant 10$. In Fig. 5.24a many of the lines appear to cross, but in reality the crossings are avoided and isolated, and so cannot yield chaotic behavior. In Fig. 5.24b no crossings are visible, but this is due to many superimposed avoided crossings which give rise to chaotic behavior.

b) *Wave Functions*

In integrable systems of two degrees of freedom the nodal lines of a wave function are regular and intersecting, as one sees in the system of a free particle in a rectangular box or in a circle. However, in nonintegrable systems the nodal lines avoid crossing each other just as in the case of avoided crossing of energy levels. Furthermore, the nodal lines are irregular in completely chaotic states. Figure 5.25 illustrates the wave functions and their contours for the system with a Hamiltonian given by (5.9.36).

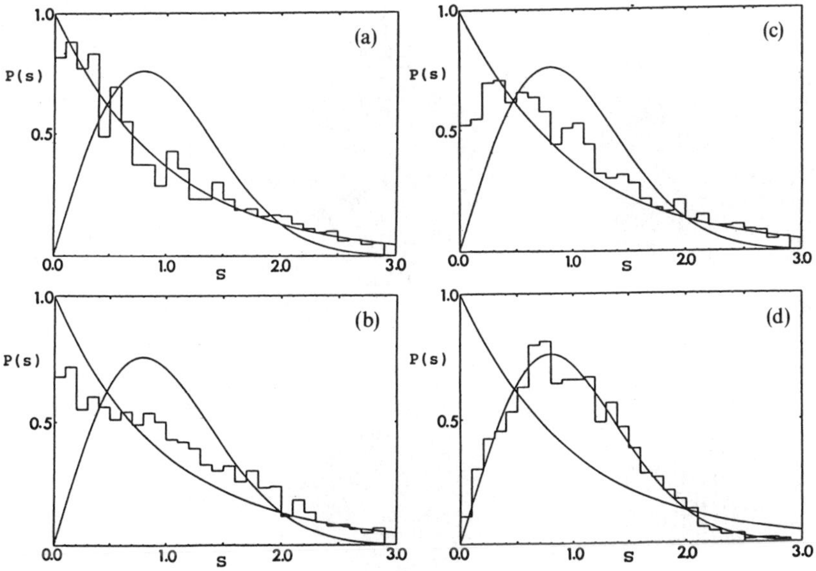

Fig. 5.23 a–d. In each panel, the smooth curve with the hump is the Wigner distribution (5.9.33) and the monotonically decreasing smooth curve is the Poisson distribution (5.9.32). The histograms show the nearest-neighbor level-spacing distributions for the homogeneous quartic system with (**a**) $C = 0$, (**b**) $C = 1$, (**c**) $C = 4$, (**d**) $C = 10$. The horizontal axis gives the normalized spacing and the vertical axis its frequency. The 600 energy levels obtained by diagonalizing 3240-dimensional matrices are used for the statistics

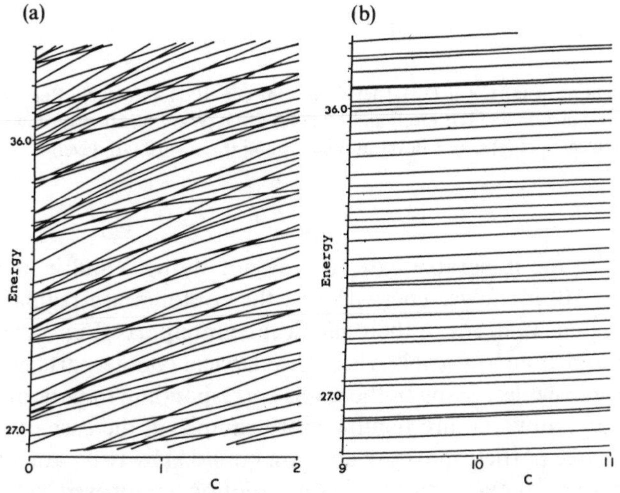

Fig. 5.24. Energy diagram in the ranges (**a**) $0 \leqslant C \leqslant 2$ and (**b**) $9 \leqslant C \leqslant 10$

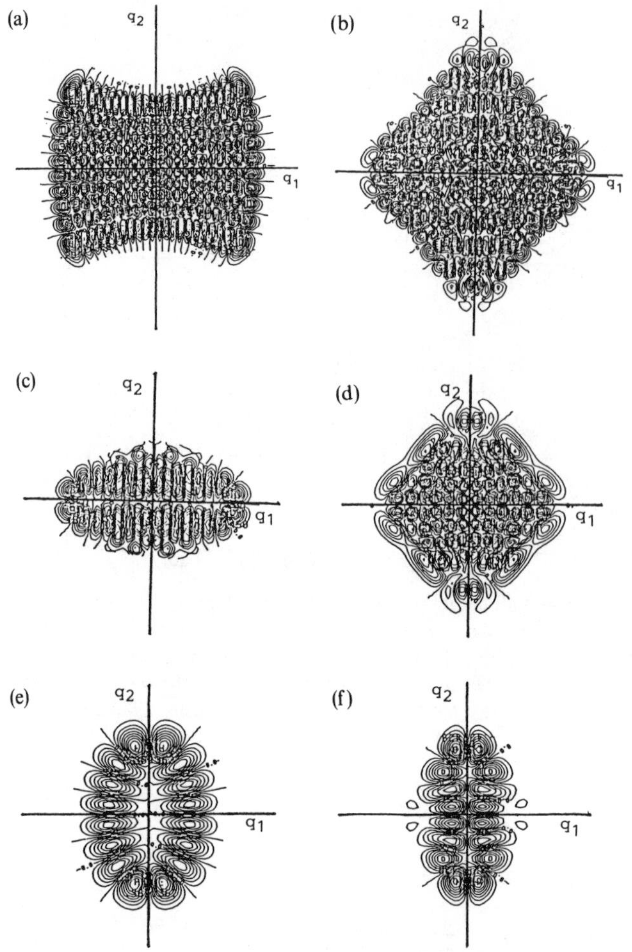

Fig. 5.25. Typical wave functions at (**a**) $C = 1$ (121st level) and (**b**) $C = 10$ (130th level). Regular wave functions at $C = 10$ are shown in (**c**) (31st level) and (**d**) (34th level). Wave functions at lower energies are also shown in the cases of (**e**) $C = 1$ (level number 7) and (**f**) $C = 10$ (7th level)

One can show that in the integrable case ($C = 0$, not shown) and in the nonintegrable case ($C = 10$) for lower energy levels the contours are regular, corresponding to the regular patterns of their classical counterparts (Fig. 5.25). On the other hand, for higher energy levels where the classical counterparts are chaotic, the contours are irregular. Some of the levels corresponding to periodic orbits in the classical case, however, are regular. For example, in the case of a stadium billiard where a free particle moves in a region bounded by two parallel edges connected by semicircular end caps, the classical motion was proved to be ergodic by *Bunimovich* [5.54] as long as the length of the parallel edges is nonzero. The nodal lines of this system are shown to be irregular [5.55], but it was pointed out that simple periodic classical orbits are retained in the profiles of the eigenfunctions [5.56]. They are called scars (Fig. 5.26). This phenomenon was discussed by *Bogomolny* [5.57] in q-space and by *Berry* [5.58] in phase space.

c) *Random Matrices*

The quantum mechanical properties of a system are determined by its Hamiltonian. The matrix elements of the Hamiltonian depend on the choice of the set of orthonormal functions describing the Hilbert space. When a Hamiltonian $\mathscr{H}$ is transformed into $\mathscr{H}' = S^{-1}\mathscr{H}S$ by a matrix S which represents the change of the sets of orthonormal functions, $\mathscr{H}'$ is equivalent to $\mathscr{H}$. In a physical system, the Hamiltonian is Hermitian and consequently the matrix S must be unitary to maintain the Hermitian property during transformation. If the elements $\mathscr{H}_{ij}$ and their transformed $\mathscr{H}'_{ij}$ are all real, S is orthogonal. Thus we have to consider in the above cases the ensemble of Hamiltonian matrices obtained by unitary or orthogonal transformations. A real symmetric Hamiltonian is found in a system with even spins invariant under time reversal, or a system (with even or odd spins) invariant under rotation. In the following we exclusively consider the ensemble of matrices obtained by orthogonal transformations of a real symmetric Hamiltonian, which will be called a Gaussian orthogonal ensemble (GOE). There is another ensemble for a system of odd spins invariant under time reversal but having no rotational symmetry. This is called a Gaussian simplectic ensemble. Further, the Gaussian unitary ensemble applies to systems without invariance under time reversal.

The Gaussian orthogonal ensemble is defined by the following requirements [5.51]:

(i) The ensemble is invariant under real orthogonal transformations:

$$\mathscr{H} \to O^{\mathrm{T}}\mathscr{H}O, \quad O^{\mathrm{T}}O = OO^{\mathrm{T}} = 1\,, \tag{5.9.37}$$

where O is any real orthogonal matrix. This implies that the probability $P(\mathscr{H})d\mathscr{H}$ that a system of this ensemble belongs to a volume element $d\mathscr{H} = \prod d\mathscr{H}_{kj}$ is invariant under the transformation O,

$$P(\mathscr{H}')d\mathscr{H}' = P(\mathscr{H})d\mathscr{H}\,. \tag{5.9.38}$$

(ii) The volume elements $\mathscr{H}_{kj}$ $(k < j)$ are statistically independent. In other words, the probability density function $P(\mathscr{H})$ is a product of the form

$$P(\mathscr{H}) = \prod f_{kj}(\mathscr{H}_{kj})\,. \tag{5.9.39}$$

These two requirements determine uniquely the functional form of $P(\mathscr{H})$:

$$P(\mathscr{H}) = C\exp(-a\,\mathrm{tr}\{\mathscr{H}^2\} + b\,\mathrm{tr}\{\mathscr{H}\})\,. \tag{5.9.40}$$

However, in actual Hamiltonian systems, we can approximate the infinite Hamiltonian matrix by finite ones to calculate energy levels. This implies that when the dimensions of the matrix are increased, the newly appearing matrix

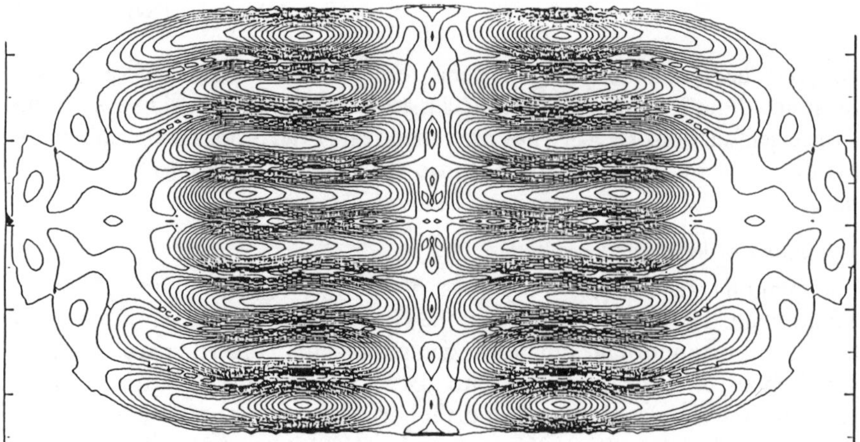

(a)

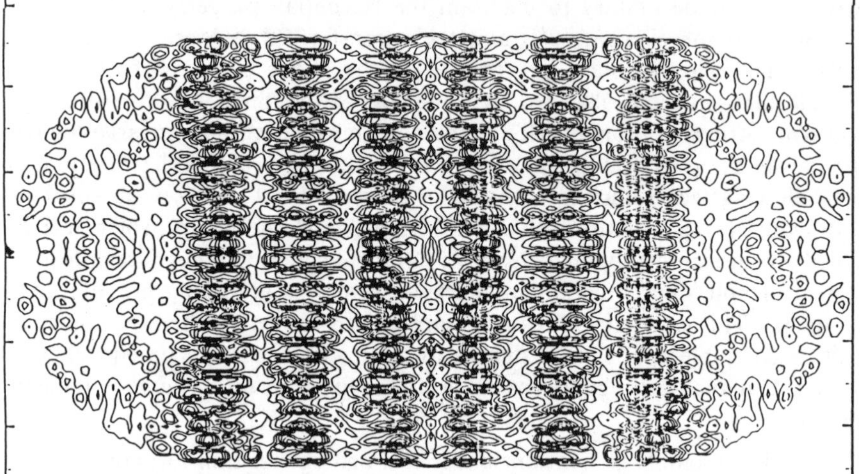

(b)

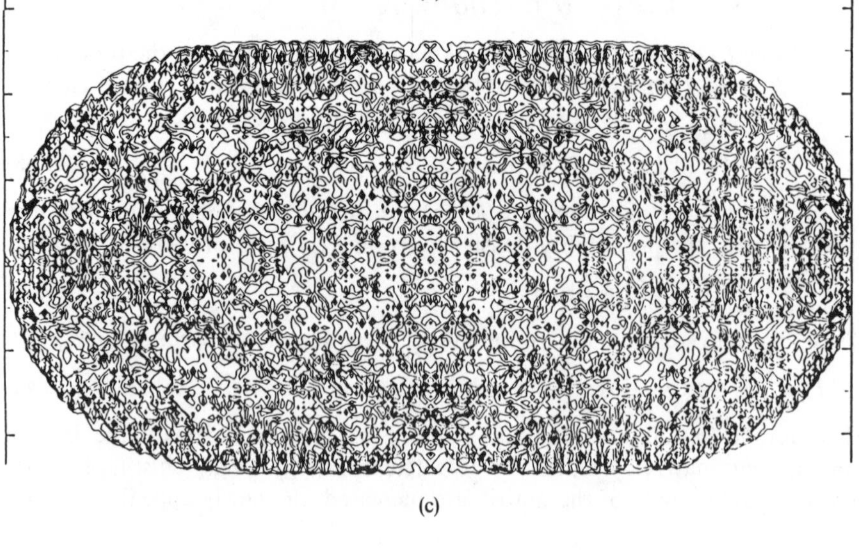

(c)

elements are usually too small to affect the calculation, and thus the matrix elements cannot be distributed uniformly, contrary to the case in a random matrix [5.59]. Nevertheless, it can derive the Wigner distribution for level spacing (see also Sect. 5.9.3). A quite different approach to derive the Wigner distribution function has been pursued by *Pechukas* [5.60] and *Yukawa* [5.61]. We do not enter this subject here.

The deviation of level spacing from a Poisson distribution towards a Wigner distribution, avoided crossings between energy levels upon changing system parameters, and the irregular pattern of nodal lines of wave functions are all considered as the manifestations of quantum chaos. In the next section, an account of the correspondence between classical and quantum chaos will be presented.

5.9.3 Correspondence Between Classical and Quantum Chaos

Chaotic behavior of classical systems usually manifests itself in the time evolution of the system. Exponential growth of the initial deviation of an orbit clealry indicates this fact. What is the quantum mechanical quantity exhibiting this exponential growth? In quantum mechanics the time evolution is described by a propagator $W(q_2, t_2|q_1, t_1)$, indicating a particle at q_2 at time t_2 with the initial condition of q_1 at time t_1, which is expressed by

$$W(q_2, t_2|q_1, t_1) = \sum_n \phi_n(q_2)\psi_n^*(q_1)\exp\left(-\frac{i}{\hbar}E_n t\right), \quad t = t_2 - t_1 \,, \tag{5.9.41}$$

where E_n and ϕ_n are the eigenvalues and the eigenfunctions, but the propagator is quasi-periodic, as can be seen in (5.9.41). $W(q_2, t_2|q_1, t_1)$ is also represented by a path integral

$$W(q_2, t_2|q_1, t_1) = \int D(\text{path})\exp\left(\frac{i}{\hbar}\int_{t_1}^{t_2} L(q, \dot{q}, t)\,dt\right), \tag{5.9.42}$$

where the action integral of the classical Lagrangian L

$$S(q_2, t_2|q_1, t_1) = \int_{t_1}^{t_2} L(q(t), \dot{q}(t), t)\,dt \tag{5.9.43}$$

is taken along a path connecting (q_1, t_1) and (q_2, t_2), and the path integral $\int D(\text{path})$ is the sum over all the paths from (q_1, t_1) to (q_2, t_2). In particular, the

Fig. 5.26 a–c. Contour lines of the wave functions for the stadium billiard. The straight edges are the same length as the diameters of the semicircles at the two ends, $1/\sqrt{(1 + \pi/4)}$, implying that the area of the stadium is 1. $\hbar/2m = 1$, $E = k^2$, where k is the wavenumber. (**a**) $k = 18.594$, (**b**) $k = 67.344$, (**c**) $k = 126.656$. (a) and (b) show the scars of periodic orbits bouncing between the parallel edges. At higher energies the scars cannot be observed (c). (After [5.65])

path integral is approximately represented by a sum of classical path's

$$W(q_1, t_1) \cong A \sum \exp\left(\frac{i}{\hbar} S_{cl}\right). \tag{5.9.44}$$

This is convenient for studying the correspondence between quantum and classical chaos. Since S_{cl} is calculated along a classical path having classical chaos, the irregular behavior of classical chaos, especially the exponential divergence of the initial deviation, reveals itself in the *phase* of the propagator in quantal systems. This behavior of the phase can affect the energy levels and the irregular pattern of nodal lines of the wave function.

This is because we have from (5.9.41),

$$f_n(q_2) \exp\left(-\frac{i}{\hbar} E_n t\right) = \int W(q_2, t_2 q_1, t_1) f_n(q_1) dq_1 ,$$

which implies that ϕ_n and $\exp(-i/\hbar E_n t)$ are respectively the eigenfunction and the eigenvalue of the Fredholm integral equation with the kernel $W(q_2 t_2 q_1 t_1)$ of chaotic nature. In a Schrodinger equation the Hamiltonian itself is usually a regular function of momenta and coordinates, but the eigenfunctions are chaotic in some cases as shown above. If we transform the basis of the Hilbert space from ϕ_n's to other orthonormal set of functions of non-chaotic nature, the transformed Hamiltonian is chaotic, and consequently it is represented by a random matrix. The eigenvalues and the distribution of level spacing remain the same, but the eigenfunctions are non-chaotic. This is the favorable point in the random matrix theory. The statistical independence of the matrix elements assumed in (5.9.39), however, is not satisfied. The random matrix theory is thus considered as an approximation [5.62] [5.66].

5.9.4 Quantum Mechanical Distribution Function

In Sect. 5.8.2 we have shown that if the distribution function $\rho(p, q, t)$ converges weakly to a constant, then the system is ergodic. The counterpart of this theorem in quantum mechanics may be described by the Wigner function [it is to be noted that the Wigner function employed here is different from (5.9.33) for level spacing], because this function can be employed like a classical distribution function in the calculation of averages of quantum mechanical physical quantities, as shown in Chap. 3. However, it was also shown there that this function oscillates with wavelength on the order of $\hbar$ and takes both positive and negative values and cannot be regarded as a true probability. The Husimi function, which is a Gaussian coarse graining of the Wigner function, is, on the other hand, always positive definite and is convenient for this purpose. These functions are defined as follows:

Wigner function:

$$\rho_W = \frac{1}{(2\pi\hbar)^N} \int d\eta \left\langle q - \frac{\eta}{2} | \hat{\rho} | q + \frac{\eta}{2} \right\rangle \exp\left(\frac{i}{\hbar} p\eta\right); \tag{5.9.45}$$

Husimi function:

$$\rho_{\mathrm{H}}(p, q) = \frac{1}{(\pi\hbar)^N} \int dq'\, dp'\, \rho_{\mathrm{W}}(p', q') \exp\left[- \sum_{i=1}^{N} \left(\frac{(q_i' - q_i)^2}{\hbar} + \frac{(p_i' - p_i)^2}{\hbar} \right) \right]$$

$$= \exp\left[\frac{\hbar}{4} \sum \left(\frac{\partial^2}{\partial q_i^2} + \frac{\partial^2}{\partial p_i^2} \right) \right] \rho_{\mathrm{W}}(p, q)$$

$$= \frac{1}{(2\pi\hbar)^2} |\langle \phi_{qp} | \phi \rangle^2| \,, \qquad (5.9.46)$$

where ϕ_{qp} is the minimum wave packet

$$\phi_{qp}(x) = \left((2\pi)^N \prod_{i=1}^{N} (\Delta q_i)^2 \right)^{-1/4} \exp\left[\sum \left(\frac{\mathrm{i}}{\hbar} p_i x_i - \frac{(x_i - q_i)^2}{2\Delta q_i^2} \right) \right] . \qquad (5.9.47)$$

The expectation value of a dynamical quantity $\hat{A}$ is given in various representations as follows:

$$\langle \hat{A} \rangle = \int \mathrm{tr}\{\hat{A}\hat{\rho}\} \qquad (5.9.48)$$

$$= \int A_{\mathrm{W}}(p, q) \rho_{\mathrm{W}}(p, q)\, dp\, dq \qquad (5.9.49)$$

$$= \int (\hat{A}\hat{\rho})_{\mathrm{H}}\, dp\, dq \,, \qquad (5.9.50)$$

where A_{W} and $(\hat{A}\hat{\rho})_{\mathrm{H}}$ are, respectively, the Wigner and Husimi representations of $\hat{A}$ and $\hat{A}\hat{\rho}$:

$$A_{\mathrm{W}} = \int d\eta \left\langle q - \frac{\eta}{2} | \hat{A} | q + \frac{\eta}{2} \right\rangle \exp\left(\frac{\mathrm{i}}{\hbar} p\eta \right) \qquad (5.9.51)$$

$$(\hat{A}\hat{\rho})_{\mathrm{H}} = \exp\left[\frac{\hbar}{4} \sum \left(\frac{\partial^2}{\partial q_i^2} + \frac{\partial^2}{\partial q_i^2} \right) \right] (\hat{A}\hat{\rho})_{\mathrm{W}}$$

$$= \langle \phi_{qp} | A | \phi_{qp} \rangle \,. \qquad (5.9.52)$$

The time evolution of ρ_{W} or ρ_{H} corresponding to the time evolution of ρ with the Hamiltonian $\mathscr{H}$ can be formulated. We are especially interested in ρ_{H}.

Now consider a system with a Hamiltonian of one degree of freedom

$$\mathscr{H}_0 = \frac{1}{2} p^2 - \frac{a_1}{2} q^2 + \frac{a_2}{4} q^4, \quad a_1 > 0, a_2 > 0 \,, \qquad (5.9.53)$$

which has a potential with a double minimum and an unstable point at the origin. Furthermore, a time-dependent perturbation

$$\mathscr{H}_1 = \frac{a_3}{4} (\sin \omega t) q^4 \qquad (5.9.54)$$

is introduced to give rise to a chaotic behavior in the classical system. In fact, the periodic perturbation is equivalent to introducing other variables $y = \sin \omega t$ and

p_y through $p_y = \dot{p}$, $\dot{p}_y = -\omega^2 y$ and consequently the system now has two degrees of freedom. We take $a_1 = 1.0$, $a_2 = 0.25$, $a_3 = 0.1$, $\omega = 0.7$ and for convenience $\hbar = 0.04$. According to a numerical calculation with an initial condition of a minimum wave packet, the Husimi function is spread over a certain region of space as time goes on, as shown in Fig. 5.27. The classical correspondence is given in Fig. 5.28 [5.63]. One can recognize the similarities

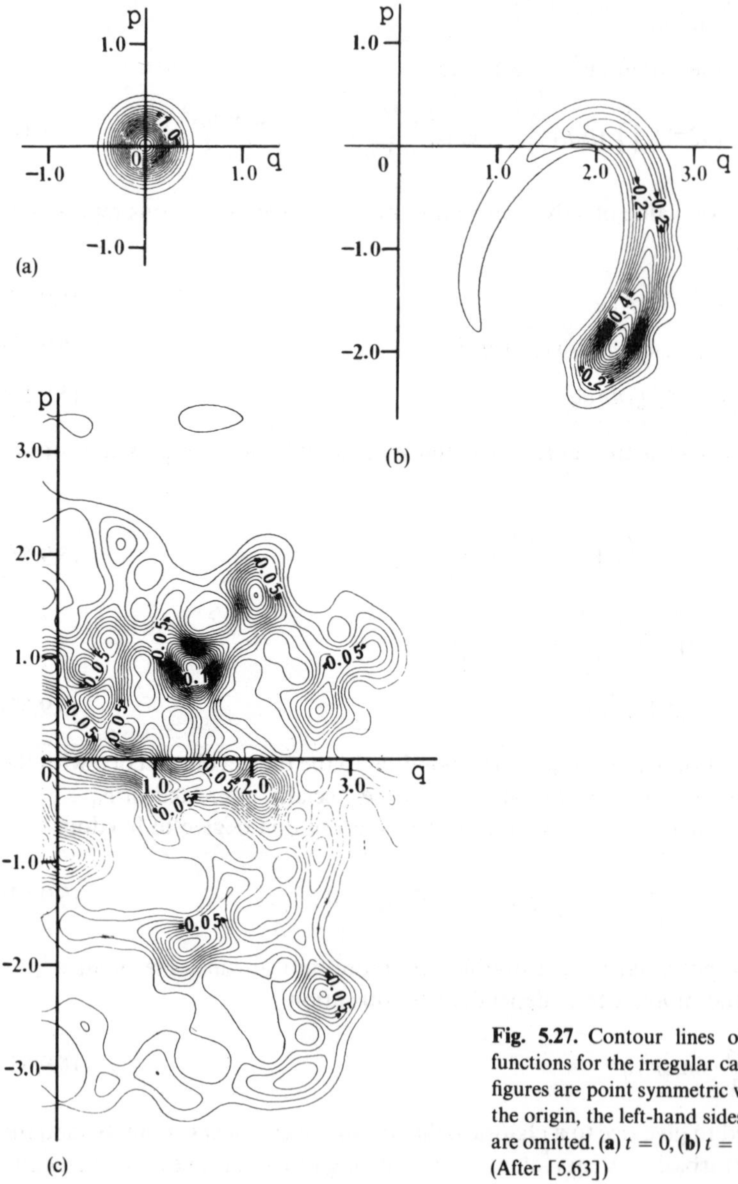

Fig. 5.27. Contour lines of the Husimi functions for the irregular case. Since these figures are point symmetric with respect to the origin, the left-hand sides of (b) and (c) are omitted. (**a**) $t = 0$, (**b**) $t = T$, (**c**) $t = 10T$. (After [5.63])

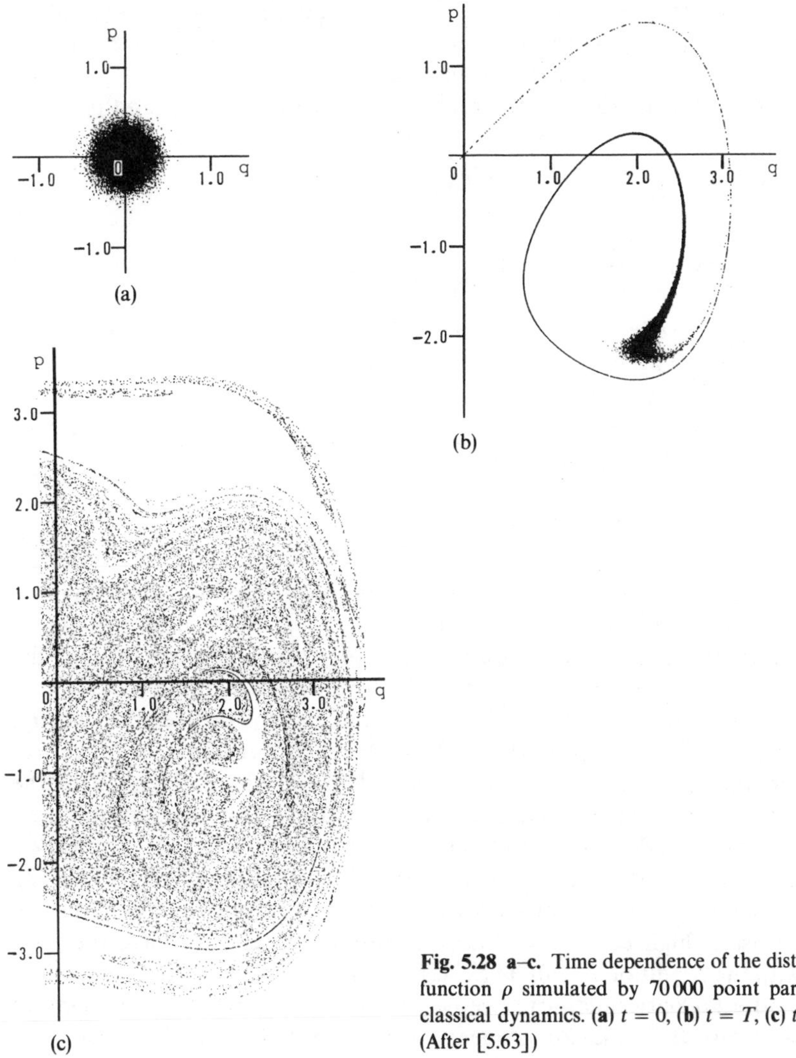

Fig. 5.28 a–c. Time dependence of the distribution function ρ simulated by 70000 point particles in classical dynamics. (**a**) $t = 0$, (**b**) $t = T$, (**c**) $t = 10\,T$. (After [5.63])

and dissimilarities between the two. One similarity is the spreading of the distribution function over a region of phase space, and one dissimilarity is the appearance and disappearance of small wave packets in the quantal case, which is obviously due to quantum mechanical interference. When small wave packets are brought about by virtue of the interference effect, the close similarity between ρ_H and ρ is lost. In this case ρ_H considered as a function of phase space always has areas of linear dimension $\sqrt{\hbar}$, corresponding to the minimum wave packet, which cannot be reduced to a uniform distribution.

From these observations it is conjectured that the Husimi distribution function has a limit of uniform distribution in the sense of weak convergence in ergodic systems which is realized in the limit $t \to \infty$ and $\hbar \to 0$.

Further, there exist the inverse transformations among $\hat{\rho}$, ρ_W and ρ_H

$$\left\langle q - \frac{\eta}{2} | \hat{\rho} | q + \frac{\eta}{2} \right\rangle = \int \rho_W(p, q) \exp\left(-\frac{i}{\hbar} p\eta \right) d\eta \,, \tag{5.9.55}$$

$$\rho_W(p, q) = \exp\left[-\frac{\hbar}{4} \sum_i \left(\frac{\partial^2}{\partial q_i^2} + \frac{\partial^2}{\partial p_i^2} \right) \right] \rho_H(p, q) \,. \tag{5.9.56}$$

Equation (5.9.56) yields $\rho_W(p, q) = \text{const.}$ if $\rho_H(p, q)$ is constant, and consequently, if $\rho_W(p, q) = \text{const}$,

$$\left\langle q - \frac{\eta}{2} | \hat{\rho} | q + \frac{\eta}{2} \right\rangle = \text{const.} \times \delta(\eta) \tag{5.9.57}$$

or

$$\langle q' | \hat{\rho} | q \rangle = \phi^*(q') \phi(q) = \text{const.} \times \delta(q - q') \,. \tag{5.9.58}$$

The last equation implies that the wave function $\phi(q)$ is rapidly oscillating and thus the product $\phi^*(q')\phi(q)$ is zero on the average, if $q' = q$. The ansatz $\rho_H(p, q) = \text{const.}$ is introduced here as a requirement of quantum ergodicity through the concept of weak convergence as mentioned above, and this ansatz is equivalent to $\rho_W = \text{const.}$ and $\hat{\rho} = 1$. In this connection one has to consider the scarring phenomena mentioned in Sect. 5.9.2b. The occurrence of scarring seems contradictory to the requirement of $\hat{\rho} = 1$ expected in the quantal version of the Bunimovich billiard, which is classically ergodic [5.54]. However, according to the theoretical considerations of *O'Connor* and *Heller* [5.64] and also *Shudo*'s numerical calculations of eigenfunctions up to high energies [5.65] (Fig. 5.28), the frequency of occurrence of scarring diminishes at higher energy levels. In other words, at high energies the scarring disappears and we have the state $\hat{\rho} = 1$ (with the condition $\hbar \to 0$). The classical billiard problem is independent of energy and ergodicity holds irrespective of energy. However, in the quantum mechanical case the behavior is energy dependent. Correspondence with the classical case is realized in the high-energy limit.

General Bibliography

Classic Literature:

G.1 L. Boltzmann: *Vorlesungen über Gastheorie*, 2 Bde. (J.A. Barth, Leipzig 1912); *Lectures on Gas Theory* [English transl. by S.B. Brush] (University of California Press, Berkeley 1964)
G.2 J.W. Gibbs: *Elementary Principles in Statistical Mechanics* (Yale University Press, New Haven, CT 1902; reprint Ox Bow, Woodbridge, CT 1981)
G.3 P. Ehrenfest, T. Ehrenfest: *The Conceptional Foundations of the Statistical Approach in Mechanics* (transl. by M.J. Moravcsik] (Cornell University Press, Ithaca, NY 1959). This was originally published in *Enzyklopädie der mathematischen Wissenschaften*, Bd. 4, Art 32 (1911)
G.4 R.H. Fowler: *Statistical Mechanics*, 2nd ed. (Cambridge University Press, Cambridge 1936)
G.5 C. Tolman: *The Principles of Statistical Mechanics* (Oxford University Press, Oxford 1938)

General Textbooks:

G.6 R. Balescu: *Equilibrium and Nonequilibrium Statistical Mechanics* (Wiley, New York 1975)
G.7 R. Becker: *Theorie der Wärme* (Springer, Berlin, Heidelberg 1955)
G.8 N.N. Bogolyubov: *Lectures on Quantum Statistics* (Gordon and Breach, New York 1967)
G.9 R.P. Feynman: *Statistical Mechanics* (Benjamin, New York 1972)
G.10 S. Flügge (ed.): *Principles of Thermodynamics and Statistical Mechanics*, Encyclopedia of Physics, Vol. 3, Part 2 (Springer, Berlin, Heidelberg 1959)
G.11 R.H. Fowler, E.A. Guggenheim: *Statistical Thermodynamics*, 2nd ed. (Cambridge University Press, Cambridge 1965)
G.12 J.I. Frenkel: *Statistische Physik* (Akademie, Berlin 1957)
G.13 T.L. Hill: *Statistical Mechanics* (McGraw-Hill, New York 1956)
G.14 K. Huang: *Statistical Mechanics* (Wiley, New York 1963)
G.15 A. Ishihara: *Statistical Physics* (Academic, New York 1971)
G.16 C. Kittel: *Elementary Statistical Physics* (Wiley, New York 1958)
G.17 R. Kubo (ed.): *Statistical Mechanics* (North-Holland, Amsterdam 1965)
G.18 L.D. Landau, E.M. Lifshitz: *Statistical Physics* (Pergamon, Oxford 1958). There are a series of revised and enlarged editions
G.19 J.E. Mayer, M.G. Mayer: *Statistical Mechanics*, (Wiley, New York 1940)
G.20 A. Münster: *Statistische Thermodynamik* (Springer, Berlin, Heidelberg 1956)
G.21 F. Reif: *Statistical and Thermal Physics* (McGraw-Hill, New York 1965)
G.22 A. Sommerfeld: *Thermodynamik und Statistik* (Dietrich 1952): *Thermodynamics and Statistical Mechanics* [transl. by J. Kestin] (Academic, New York 1956)
G.23 D. terHarr: *Elements of Statistical Mechanics* (Holt, Rinehart & Winston, New York 1961)
G.24 G.H. Wannier: *Statistical Physics* (Wiley, New York 1966)

Mathematical Aspects of Fundamental Problems:

G.25 A.I. Khinchin: *Mathematical Foundations of Statistical Mechanics* [transl. by G. Gamow] (Dover, New York 1949)
G.26 R. Kurth: *Axiomatics of Classical Statistical Mechanics* (Pergamon, Oxford 1960)
G.27 O. Penrose: *Foundations of Statistical Mechanics* (Pergamon, Oxford 1970)
G.28 R. Ruelle: *Statistical Mechanics* (Benjamin, New York 1969)

References

Chapter 1

1.1 L. Boltzmann: *Vorlesungen über Gastheorie*, 2 Bde. (J.A. Barth, Leipzig 1912); *Lectures on Gas Theory* [English transl. by S.G. Brush] (University of California Press, Berkeley 1964)

1.2 J.W. Gibbs: *Elementary Principles in Statistical Mechanics* (Yale University Press, New Haven, (T 1902; reprint Ox Bow, Woodbridge, CT 1981)

1.3 P. Ehrenfest, T. Ehrenfest: *The Conceptional Foundations of the Statistical Approach in Mechanics* [transl. by M.J. Moravcsik] (Cornell University Press, Ithaca, NY 1959). Originally published in *Enzyklopädie der mathematischen Wissenschaften*, Bd. 4, Art 32 (1911)

1.4 A. Einstein: *Investigations on The Theory of the Brownian Motion*, ed. by R. Furth [transl. by A. D. Cowper] (Dover, New York 1956)

1.5 P. Jordan: *Statistische Mechanik auf Quantenmechanischer Grundlage* (Vieweg, Braunschweig 1933)

1.6 R.C. Tolman: *The Principles of Statistical Mechanics* (Oxford University Press, Oxford 1938)

1.9 A.I. Khinchin: *Mathematical Foundations of Statistical Mechanics* [transl. by G. Gamow] (Dover, New York 1949)

1.10 R. Kurth: *Axiomatics of Classical Statistical Mechanics* (Pergamon, Oxford 1960)

1.11 O. Penrose: *Foundations of Statistical Mechanics* (Pergamon, Oxford 1970)

1.12 R. Ruelle: *Statistical Mechanics* (Benjamin, New York 1969)

1.13 E.H. Kennard: *Kinetic Theory of Gases* (McGraw-Hill, New York 1938)

Chapter 2

2.1 L.D. Landau, E.M. Lifshitz: *Statistical Physics* [transl. by D. Shoenberg] (Clarendon,Oxford 1938) Various revised and enlarged editions have been published by Pergamon, Oxford

2.2 J.E. Mayer, M.G. Mayer: *Statistical Mechanics* (Wiley, New York 1940)

2.3 D. terHaar: *Elements of Statistical Mechanics* (Holt, Rinehart and Winston, New York 1961)

2.4 D. terHaar: *Elements of Thermostatics* (Holt, Rinehart and Winston, New York 1966)

2.5 R. Kubo, H. Ichimura, T. Usui, N. Hashitsume: *Statistical Mechanics* (North-Holland, Amsterdam 1965)

2.6 C. Kittel: *Elementary Statistical Mechanics* (Wiley, New York 1958)

2.7 R. W. Gurney: *Introduction to Statistical Mechanics* (McGraw-Hill, New York 1949)

2.8 G.S. Rushbrooke: *Introduction to Statistical Mechanics* (Oxford 1951)

2.9 A. Sommerfeld: *Thermodynamik und Statistik* (Dietrich 1952); *Thermodynamics and Statistical Mechanics* [English transl. by J. Kestin] (Academic, New York 1956)

2.10 R. Becker: *Theorie der Wärme* (Springer, Berlin, Heidelberg 1955)

2.11 A. Münster: *Statistische Thermodynamik* (Springer, Berlin, Heidelberg 1956); English transl.: *Statistical Thermodynamics* (Springer, Berlin, Heidelberg 1969)

2.12 T.L. Hill: *Statistical Mechanics* (McGraw-Hill, New York 1956)

2.13 S. Flügge (ed.): *Principles of Thermodynamics and Statics*, Encyclopedia of Physics, Vol. 3, Part 2 (Springer, Berlin, Heidelberg 1959)

2.14 K. Huang: *Statistical Mechanics* (Wiley, New York 1963)

2.15 F. Reif: *Statistical and Thermal Physics* (McGraw-Hill, New York 1965)
2.16 G.H. Wannier: *Statistical Physics* (Wiley, New York 1966)
2.17 A. Isihara: *Statistical Mechanics* (Academic, New York 1971)
2.18 K. Husimi: Proc. Phys.-Math. Soc. Jpn. **22**, 246 (1940)

Chapter 3

3.1 M. Toda: J. Phys. Soc. Jpn, **7**, 230 (1952)
3.2 The method of steepest descent (the saddle point method) was applied to statistical mechanics by C.G. Darwin and R.H. Fowler in many applications: F.G. Fowler: *Statistical Mechanics*, 2nd ed. (Cambridge University Press, Cambridge 1936)
3.3 R.H. Fowler, E.A. Guggenheim: *Statistical Thermodynamics* (Cambridge University Press, Cambridge 1939, 1965)
3.4 J.O. Hirschfelder, C.F. Curtiss, R.B. Bird: *Molecular Theory of Gases and Liquids* (Wiley, New York 1954)
3.5 J. de Boer: Physica **14**, 139 (1948)
3.6 M. Toda: *Recent Problems in Physics* (Iwanami Shoten, Tokyo 1948) [in Japanese]
3.7 J. G. Kirkwood, F.P. Buff: J. Chem. Phys. **17**, 338 (1949)
3.8 A. Harasima: J. Phys. Soc. Jpn. **8**, 343 (1953)
3.9 M. Toda: J. Phys. Soc. Jpn. **10**, 512 (1955)
3.10 S. Ono, S. Kondo: *Molecular Theory of Surface Tension in Liquids*, Handbuch der Physik, Bd. 10, Struktur der Flüssigkeiten (Springer, Berlin, Heidelberg 1960)
3.11 J.E. Mayer, M.G. Mayer: *Statistical Mechanics* (Wiley, New York 1940)
3.12 K. Husimi: J. Chem. Phys. **18**, 682 (1950)
3.13 J. de Boer, G.E. Uhlenbeck (eds.): *Studies in Statistical Mechanics*, Vol. 1 (North-Holland, Amsterdam 1962). Continuing series
3.14 P. Debye, E. Hückel: Phys. Z. **24**, 185 (1923)

Chapter 4

[4.1–13] may serve as general references to Chap. 4.
4.1 R. Brout: *Phase Transitions* (Benjamin, New York 1965)
4.2 M. Fisher: Rep. Prog. Phys. **30**, Part 2, 615 (1967)
4.3 M. Fisher: *The Nature of Critical Points*, ed. by W.E. Britten, *Lectures on Theoretical Physics* (University of Colorado Press 1965)
4.4 H.S. Green, C.A. Hurst: *Order-Disorder Phenomena* (Interscience, New York 1964)
4.5 G.F. Newell, E.W. Montroll: Rev. Mod. Phys. **25**, 353 (1953)
4.6 H.E. Stanley: *Introduction to Critical Phenomena* (Oxford University Press, Oxford 1971)
4.7 L.P. Kadanoff, W. Götze, D. Hamblen, R. Hecht, E.A.S. Lewis, V.V. Palciauskas, M. Rayl, J. Swift: Rev. Mod. Phys. **39**, 395 (1967)
4.8 Shang-Keng Ma: *Modern Theory of Critical Phenomena* (Benjamin, New York 1976)
4.9 P. Pfeuty, G. Toulouse: *Introduction to the Renormalization Group and Critical Phenomena* [transl. by G. Barton] (Wiley, New York 1977)
4.10 K.G. Wilson, G. Kogut: The renormalization group and the ε expansion. Phys. Rep. C12, No. 2, 75 (1974)
4.11 K.G. Wilson: The renormalization group: critical phenomena and the Kondo problem. Revs. Mod. Phys. **47**, 773 (1975)
The above two articles were written by Wilson, who introduced the renormalization method to critical phenomena.
4.12 C. Domb, M.E. Green (eds.): *Phase Transitions and Critical Phenomena*, Vols. 1–4, 5A, 5B (Academic, New York 1972–1976).
This series contains many review articles on experimental and theoretical approaches to the studies of phase transitions.
4.13 H.E. Stanley (ed.): *Cooperative Phenomena near Phase Transitions, a Bibliography with Selected Readings* (MIT, Cambridge, MA 1973)

4.14 E. Leib, T. Schultz, D. Mattis: Ann. Phys. **16**, 407 (1961);
S. Katsura: Phys. Rev. **127**, 1508 (1962); ibid. **129**, 2835 (1963)
4.15 C.N. Yang: Phys. Rev. **85**, 808 (1952)
4.16 L. Onsager: Phys. Rev. **65**, 117 (1944)
4.17 T.D. Lee, C.N. Yang: Phys. Rev. **87**, 410 (1952)
4.18 J.E. Mayer, M.G. Mayer: *Statistical Mechanics* (Wiley, New York 1940)
4.19 C.N. Yang, T.D. Lee: Phys. Rev. **87**, 404 (1952)
4.20 T. Asano: J. Phys. Soc. Jpn. **29**, 350 (1970)
4.21 E.H. Lieb, D.C. Mattis (eds.): *Mathematical Physics in One Dimension* (Academic, New York 1966);
H. Takahashi: Proc. Phys. Math. Soc. Jpn. **24**, 60 (1942)
4.22 R. Ruelle: Commun. Math. Phys. **9**, 267 (1968)
4.23 F.J. Dyson: Commun, Math. Phys. **12**, 91, 212 (1969)
4.24 G.A. Baker: Phys. Rev. **126**, 2071 (1962)
4.25 M. Kac, G.E. Uhlenbeck, P.C. Hemmer: J. Math. Phys. **4**, 216 (1963)
4.26 N.G. van Kampen: Phys. Rev. **135A**, 362 (1964); Physica **48**, 313 (1970)
4.27 O. Penrose, J.L. Lebowitz: J. Stat. Phys. **3**, 211 (1971);
see also P.C. Hemmer, J.L. Lebowitz in [4.12]: **5B**, 108 (1976)
4.28 D. Poland, H.A. Scherega: *Theory of Helix-coil Transitions in Biopolymers* (Academic, New York 1970)
4.29 H.A. Kramers, G.H. Wannier: Phys. Rev. **60**, 252 (1941)
4.30 R.J. Baxter: Ann. Physics **70**, 193 (1972)
4.31 M. Suzuki: J. Phys. Soc. Jpn **55**, 4205 (1986)

Chapter 5

[5.1–9] may serve as general references to Chap. 5.
5.1 L.D. Landau, E.M. Lifshitz: *Statistical Physics*, transl. by D. Shoenberg (Clarendon, Oxford 1938, Pergamon 1958)
5.2 A.I. Khinchin: *Mathematical Foundations of Statistical Mechanics* (Dover, New York 1949)
5.3 D. ter Haar: Rev. Mod. Phys. **27**, 289 (1955)
5.4 Proceedings of the International School of Physics "Enrico Fermi" XIV, *Ergodic Theories* (Academic, New York 1961)
5.5 I.E. Farquhar: *Ergodic Theory in Statistical Mechanics* (Interscience, New York 1964)
5.6 R. Jancel: *Foundations of Classical and Quantum Statistical Mechanics* (Pergamon, Oxford 1963)
5.7 E. Hopf: *Ergodentheorie* (Chelsea, New York 1948)
5.8 V.I. Arnold, A. Avez: *Ergodic Problems of Classical Mechanics* (Benjamin, New York 1968)
5.9 P. and T. Ehrenfest: *The Conceptual Foundations of the Statistical Approach in Mechanics* [Transl. by M.J. Moravcsik] (Cornell University Press, Ithaca 1959)
5.10 R. Abraham, J.E. Marsden: *Foundation of Mechanics*, 2nd ed. (Benjamin, New York 1978)
5.11 G.D. Birkhoff: Proc. Nat. Acad. Sci. **17**, 656 (1931)
5.12 B.O. Koopman: Proc. Nat. Acad. Sci. **17**, 315 (1931)
5.13 G.D. Birkhoff, B.O. Koopman: Proc. Nat. Acad. Sci. **18**, 279 (1932)
5.14 P.A. Smith: J. Math. **7**, 365 (1928)
5.15 L. Van Hove: Math. Rev. **17**, 812 (1956)
5.16 D.S. Ornstein: Adv. Math. **4**, 337 (1970)
5.17 Y. Aizawa: J. Phys. Soc. Jpn. **33**, 1693 (1972)
5.18 J.G. Sinai: Ergodicity of Boltzmann's gas model. In *Statistical Mechanics, Foundations and Applications*, Proc. of the I.U.P.A.P. Meeting, Copenhagen, 1966, ed. by T.A. Bak (Benjamin, New York 1967) p. 559
[5.19–28] may serve as reviews of the material in Sects. 5.4–7.
5.19 S. Jorna (ed.): *Topics in Nonlinear Dynamics*, AIP Conf. Proc., No. 46 (American Institute of Physics, New York 1978). Especially, see papers by J. Moser, M.V. Berry, J. Ford, Y.M. Treve

5.20 J. Ford: *Fundamental Problems in Statistical Mechanics 1*, ed. by E.D.G. Cohen (North-Holland, Amsterdam 1975)

5.21 G.E.O. Giacaglia: *Perturbation Methods in Nonlinear Systems* (Springer, Berlin, Heidelberg 1972)

5.22 C.L. Siegel, J.K. Moser: *Lectures in Celestial Mechanics* (Springer, Berlin, Heidelberg 1971)

5.23 J. Moser: *Stable and Random Motions in Dynamical Systems* (Princeton University Press, Princeton 1973)

5.24 Y. Hagiwara: *Celestial Mechanics*, Vol. 4, Part 2 (Japan Soc. for the Promotion of Science, Tokyo 1975)

5.25 G. Casati, J. Ford (eds.): *Stochastic Behavior in Classical and Quantum Hamiltonian Systems*, Como, 1977, Lecture Notes Phys., Vol. 93 (Springer, Berlin, Heidelberg 1979)

5.26 I.C. Percival: *Semiclassical Theory of Bound States*, ed. by I. Prigogine, S.A. Rice, Advances in Chemical Physics, Vol. 36 (Wiley, New York 1977)

5.27 B.V. Chirikov: Phys. Rep. **52**, 263 (1979)

5.28 R.H.G. Helleman: In *Fundamental Problems in Statistical Mechanics V*, ed. by E.G.D. Cohen (North-Holland, Amsterdam 1980)

5.29 H. Bruns: Acta Math. **11**, 25 (1887)

5.30 H. Poincaré: *Les Methodes Nouvelles de la Mécanique Céleste* (Dover, New York 1957)

5.31 E. Fermi: Phys. Z. **24**, 261 (1923)

5.32 E. Fermi, J. Pasta, S. Ulam: Los Alamos Rpt. LA-1940 (1955); *Collected Papers of Enrico Fermi*, Vol. II, University of Chicago press, Chicago 1965) p. 977

5.33 P.C. Hemmer: Dynamic and Stochastic Types of Motion in the Linear Chain, Thesis (Trondheim 1959)

5.34 J. Ford, J. Waters: J. Math. Phys. **4**, 1293 (1963)

5.35 N. Saitô, N. Hirotomi, A. Ichimura: J. Phys. Soc. Jpn. **39**, 1931 (1975)

5.36 M. Hénon, C. Heiles: Astron. J. **69**, 73 (1964)

5.37 J.N. Greene: J. Math. Phys. **9**, 760 (1968)

5.38 A.N. Kolmogorov: Intern. Congress Mathematician, Amsterdam **1**, 315 (1954)

5.39 J. Moser: Commun. Pure Appl. Math. **9**, 673 (1956)

5.40 N. Saitô, H. Hirooka, J. Ford, F. Vivaldi, G.H. Walker: Physica **5D**, 273 (1983); H. Hirooka, J. Ford, N. Saitô: J. Phys. Soc. Jpn. **53**, 385 (1984)

5.41 N. Saitô: J. Phys. Soc. Jpn. **51**, 374 (1982)

5.42 N. Saitô, Y. Matsunaga: J. Phys. Soc. Jpn. **58**, 3089 (1989)

5.43 G. Ludwig: In Ref. [5.4] and *Die Grundlagen der Quantenmechanik* (Springer, Berlin, Heidelberg 1954)

5.44 N.G. van Kampen: Physica **20**, 603 (1954)

5.45 M. Fierz: Helv. Phys. Acta **28**, 705 (1955)

5.46 J.E. Farquhar, P.T. Landsberg: Proc. R. Soc. London **239**, 134 (1957)

5.47 P. Bocchieri, A. Loinger: Phys. Rev. **111**, 668 (1958); ibid. **114**, 948 (1959); G.M. Prosperi, A. Scotti: Nuovo Cimento **13**, 1007 (1959); ibid. **17**, 267 (1960); P. Bocchieri, G.M. Prosperi: Ref. [5.17, p. 17]

5.48 S. Golden, H.C. Longuet-Higgins: J. Chem. Phys. **33**, 1479 (1960)

5.49 B. Eckhardt: Phys. Rep. **163**, 205 (1988). Les Houches Lectures **34**, *Chaotic Behaviors of Deterministic Systems* (1983, *ibid.* **52**, *Chaos and Quantum Physics* (1991) (North Holland Pub. Co.)

5.50 M.V. Berry, M. Tabor: Proc. R. Soc. London **A 356**, 375 (1977)

5.51 M.L. Mehta: *Random Matrices and the Statistical Theory of Energy Levels* (Academic, New York 1967) Revised ed. (1991)

5.52 T.A. Brody, J.B. Flores, P.A. Mello, A. Pandey, S.S.M. Wong: Rev. Mod. Phys. **53**, 385 (1981)

5.53 A. Shudo, N. Saitô: J. Phys. Soc. Jpn. **56**, 2641 (1987)

5.54 L.A. Bunimovich: Funct. Anal. Appl. **8**, 254 (1974)

5.55 S.W. McDonald, A.N. Kaufman: Phys. Rev. Lett. **42**, 1189 (1979)

5.56 E.J. Heller: Phys. Rev. Lett. **53**, 1515 (1984); In *Quantum Chaos and Statistical Nuclear Physics. Proc. Cuernavaca, Mexico 1986*, ed. by T.H. Seligman, H. Nishioka, Lect. Notes Phys., Vol. **263** (Springer, Berlin, Heidelberg 1986)

5.57 E.B. Bogomolny: Physica D **31**, 405 (1988)
5.58 M.V. Berry: Proc. R. Soc. London A **423**, 219 (1989)
5.59 A. Shudo: Prog. Theor. Phys. Suppl. **98**, 173 (1989)
5.60 P. Pechukas: Phys. Rev. Lett. **51**, 943 (1983)
5.61 T. Yukawa, T. Ishikawa: Prog. Theor. Phys. Suppl. **98**, 157 (1989)
5.62 N. Saitô: Prog. Theor. Phys. Suppl. **98**, 376 (1989)
5.63 K. Takahashi: Prog. Theor. Phys. Suppl. **98**, 1109 (1989)
5.64 P.W. O'Connor, E.J. Heller: Phys. Rev. Lett. **61**, 2288 (1988)
5.65 N. Saitô and A. Shudo: J. Phys. Soc. Jpn. **62**, 53 (1993)
5.66 N. Saitô: to be published

Subject Index

Springer Series in Solid-State Sciences

Editors: M. Cardona P. Fulde K. von Klitzing H.-J. Queisser

1 **Principles of Magnetic Resonance** 3rd Edition By C. P. Slichter

2 **Introduction to Solid-State Theory** By O. Madelung

3 **Dynamical Scattering of X-Rays in Crystals** By Z. G. Pinsker

4 **Inelastic Electron Tunneling Spectroscopy** Editor: T. Wolfram

5 **Fundamentals of Crystal Growth I** Macroscopic Equilibrium and Transport Concepts By F. E. Rosenberger

6 **Magnetic Flux Structures in Superconductors** By R. P. Huebener

7 **Green's Functions in Quantum Physics** 2nd Edition By E. N. Economou

8 **Solitons and Condensed Matter Physics** Editors: A. R. Bishop and T. Schneider

9 **Photoferroelectrics** By V. M. Fridkin

10 **Phonon Dispersion Relations in Insulators** By H. Bilz and W. Kress

11 **Electron Transport in Compound Semiconductors** By B. R. Nag

12 **The Physics of Elementary Excitations** By S. Nakajima, Y. Toyozawa, and R. Abe

13 **The Physics of Selenium and Tellurium** Editors: E. Gerlach and P. Grosse

14 **Magnetic Bubble Technology** 2nd Edition By A. H. Eschenfelder

15 **Modern Crystallography I** Fundamentals of Crystals Symmetry, and Methods of Structural Crystallography 2nd Edition By B. K. Vainshtein

16 **Organic Molecular Crystals** Their Electronic States By E. A. Silinsh

17 **The Theory of Magnetism I** Statics and Dynamics By D. C. Mattis

18 **Relaxation of Elementary Excitations** Editors: R. Kubo and E. Hanamura

19 **Solitons** Mathematical Methods for Physicists By. G. Eilenberger

20 **Theory of Nonlinear Lattices** 2nd Edition By M. Toda

21 **Modern Crystallography II** Structure of Crystals 2nd Edition By B. K. Vainshtein, V. L. Indenbom, and V. M. Fridkin

22 **Point Defects in Semiconductors I** Theoretical Aspects By M. Lannoo and J. Bourgoin

23 **Physics in One Dimension** Editors: J. Bernasconi and T. Schneider

24 **Physics in High Magnetics Fields** Editors: S. Chikazumi and N. Miura

25 **Fundamental Physics of Amorphous Semiconductors** Editor: F. Yonezawa

26 **Elastic Media with Microstructure I** One-Dimensional Models By I. A. Kunin

27 **Superconductivity of Transition Metals** Their Alloys and Compounds By S. V. Vonsovsky, Yu. A. Izyumov, and E. Z. Kurmaev

28 **The Structure and Properties of Matter** Editor: T. Matsubara

29 **Electron Correlation and Magnetism in Narrow-Band Systems** Editor: T. Moriya

30 **Statistical Physics I** Equilibrium Statistical Mechanics 2nd Edition By M. Toda, R. Kubo, N. Saito

31 **Statistical Physics II** Nonequilibrium Statistical Mechanics 2nd Edition By R. Kubo, M. Toda, N. Hashitsume

32 **Quantum Theory of Magnetism** 2nd Edition By R. M. White

33 **Mixed Crystals** By A. I. Kitaigorodsky

34 **Phonons: Theory and Experiments I** Lattice Dynamics and Models of Interatomic Forces By P. Brüesch

35 **Point Defects in Semiconductors II** Experimental Aspects By J. Bourgoin and M. Lannoo

36 **Modern Crystallography III** Crystal Growth By A. A. Chernov

37 **Modern Chrystallography IV** Physical Properties of Crystals Editor: L. A. Shuvalov

38 **Physics of Intercalation Compounds** Editors: L. Pietronero and E. Tosatti

39 **Anderson Localization** Editors: Y. Nagaoka and H. Fukuyama

40 **Semiconductor Physics** An Introduction 6th Edition By K. Seeger

41 **The LMTO Method** Muffin-Tin Orbitals and Electronic Structure By H. L. Skriver

42 **Crystal Optics with Spatial Dispersion, and Excitons** 2nd Edition By V. M. Agranovich and V. L. Ginzburg

43 **Structure Analysis of Point Defects in Solids** An Introduction to Multiple Magnetic Resonance Spectroscopy By J.-M. Spaeth, J. R. Niklas, and R. H. Bartram

44 **Elastic Media with Microstructure II** Three-Dimensional Models By I. A. Kunin

45 **Electronic Properties of Doped Semiconductors** By B. I. Shklovskii and A. L. Efros

46 **Topological Disorder in Condensed Matter** Editors: F. Yonezawa and T. Ninomiya

Springer Series in Solid-State Sciences

Editors: M. Cardona P. Fulde K. von Klitzing H.-J. Queisser

47 **Statics and Dynamics of Nonlinear Systems**
Editors: G. Benedek, H. Bilz, and R. Zeyher

48 **Magnetic Phase Transitions**
Editors: M. Ausloos and R. J. Elliott

49 **Organic Molecular Aggregates**
Electronic Excitation and Interaction Processes
Editors: P. Reineker, H. Haken, and H. C. Wolf

50 **Multiple Diffraction of X-Rays in Crystals**
By Shih-Lin Chang

51 **Phonon Scattering in Condensed Matter**
Editors: W. Eisenmenger, K. Laßmann, and S. Döttinger

52 **Superconductivity in Magnetic and Exotic Materials** Editors: T. Matsubara and A. Kotani

53 **Two-Dimensional Systems, Heterostructures, and Superlattices**
Editors: G. Bauer, F. Kuchar, and H. Heinrich

54 **Magnetic Excitations and Fluctuations**
Editors: S. W. Lovesey, U. Balucani, F. Borsa, and V. Tognetti

55 **The Theory of Magnetism II** Thermodynamics and Statistical Mechanics By D. C. Mattis

56 **Spin Fluctuations in Itinerant Electron Magnetism** By T. Moriya

57 **Polycrystalline Semiconductors**
Physical Properties and Applications
Editor: G. Harbeke

58 **The Recursion Method and Its Applications**
Editors: D. G. Pettifor and D. L. Weaire

59 **Dynamical Processes and Ordering on Solid Surfaces** Editors: A. Yoshimori and M. Tsukada

60 **Excitonic Processes in Solids**
By M. Ueta, H. Kanzaki, K. Kobayashi, Y. Toyozawa, and E. Hanamura

61 **Localization, Interaction, and Transport Phenomena** Editors: B. Kramer, G. Bergmann, and Y. Bruynseraede

62 **Theory of Heavy Fermions and Valence Fluctuations** Editors: T. Kasuya and T. Saso

63 **Electronic Properties of Polymers and Related Compounds**
Editors: H. Kuzmany, M. Mehring, and S. Roth

64 **Symmetries in Physics** Group Theory Applied to Physical Problems 2nd Edition
By W. Ludwig and C. Falter

65 **Phonons: Theory and Experiments II**
Experiments and Interpretation of Experimental Results By P. Brüesch

66 **Phonons: Theory and Experiments III**
Phenomena Related to Phonons
By P. Brüesch

67 **Two-Dimensional Systems: Physics and New Devices**
Editors: G. Bauer, F. Kuchar, and H. Heinrich

68 **Phonon Scattering in Condensed Matter V**
Editors: A. C. Anderson and J. P. Wolfe

69 **Nonlinearity in Condensed Matter**
Editors: A. R. Bishop, D. K. Campbell, P. Kumar, and S. E. Trullinger

70 **From Hamiltonians to Phase Diagrams**
The Electronic and Statistical-Mechanical Theory of sp-Bonded Metals and Alloys By J. Hafner

71 **High Magnetic Fields in Semiconductor Physics**
Editor: G. Landwehr

72 **One-Dimensional Conductors**
By S. Kagoshima, H. Nagasawa, and T. Sambongi

73 **Quantum Solid-State Physics**
Editors: S. V. Vonsovsky and M. I. Katsnelson

74 **Quantum Monte Carlo Methods in Equilibrium and Nonequilibrium Systems** Editor: M. Suzuki

75 **Electronic Structure and Optical Properties of Semiconductors** 2nd Edition
By M. L. Cohen and J. R. Chelikowsky

76 **Electronic Properties of Conjugated Polymers**
Editors: H. Kuzmany, M. Mehring, and S. Roth

77 **Fermi Surface Effects**
Editors: J. Kondo and A. Yoshimori

78 **Group Theory and Its Applications in Physics**
2nd Edition
By T. Inui, Y. Tanabe, and Y. Onodera

79 **Elementary Excitations in Quantum Fluids**
Editors: K. Ohbayashi and M. Watabe

80 **Monte Carlo Simulation in Statistical Physics**
An Introduction 3rd Edition
By K. Binder and D. W. Heermann

81 **Core-Level Spectroscopy in Condensed Systems**
Editors: J. Kanamori and A. Kotani

82 **Photoelectron Spectroscopy**
Principle and Applications 2nd Edition
By S. Hüfner

83 **Physics and Technology of Submicron Structures**
Editors: H. Heinrich, G. Bauer, and F. Kuchar

84 **Beyond the Crystalline State** An Emerging Perspective By G. Venkataraman, D. Sahoo, and V. Balakrishnan

85 **The Quantum Hall Effects**
Fractional and Integral 2nd Edition
By T. Chakraborty and P. Pietiläinen

86 **The Quantum Statistics of Dynamic Processes**
By E. Fick and G. Sauermann

87 **High Magnetic Fields in Semiconductor Physics II**
Transport and Optics Editor: G. Landwehr

88 **Organic Superconductors** 2nd Edition
By T. Ishiguro, K. Yamaji, and G. Saito

89 **Strong Correlation and Superconductivity**
Editors: H. Fukuyama, S. Maekawa, and A. P. Malozemoff

Printing: Druckhaus Beltz, Hemsbach
Binding: Buchbinderei Schäffer, Grünstadt